Monte Miles
1055 Aber Hall

The Science of
Entomology

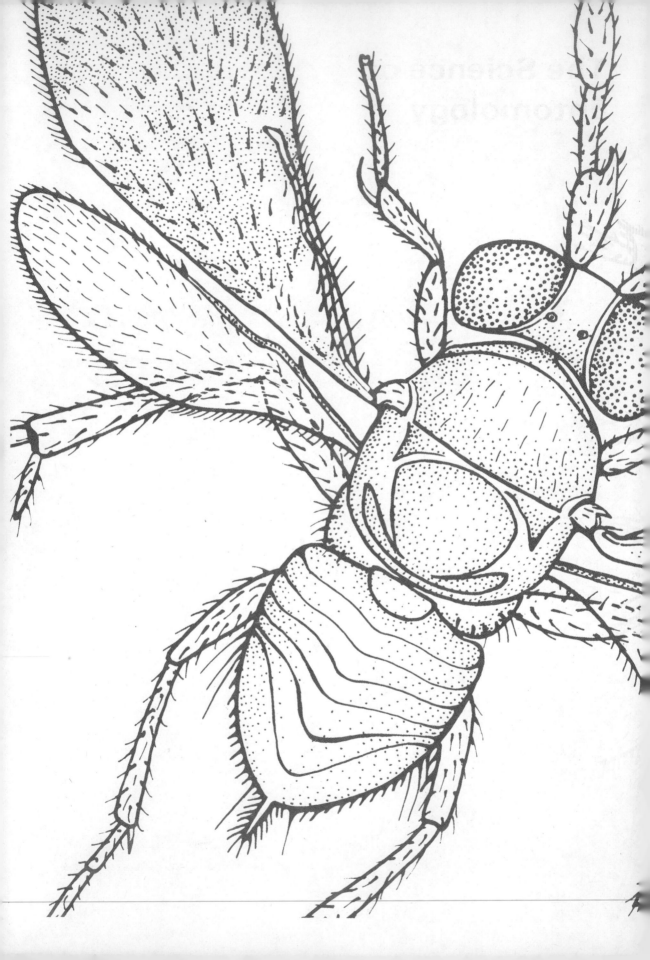

The Science of Entomology

William S. Romoser
Ohio University

Macmillan Publishing Co., Inc.
New York
Collier Macmillan Publishers
London

To Professor Carl E. Venard

Macmillan Publishing Co., Inc.
866 Third Avenue, New York, New York 10022

Collier-Macmillan Canada, Ltd., Toronto, Ontario

Library of Congress Cataloging in Publication Data
Romoser, William S.
  The science of entomology.
  Bibliography: p.
  1. Entomology.   I. Title.
QL463.R65   595.7   72–80074
ISBN 0–02–403420–7

Printing: 4 5 6 7 8   Year: 6 7 8 9

Acknowledgments

For permission to include materials from the following publications,
as indicated in the figure and table captions.

ACADEMIC PRESS, INC.
Clark and Rockstein, 1964: Aging in insects, by A. M. Clark and M. Rockstein, in
    The Physiology of Insecta, Vol. 1, pp. 227–81, M. Rockstein, ed. Copyright 1964 by
    Academic Press, Inc., New York.
Gilbert, 1964: Physiology of growth and development: endocrine aspects, by L. I.
    Gilbert, in *The Physiology of Insecta*, Vol. 1, pp. 149–225, M. Rockstein, ed. Copy-
    right 1964 by Academic Press, Inc., New York.
Hoyle, 1965: Neural control of skeletal muscles, by G. Hoyle, in *The Physiology of
    Insecta*, Vol. 2, pp. 407–49, M. Rockstein, ed. Copyright 1965 by Academic Press,
    Inc., New York.
Miller, 1964: Respiration—aerial gas transport, by P. L. Miller, in *The Physiology of
    Insecta*, Vol. 3, pp. 557–615, M. Rockstein, ed. Copyright 1964 by Academic Press,
    Inc., New York.

AMERICAN ASSOCIATION FOR THE ADVANCEMENT OF SCIENCE
Steyskal, 1965: Trend curves of the rate of species description in zoology, by G. C.
    Steyskal, Science, 149: 880–82. Copyright 1965 by the American Association for
    the Advancement of Science.
Locke, 1965: Permeability of the insect cuticle to water and lipids, by M. Locke,
    Science, 147: 295–98. Copyright 1965 by the American Association for the Advance-
    ment of Science.

THE AMERICAN ENTOMOLOGICAL SOCIETY
Breland, Eddleman, and Biesele, 1968: Studies of insect spermatozoa I, by O. P.
    Breland, C. D. Eddleman, and J. J. Biesele, Entomol. News, 79(8): 197–216, Phila-
    delphia, 1968.

Acknowledgments

AMERICAN SCIENTIST

Harcourt and Leroux, 1967: Population regulation in insects and man, by D. G. Harcourt and E. J. Leroux, Amer. Sci., 55(4): 400–15, New Haven, Conn., 1967.

ANNUAL REVIEWS, INC.

Jander, R. 1963. Insect orientation, by R. Jander, Ann. Rev. Entomol., 8: 94–114, Palo Alto, Calif., 1963.

ASSOCIATED BOOK PUBLISHERS LTD.

Clark et al., 1967: *The Ecology of Insect Populations in Theory and Practice*, by L. R. Clark, P. W. Geier, R. D. Hughes, and R. F. Morris, London, 1967.

Wigglesworth, 1965: *The Principles of Insect Physiology,* 6th ed., by V. B. Wigglesworth, London, 1965.

CCM: GENERAL BIOLOGICAL, INC.

Selected portions of key cards.

THE COMPANY OF BIOLOGISTS LTD.

Beament, 1958: The effect of temperature on the waterproofing mechanism of an insect, by J. W. L. Beament, J. Exp. Biol., 35: 494–519, Cambridge, 1958.

Wigglesworth, 1930: The formation of the peritrophic membrane in insects with special reference to the larvae of mosquitoes, by V. B. Wigglesworth, Quart. J. Microscop. Sci., 73: 593–616, Cambridge, 1930.

Wigglesworth, 1945: Transpiration through the cuticle of insects, by V. B. Wigglesworth, J. Exp. Biol., 21: 97–114, Cambridge, 1945.

CORNELL UNIVERSITY PRESS

Comstock. 1918: *The Wings of Insects*, by John Henry Comstock. Copyright 1918 by the Comstock Publishing Company, Inc. Used by permission of Cornell University Press.

Comstock, 1940: *An Introduction to Entomology*, by John Henry Comstock. Copyright 1933, 1936, 1940 by Comstock Publishing Company, Inc. Used by permission of Cornell University Press.

von Frisch, 1950: *Bees: Their Vision, Chemical Senses, and Language,* by Karl von Frisch. Copyright © 1950 by Cornell University. Used by permission of Cornell University Press.

Matheson, 1944: *Handbook of the Mosquitoes of North America,* Second Edition, by R. Matheson. Copyright 1944 by Comstock Publishing Company, Inc. Used by permission of Cornell University Press.

Maynard, 1951: *The Collembola of New York State,* by Elliott A. Maynard. Copyright 1951 by Comstock Publishing Company, Inc. Used by permission of Cornell University Press.

Snodgrass, 1956: *Anatomy of the Honeybee,* by R. E. Snodgrass. © 1956 by Cornell University. Used by permission of Cornell University Press.

West, 1951: *The Housefly: Its Natural History, Medical Importance, and Control,* by Luther S. West. Copyright 1951 by Comstock Publishing Company, Inc. Used by permission of Cornell University Press.

Wigglesworth, 1959a: *The Control of Growth and Form: A Study of the Epidermal Cell in an Insect,* by V. B. Wigglesworth. © 1959 by Cornell University. Used by permission of Cornell University Press.

DADANT & SONS, INC.

Snodgrass, 1963b: The anatomy of the honey bee, in *The Hive and the Honey Bee,* pp. 141–90, R. A. Grout, ed., Hamilton, Ill., Dadant & Sons, Inc.

ENTOMOLOGICAL SOCIETY OF AMERICA

Claassen, 1931: *Plecoptera Nymphs of America (North of Mexico)*, by P. W. Claussen, The Thomas Say Foundation, Publ. 3.

Ewing, 1940: *The Protura of North America,* by H. E. Ewing, Ann. Entomol. Soc. Amer., 33(3): 495–551.

Froeschner, 1947: Notes and keys to the Neuroptera of Missouri, by R. C. Froeschner, Ann. Entomol. Soc. Amer., 40(1): 123–36, College Park, Md., 1947.

Acknowledgments

Hess, 1917: The chordotonal organs and pleural discs of cerambycid larvae, by W. N. Hess, Ann. Entomol. Soc. Amer., 10: 63–78, College Park, Md., 1917.

Wolglum and McGregor, 1958: Observations on the life history and morphology of *Agulla bractea* Carpenter (Neuroptera: Raphidiodea: Raphidiidae), by R. S. Wolglum and F. A. McGregor, Ann. Entomol. Soc. Amer., 51: 129–41, College Park, Md., 1958.

ENTOMOLOGICAL SOCIETY OF WASHINGTON

Caudell, 1918: Zoraptera not an apterous order, by A. N. Caudell, Proc. Entomol. Soc. Wash., 22: 84–97, Washington, D.C., 1918.

GUSTAV FISCHER VERLAG

Barth, 1937: Muskulatur und Bewegungsart der Raupen, by R. Barth, Zool. Jahrb., 62: 507–66, Jena, Germany, 1937.

Vogel, 1921: Zur Kenntnis des Baues und der Funktion des Stachels und des Vorderdarmes der Kleiderlaus, by R. Vogel, Zool. Jahrb., Anat., 42: 229–58, 1921.

HOLT, RINEHART AND WINSTON, INC.

Borror and DeLong, 1971: *An Introduction to the Study of Insects,* third edition, by Donald J. Borror and Dwight M. DeLong. Copyright © 1964, 1971 by Holt, Rinehart and Winston, Inc. Copyright 1954 by Donald J. Borror and Dwight M. DeLong. Reprinted by permission of Holt, Rinehart and Winston, Inc.

LONGMAN GROUP LTD.

Skaife, 1961: *Dwellers in Darkness,* by S. H. Skaife, London, 1961.

MCGRAW-HILL BOOK COMPANY

Folsom and Wardle, 1934: *Entomology, with Special Reference to Its Ecological Aspects,* 4th ed., by J. W. Folsom and R. A. Wardle, copyright 1934 by McGraw-Hill, Inc. Used with permission of McGraw-Hill Book Company.

Graham and Knight, 1965: *Principles of Forest Entomology,* 4th ed., by S. A. Graham and F. B. Knight. Copyright 1965 by McGraw-Hill Book Company. Used with permission of McGraw-Hill Book Company.

Johannsen and Butt, 1941: *Embryology of Insects and Myriapods,* by O. A. Johannsen and F. H. Butt. Copyright 1941 by McGraw-Hill, Inc. Used with permission of McGraw-Hill Book Company.

Metcalf, Flint, and Metcalf, 1962: *Destructive and Useful Insects,* by C. L. Metcalf, W. P. Flint, and R. L. Metcalf. Copyright 1962 by McGraw-Hill, Inc. Used with permission of McGraw-Hill Book Company.

Snodgrass, 1925: *Anatomy and Physiology of the Honey Bee,* by R. E. Snodgrass. Copyright 1925 by McGraw-Hill, Inc. Used with permission of McGraw-Hill Book Company.

Snodgrass, 1935: *Principles of Insect Morphology,* by R. E. Snodgrass. Copyright 1935 by McGraw-Hill, Inc. Used with permission of McGraw-Hill Book Company.

MACMILLAN PUBLISHING CO., INC.

Essig, 1942: *College Entomology,* by E. O. Ossig. Copyright 1942 by Macmillan Publishing Co., Inc., New York.

Essig, 1958: *Insects and Mites of Western North America,* by E. O. Essig. Copyright 1958 by Macmillan Publishing Co., Inc., New York.

Herms and James, 1961: *Medical Entomology,* 5th ed. Copyright 1961 by Macmillan Publishing Co., Inc., New York.

James and Harwood, 1969: *Herm's Medical Entomology,* by M. T. James and R. F. Harwood. Copyright 1969 by Macmillan Publishing Co., Inc., New York.

Stiles, Hegner, and Boolootian, 1969: *College Zoology,* 8th ed., by K. A. Stiles, R. W. Hegner, and R. Boolootian. Copyright 1969 by Macmillan Publishing Co., Inc., New York.

THE MARINE BIOLOGICAL LABORATORY

Schneiderman and Williams, 1955; An experimental analysis of the discontinuous respiration of the cecropia silkworm, by H. A. Schneiderman and C. M. Williams, Biol. Bull., 109: 123–43, Woods Hole, Mass., 1955.

Acknowledgments

NATIONAL ACADEMY OF SCIENCES

Committee on Plant and Animal Pests, 1969: *Insect Pest Management and Control*, Publication 1695, by the Committee on Plant and Animal Pests, National Academy of Sciences. National Research Council, Washington, D.C., 1969.

OLIVER & BOYD LTD.

Davey, 1965: *Reproduction in the Insects*, by K. G. Davey, Edinburgh, 1965.

Wigglesworth, 1970: *Insect Hormones*, by V. B. Wigglesworth, Edinburgh, 1970.

PAUL PAREY

Sturm, 1956: Die Paarung beim Silberfischen, *Lepisma saccharina*, by H. Sturm, Z. Tierpsychol., 13: 1–2, Berlin, 1956.

PERGAMON PRESS, INC.

Engelmann, 1970: *The Physiology of Insect Reproduction*, by F. Engelmann, New York, 1970. Reprinted with permission.

ALVAH PETERSON

Peterson, 1951: *Larvae of Insects*, part II, Coleoptera, Diptera, Neuroptera, Siphonaptera, Mecoptera, Trichoptera, by A. Peterson, Columbus, Ohio, 1951.

Peterson, A. 1948: *Larvae of Insects*, part I, Lepidoptera and plant infesting Hymenoptera, by A. Peterson, Columbus, Ohio, 1948.

THE RONALD PRESS COMPANY

Hagen, 1951: *Embryology of the Viviparous Insects*, by H. R. Hagen. Copyright 1951, The Ronald Press Company, New York.

Rolston and McCoy, 1966: *Introduction to Applied Entomology*, by L. H. Rolston and L. E. McCoy. Copyright 1966, The Ronald Press Company, New York.

THE ROYAL SOCIETY

Gillett and Wigglesworth, 1932: The climbing organ of an insect, *Rhodnius prolixus* (Hemiptera: Reduviidae). Proc. Roy. Soc. London, Series B, 111: 364–76, 1932.

W. B. SAUNDERS COMPANY

Patton, 1963: *Introductory Insect Physiology*, by R. L. Patton, Philadelphia, 1963.

SPRINGER-VERLAG

Baldus, 1926: Experimentelle Untersuchungen über die Entfernungslokalisation der Libellen *(Aeschna cyanea)*, by K. Baldus, Z. vergl. Physiol., 3: 375–505, Berlin, 1926.

Fraenkel, 1932: Untersuchungen über die Koordination von Reflexen und automatisch-nervösen Rhythmen bei Insekten III, by G. Fraenkel, Z. vergl. Physiol., 16: 394–460, Berlin, 1932.

Hertz, 1929: Die Organization des Optischen Feldes bei der Biene, I, by M. Hertz, Z. vergl. Physiol., 8: 693–748, Berlin, 1929.

Markl, 1962: Schweresinnesorgane bei Ameisen und anderen Hymenopteren, by H. Markl, Z. verfl. Physiol., 44: 475–569, Berlin, 1962.

Nachtigall, 1960: Über kinematik, Dynamik und Emergetik des schwimmens einheimischen Dytisciden, by W. Nachtigall, Z. vergl. Physiol., 43: 48–118, Berlin, 1960.

Nachtigall, 1963: Zur Lokomotionsmechanik schwimmender Dipterenlarven, by W. Nachtigall, Z. vergl. Physiol., 46: 449–66, Berlin, 1963.

UNIVERSITY OF CHICAGO PRESS

Andrewartha and Birch, 1954: *The Distribution and Abundance of Animals*, by H. G. Andrewartha and L. C. Birch. Copyright 1954 by The University of Chicago. All rights reserved. Published 1954.

UNIVERSITY OF MICHIGAN PRESS

Goetsch, 1957: *The Ants*, by W. Goetsch. Copyright © by the University of Michigan, 1957.

UNIVERSITY OF MINNESOTA PRESS

Richards, 1951: *The Integument of Arthropods*, by A. Glenn Richards. University of Minnesota Press, Minneapolis, © 1951 University of Minnesota.

Acknowledgments

VAN NOSTRAND REINHOLD COMPANY

Cummins, Miller, Smith, and Fox, 1965: *Experimental Entomology,* by K. W. Cummins, L. D. Miller, Ned A. Smith, and R. M. Fox, New York, 1965.

DuPorte, 1961: *Manual of Insect Morphology,* by E. M. DuPorte, New York, 1961.

VERLAG CHEMIE GMBH

Schildknecht and Holoubek, 1961: Die Bombardierkäfer und ihre Explosionschemie V. Mitteilung über insekten. Abwehrstoffe, by H. Schildknecht and K. Holoubek, Angew. Chem., 73(1): 1–6, Weinheim/Bergstr., Germany, 1961.

GEORGE WEIDENFELD & NICOLSON LTD.

Chauvin, 1967: *The World of an Insect,* by R. Chauvin, London, World University Library.

JOHN WILEY & SONS, INC.

Ross, 1965: *A Textbook of Entomology,* 3rd ed., by H. H. Ross, New York, 1965. By permission of John Wiley & Sons, Inc.

Schneirla, 1953: Insect behavior in relation to its setting, by T. C. Schneirla, in *Insect Physiology,* pp. 685–722, K. D. Roeder, ed., New York, 1953.

Preface

The magnitude of the role played by insects in the scheme of life is undisputed. In adaptive diversity and number of species, they are among the most successful of all organisms. The relatively few that are problems for man have been very effective in taxing his ingenuity to its fullest throughout history and the battle is far from over. Thus the science of entomology is one of the major areas of basic biology as well as a vital applied science.

My objective in writing this text has been to provide a broad, balanced introduction to the field of entomology. What constitutes a broad, balanced coverage of a field as vast as entomology is largely a matter of opinion. Perhaps a more realistic statement would be that this text represents what I view as a broad, balanced coverage of entomology at the time of this writing.

I have treated entomology as a branch of biology that has applied aspects, but which is not strictly an applied science. The major portion of the text is developed around two concepts: *structure* and *function* at the various levels of biological organization, and *unity* and *diversity* as the result of organic evolution. This approach should give the reader a realistic view of insects in the scheme of life as opposed to a purely negative "insects as pests" point of view. I have included a discussion of the literature of entomology with the hope that the student will be encouraged to make full use of the vast information available.

Under the heading "Structure and Function," I have included a consideration of insect ecology as well as those topics which are more commonly included. This approach is based on the idea that insects, in addition to being functional units, are also parts of higher orders of structure and function—that is, parts of populations and the various ecosystems. In the section "Unity and Diversity" I consider in a general way the origin of Insecta and the basic principles upon which the groupings of contemporary insects are based. I have included the discussion of insect systematics because I feel that it is important for the student to appreciate how the various groupings of insects have been developed; undergraduates in these days of molecular and cellular biology need exposure to the practical aspects of systematics. The chapter entitled "Survey of Class Insecta" provides a brief overview of the orders of insects. I have tried to show how the various orders relate to one another as well as provide information regarding their biology and medical and economic significance. The final chapter of this text is a consideration of the applied aspects of entomology. It begins with a discussion of how certain insects have become problems for man and the various methods which have been used or may be used in the future to combat them. The many ways insects are beneficial are also discussed in this chapter.

I have arranged the sequence of topics in the way I think most appropriate for dealing with the various aspects of entomology. How-

ever, each chapter is so designed that it can be read and understood
with minimal reference to other chapters. Thus this text should be
useful in any organizational framework a given instructor may
choose to develop. A bibliography consisting mainly of major review
papers, monographs, and specialized textbooks is given at the end of
each chapter.

As I have stated, this text has been written in response to the
realization of the need for a broadly based treatment of general
entomology. Therefore, its major role should be in the one-quarter
or one-semester course in general entomology. However, I also hope
that the professional entomologist or zoologist who is engaged in one
of the many specialized areas will find it to be useful as a reasonably
up-to-date review.

Several persons have been of assistance in the preparation of this
text and I wish to express my sincere appreciation for their efforts.

Wayne P. Aspey, Irving J. Cantrall, Woodbridge Foster, Arden O.
Lea, Jerome S. Rovner, Charles A. Triplehorn, James A. Wilson,
Frank N. Young, and James R. Zimmerman critically reviewed
selected parts of the manuscript.

Clifford S. Crawford provided constructive criticism throughout the
development of the manuscript.

Richard F. Ashley, Harold C. Chapman, Alfred Dietz, John D.
Edman, Woodbridge Foster, Walter Humphreys, Arden O. Lea, and
Shirlee Meola each provided one or more photographs.

Elisabeth Belfer of Macmillan Publishing Co., Inc., expertly guided
the manuscript from the final typed form to the finished product.

I also wish to express my gratitude to many of my colleagues at
Ohio University who freely offered encouragement and advice
throughout the period of this project.

W. S. R.

Contents

PART ONE

Structure and Function

1

Introduction

Insects are members of the class Insecta in the invertebrate phylum Arthropoda, the largest in the Animal Kingdom. The members of this phylum are characterized by a segmented body that bears a varying number of paired and segmented appendages; bilateral symmetry; an exoskeleton that contains the nitrogenous polysaccharide, chitin; and various internal features, such as an open circulatory system, Malpigian tubules (generally), and in most a system of ventilatory tubules, the tracheae and tracheoles. Present-day arthropods are often divided into two large subphyla, Chelicerata and Mandibulata. The trilobites, a long-extinct group of organisms, comprise a third arthropod subphylum, the Trilobita. The chelicerates bear a pair of appendages called *chelicerae* near the oral opening; the mandibulates are characterized by a pair of grinding structures associated with the mouthparts, the *mandibles*. Both chelicerae and mandibles are subject to considerable variation in structure; consequently, there are rather profound deviations from the "typical" forms. Although both structures usually function as parts of the feeding apparatus, they are not homologous. Spiders, ticks, scorpions, and horseshoe crabs are examples of chelicerates. Insects, together with millipedes, centipedes, and others, make up the subphylum Mandibulata.

Insects can be differentiated from the vast majority of other arthropods by several rather distinct traits. Among these are three well-defined body regions or tagmata: a head, a thorax, and an abdomen; three pairs of legs in the adult stage; commonly one or two pairs of wings; a single pair of segmented antennae on the head; and several, less obvious, but equally distinctive characteristics which will become apparent as the reader proceeds through this text. The name *Hexapoda* (six legs) is commonly applied to insects. However, the name *Insecta* is preferable since there is some question as to whether all arthropods with six legs in the adult stage actually belong in the same class (Sharov, 1966). Insecta literally means "in-cut," which describes the segmented appearance of the members of this class. The phylum Arthropoda will be discussed in more detail when we consider the evolution of insects.

Significance of Insects

Insects as a group are highly successful organisms. Their significance can be looked upon from two standpoints: (1) their tremendous success relative to organisms other than man, and (2) their extreme importance from man's point of view.

One useful measure of the success of insects is the number of extant species. Estimates based on the current rates of description of new species of insects run from one to several million. It has been pointed out by a number of entomologists that several large groups of insects have hardly been studied at all. Brues, Melander, and Carpenter (1954) explain that approximately 750,000 species have been described and named and suggest that this figure is possibly only one fifth to one tenth of the insectan species that exist. Insects have been said to outnumber all the other species of animals and the plants combined.

Other important criteria for success include the span of geologic time traversed by a group of organisms and the adaptability to various environmental situations. Insects are thought to have evolved in the Silurian era, approximately 360 million years ago. Mammals as a group are approximately 230 million years old; modern man has been around for perhaps 1 million years. In this sense, insects have not invaded man's world; man has invaded theirs!

The adaptability of the basic insectan plan has been phenomenal. Insects can be found in nearly every conceivable situation. As you proceed through this text, you will come to realize the seemingly unlimited adaptability of insects and gain some insight as to how they have reached their position of success.

From man's point of view a number of insect species have been considered arch enemies from the earliest times. Although the number of pest species compared to the total number of insect species is probably very small, the troubles for man wrought by this group have in the past and still do reach astonishing proportions. Insects destroy annually millions of dollars worth of agricultural crops, fruits, shade trees and ornamental plants, stored products of various sorts, household items, and other material goods valued by man. They serve as vectors of the causative agents of a sizable number of diseases of man and domestic animals. In addition, they attack man and other animals, directly causing irritation, blood loss, and, in some instances, death. However, there are two sides to the picture. Insects have provided over the years, and still continue to provide, many goods and services, so to speak, looked upon very favorably by man. Such insect products as honey and beeswax, silk, shellac, and cochineal are utilized for a variety of applications, ranging from sweetening biscuits to furnishing one of the basic components of many cosmetics. In addition, there are many indirect benefits of insect activities, such as plant pollination. A more detailed consideration of these subjects will be found in Chapter 12.

Although there is much that can be said both for and against insects as they relate to man, the vast majority are quite neutral, neither bestowing any great benefit nor causing any great harm.

Entomology as a Science

What Is Science?

Before we can intelligently consider entomology as a science, we must first discuss exactly what we mean by science. Science is a body of knowledge obtained in a very special way, which we call the *scientific method.* Scientists assume that entities of the universe interact with one another in a predictable pattern—in other words, follow a very definite set of rules. We might say then that the major objective of the scientist is to discover and seek to understand these rules. From this understanding comes the ability to predict events and in many instances exert controlling or modifying influences on these events. Therefore, a knowledge of the life cycle and biology of a given insect enables us to predict the insect's activities at a given time and may allow us to act in such a way that we can either effectively limit the insect or make effective use of its favorable points. For example, understanding of the reproductive processes of screw-worm flies, serious pests of both wild and domestic animals in the southern United States and elsewhere, has enabled investigators to develop the comparatively new, quite successful sterile-male method of insect eradication. Using this method, the screw-worm has apparently been eradicated from the southeastern United States. Likewise, extensive knowledge of the behavior of the honey bee has enabled apiarists over the years to harvest greater and greater amounts of honey per hive.

The difference between scientific and other approaches to learning of the universe lies in the rigid application of objective or rational thinking and the scientific method. Basically, this method consists of the following steps:

1. *Observation and description:* Critical observation and accurate description of events are essential prerequisites to the remaining steps.
2. *Hypothesis:* Once an event or series of events has been adequately described, tentative explanations or hypotheses are formulated.
3. *Experiment:* It is the purpose of experimentation to determine which, if any, of the hypotheses is the correct or the "most" correct one. Experimentation may also suggest hypotheses that had not been apparent earlier in the process. An experiment is a manipulation of the entities being investigated in such a way that the influence of changing a single factor or variable can be evaluated. Of course, it is usually not feasible, or, for that matter, possible, to achieve a sufficient degree of constancy of all pertinent variables. It then becomes necessary to introduce "controls" into the experiment. A control is carried out concurrently with and is essentially identical in all respects to the experimental situation except that the variable being evaluated is unchanged. This helps to correct for the effects of variables that may change in an unknown and/or uncontrollable way during the course of the experiment.

The introduction of controls is especially important in biological research where one is commonly dealing with a large number of uncontrollable variables. Hypotheses borne out by repeated experiments over a long period of time often come to be considered laws, but even these well-supported hypotheses may fall in the light of new, more critical or sophisticated experimentation. At some point between hypotheses and laws lie theories, which may be considered to be hypotheses that have survived some experimentation but still remain more within the realm of conjecture than laws. Thus there may be a number of current theories explaining a given natural phenomenon, all of which are supported by a degree of experimental evidence, but none of which have, as yet, been established as the "correct" one.

Every step of the scientific method outlined above is not necessarily a part of every scientist's work. A given investigator may be involved only with the observational and descriptive level and for a variety of possible reasons may not carry on experiments. Another may take the observations and hypotheses of another investigator and put them through rigorous experimentation. Both these men are scientists. The essential requirement is that they work within the framework of the scientific method and adhere closely to its basic assumptions.

Basic Components of Science

The application of the scientific method to the entities and events of the universe is a progressive and accumulative activity. New information modifies, clarifies, and supplements the old. Old information serves as an ever-increasing foundation for the new. Basically, there are two major components of science: the scientists and their students, and the great body of printed words, the literature. One often wonders how much subtle, unrecorded information that has developed as a result of years of professional experience is stored within the minds of scientists and how much lost with them or is passed on to their students. Although probably a significant amount of information is, in fact, passed from scientist-teacher to student, the true, long-term memory bank of science is the literature. This topic will be pursued in greater detail later in this chapter.

Acquisition of Scientific Information

We are now in a position to consider how scientific information can be acquired. The most primitive and still one of the most important means is by word of mouth. This is the method employed in the classroom, at scientific meetings, conferences, and so on. Perhaps its most valuable aspect is the opportunity for a two-way exchange of information. Ideally, in this situation the lines of communication are wide open and a minimal amount of ambiguity should be the result. Another, equally important—in fact, essential—means of acquiring scientific information is consultation of the literature. This means has the advantage that one can go back in time as far as one wishes; however, the two-way exchange of information becomes impossible if the author of a piece of recorded work is no longer living. Both these means of information acquisition should be considered to be prerequisites for a third means—personal investigation. This method is, of course, the source of new information. All three means are essential to the existence of science, and certainly no one could take precedence over the other two.

In applying the various means of information acquisition, a given scientist may have a variety of motivations. These motivations can be looked upon as placing this scientist somewhere between the two end

points of a continuum. On the one end, the scientist's motivations
would be based entirely on the desire to satisfy a direct human need;
at the other end, his motivations would be based purely on human
curiosity or academic interest. Individual scientists, of course, vary a
great deal with regard to position on this continuum.

Where Does Entomology Fit into the Total Framework?

Entomology is, first of all, one of the biological sciences, since it is
concerned with "living" systems. The biological sciences can be
divided into basic divisions, such as morphology, physiology, genetics,
and ecology, and into taxonomic divisions: ornithology, mycology,
and so on. Entomology, since it deals with the study of a very specific
group of organisms is, of course, a taxonomic division. This being the
case, we can logically approach the science of entomology by considering the "basic" divisions as they apply to insects: insect morphology, physiology, ecology, and so on.

The Literature of Entomology

The literature of entomology consists of a wide variety of publications, ranging from the rather popularized accounts intended for the
layman to the highly technical treatises on very specific aspects of the
science. The information available on insects is vast and is growing so
rapidly that no one person can stay abreast of and learn more than
comparatively small portions of it. This situation has resulted in
specialization, the concentration of effort on a single topic or a group
of closely related topics. Thus, among entomologists there are, for
example, insect physiologists, ecologists, morphologists, systematists,
toxicologists, and economic entomologists. Generally, the specialization goes even further; a man in one of the preceding groups may concentrate on a particular insect species or group of insects, or on a
specific topic, or both. For example, there are specialists in the systematics of a particular family of beetles, in mosquito physiology, and so on.

The tremendous number of entomological publications makes a
comprehensive review inappropriate within the context of this book.
However, a brief discussion of some of the different types of entomological literature will at least broaden the reader's perspective. More
extensive treatments of the literature of entomology and zoology in
general can be found in Smith and Painter (1966), Chamberlin (1952),
and Blackwelder (1967). Arnett (1970) provides insight into the problems involved with the storage and retrieval of entomological information and discusses the current and proposed solutions.

Publications pertinent to entomology (for that matter, the entire
field of zoology) can be divided into two basic groups. The first group
includes all those publications, irrespective of type, which contain the
actual information about animals (insects). The members of this
group contain all the products of zoological (entomological) research.
The second group includes all the publications that attempt to
coordinate the vast information contained in the first group, making it
more readily available to investigators. Examples of publications

from each of these groups are presented below. This method of classification, as with most, is certainly not without exception, as will be shown.

Publications That Contain the Actual Information of Entomology

Textbooks.   A textbook is generally designed to give the reader an understanding of the basic principles involved in a given subject. A textbook may be quite general in scope, presenting a survey or overview of an entire field—for example, a general entomology textbook such as the one you are reading. On the other hand, textbooks commonly deal with a particular area within a given field—for example, a text devoted to insect physiology. Books of this type, of course, handle a subject in considerably more depth than the general text and are more likely to be used in advanced courses.

Monographs.   A monograph is limited in scope, dealing only with a very small area within a science. However, monographs are usually comprehensive in their coverage of pertinent literature and handling of subject matter.

Symposia.   Symposia are published collections of the presentations of several experts in a given area who have met to consider a specific aspect of science.

Lectures.   A lecture is an oral discourse presented to an audience or class, particularly for instructional purposes. Good entomological examples of lectures are the talks presented by outstanding scientists to the general sessions usually held several times during each annual meeting of the various entomological societies throughout the world. These talks are commonly published in bulletins issued periodically by these societies.

Essays.   Essays are analytic or interpretative expositions that usually deal with a given topic from a rather personal or limited point of view. A well-known example is the book *Silent Spring* by Rachel Carson.

Reference Works.   Reference books generally attempt to offer comprehensive coverage of a specific area and are designed to be consulted as needed rather than read and digested in their entirety. This category may seem rather arbitrary, since any piece of literature that is consulted can be classified as a reference, and rigorously documented and highly technical entomological texts, such as Imms' classic *A General Textbook of Entomology* (Richards and Davies, 1957) can as easily be looked upon as reference works as they can textbooks. A particularly good example of a publication designed for reference is *The Encyclopedia of the Biological Sciences* edited by Peter Gray.

Pamphlets.   Pamphlets are usually rather brief writings on a specific topic, commonly geared to the persons who put many of the findings of entomological research into practice: farmers, exterminators, and so on. Examples of pamphlets are the *Farmer's Bulletins* published by the U.S. Department of Agriculture, which pertain to the biology and control of economically important species of insects.

Reports. From time to time, groups of experts are called together to investigate or to discuss an issue or problem of particular significance. For example, the World Health Organization (WHO) periodically sponsors meetings of expert committees on various problems pertaining to international health matters. These committees usually submit reports describing the problems discussed and conclusions reached in the course of the meeting.

Series. Serial publications are issued periodically (sometimes at irregular intervals) and have a certain unity of subject matter; that is, a particular series deals, volume after volume, with more or less the same general subject matter. Series are published by most professional societies and many private and governmental concerns in the form of journals, bulletins, miscellaneous publications, yearbooks, and so on. Some may be published as a series, for example the *Farmer's Bulletins*. Hammack (1970) provides a useful and comprehensive description of the serial literature pertinent to entomology.

Publications That Attempt to Coordinate the Literature

From the time scientists realized the fantastic rate of growth of the literature of science, many very useful attempts to coordinate and integrate the works in various areas have been made. In this section we want to discuss briefly some of the more common publications in this category. Most of these publications are quite expensive and are seldom purchased by an individual.

Bibliographies. A bibliography, as Smith and Painter (1966) aptly point out, is to the literature of science as an index is to a book, the basic difference being that a bibliography lists only publications of various sorts on a single topic, instead of the subjects, authors, and so on, to be found in the index of a single book. Bibliographies appear in different forms. For example, one form is the type that cites pertinent references at the end of a chapter of a book or a scientific paper. Another type of bibliography is issued periodically and lists current publications in a particular field. Important examples of this type are *Zoological Record, Bioresearch Index, Bibliography of Agriculture, Cumulated Index Medicus,* and *Current Contents.* The *Zoological Record* has a very broad coverage, both foreign and domestic, and is arranged according to taxonomic groups and by subjects. The section on insects is quite extensive and contains references to papers of interest to most entomologists, not just taxonomists. *Bioresearch Index* is a monthly publication that furnishes bibliographies by journals and contains citations of research papers not covered by *Biological Abstracts.* The *Bibliography of Agriculture,* a monthly publication, generally contains a large number of references to entomological papers. The last issue of each year is a cumulative subject index.

Cumulated Index Medicus is issued four times yearly, covers much of the foreign and domestic medical literature, and may contain references of interest particularly to medical entomologists. *Current Contents* is a publication in which are reproduced the tables of contents of many journals, several entomologically oriented ones included. It is issued weekly and is probably one of the best ways to keep abreast of the most current literature. This is especially important when one is working in a very active area in which a number of researchers are publishing extensively.

Abstracting Journals. Abstracting journals contain brief descriptions or abstracts of the results reported in the journals that fall within their scope. Abstracts are extremely useful since they give an investigator a better idea of the content of a given reference than does a mere listing of titles, although they do serve also as bibliographies. This kind of information helps one decide whether or not he will find it necessary to consult the actual reference and compensates somewhat for those publications which may be extremely difficult to obtain or translate from a foreign language. Abstracting journals are usually extensively cross-indexed, which makes them efficient to use. One of the most significant examples of this type of publication is *Biological Abstracts*. This bimonthly publication is quite comprehensive in its coverage of the literature of theoretical and applied biology. In addition to the volumes containing the abstracts, a semimonthly publication, *B.A.S.I.C.,* provides a very elaborate computerized subject index to the issues of *Biological Abstracts*. A cumulative subject index based on all the issues of *B.A.S.I.C.* is published for each year's volume of *Biological Abstracts*. Since January 1970 the abstracts and citations of research papers pertaining to insects and arachnids in *Biological Abstracts* and *Bioresearch Index* have been compiled into a separate publication, *Abstracts of Entomology*. One issue of this publication corresponds to two issues of *Biological Abstracts* and one issue of *Bioresearch Index*.

An additional abstracting journal that specializes in covering research literature in entomology is *Entomology Abstracts*. This periodical is published monthly and covers a wide variety of entomological topics. Another very useful monthly publication is *Dissertation Abstracts*. This contains abstracts of all dissertations by contributing institutions in the United States arranged by the type of subject matter. These abstracts are quite useful since there is commonly a hiatus between the writing of a dissertation and the publication of a research paper or papers based on it. If one decides on the basis of a given abstract that he needs more information, he may readily obtain, for a fee, a microfilm of an entire dissertation.

An abstracting journal of particular importance to entomologists is the *Review of Applied Entomology*. It is published in two series: Series A, Agricultural, and Series B, Medical and Veterinary. It is well indexed, both by subject and author, and contains abstracts covering a wide variety of entomological topics. Other periodical publications that are at least partly abstracting journals are *Biologisches Zentralblatt, Physiological Abstracts, Tropical Diseases Bulletin, Apicultural Abstracts,* and several others from various countries.

Review Journals. Review journals contain papers that discuss the literature on a rather specific topic in a given field. Review papers not only bring together information from the pertinent literature on a given topic but also commonly contain useful syntheses of information that may not occur in any other type of publication. In this sense they may be classed in either of the two rather arbitrary categories we have used to discuss entomological literature. Two very important review journals in the field of entomology are the *Annual Review of Entomology* and the *Annual Review of Insect Physiology.* Other review journals that may contain reviews of interest to entomologists are the *Annual Review of Physiology* and the *Annual Review of Medicine.* In addition, review papers may appear in journals, bulletins, and so on, which contain other types of publications.

Taxonomic Indexes and Catalogs. Taxonomic indexes include literature references to such items as the original description of a given genus or species, and revisions of genera. These publications are quite useful for tracing the taxonomic literature pertinent to a given group and for determining the systematic position of a given genus or species. Especially important indexes are *Nomenclator Zoologicus* edited by A. A. Neave, *Zoological Record,* and *Biological Abstracts. Nomenclator Zoologicus* lists the names of genera and subgenera of all zoological groups from 1758, the year of the publication of the 10th edition of Carl Linne's (Linnaeus) *Systema Naturae,* to 1950. *Zoological Record* contains the names of all new genera described each year and pertinent literature references from 1864. *Biological Abstracts,* in the section "Systematic Zoology," provides references to the original descriptions of genera and subgenera of animals since 1935. Smith (1958) points out that "Neave's *Nomenclator Zoologicus* and *Biological Abstracts* serve admirably as a complete generic index from 1758 to the present." He further suggests that ". . . for names published since the most recent issues of *Biological Abstracts,* journals in which new genera of the various groups might be expected to occur must be consulted."

For species indexes and catalogs similar to the generic ones just described, one must refer to one or more of several currently available. Sherborn's *Index Animalium* is the only general species index available and covers all the specific names proposed for animals from 1758 through 1800. Smith and Painter (1966), Chamberlin (1952), and Blackwelder (1967) each contain lists of catalogs for various insectan and other groups.

Science Citation Index. *The Science Citation Index* is a recent addition to the literature coordinating publications. It is published by the organization that publishes *Current Contents* and is composed of two sets of indexes, a citation index and a source index, both of which are cumulative. Its objective is to list and index the current and past research papers that cite a given reference. It enables an investigator to begin with a given reference and find other references that have cited the "starting reference." Since both the "starting reference" and "citing reference" are likely to pertain to the same or very closely related topics, one is able to proceed forward or backward in time, using "citing references" as "starting references" in a cyclical manner and by doing so accumulate a bibliography of the literature on a given topic.

Approach of This Text

As scientists we are trying to understand the nature of the universe by applying the scientific method. As biologists we are interested in those fragments of the universe that we consider to be "alive." As entomologists we are interested in those living organisms classified as insects. We shall approach this group of organisms by attempting to answer in general terms and to the extent of presently available knowledge four basic questions:

1. How do insects function?
2. How do insects reproduce, grow, and develop?
3. How did insects arise and diversify?
4. What are the relationships between man and insects?

These four questions have served as the basis for the organization of this book.

Selected References

ENTOMOLOGICAL LITERATURE
Arnett (1970); Blackwelder (1967); Chamberlin (1952); Hammack (1970); Smith and Painter (1966).

HISTORY OF ENTOMOLOGY
Cushing (1957); Essig (1931); Howard (1930); Osborn (1937).

The Integumentary System

The general body covering, or *integument,* of insects is a truly remarkable, complex organ of diverse structure and function. Probably its most obvious role is that of forming the supportive shell, the *skeleton.* The insect skeleton is quite different from the internal skeleton, or endoskeleton, of vertebrates in that it is located on the outside of the body, forming an *exoskeleton.* Even the internal processes referred to by some authors as endoskeletal structures are continuous with the external shell. In addition to forming the outer covering and internal processes, a modified form of the integument lines a major part of the alimentary canal, the tracheal system, genital ducts, and the ducts of the various dermal glands.

As in the vertebrates, the insect skeleton affords many points for muscle attachment. In many cases the elastic nature of certain parts of the skeleton may actually oppose muscular contraction in a manner similar to the mutual opposition of antagonistic muscles in vertebrates. In addition, this elasticity may, in some cases, aid in the inspiration of gases during ventilatory movements. The insect skeleton is hardened, or *sclerotized,* to varying degrees and thus serves as a protective armor for the organs encased by it. Properties of the integument other than hardness also play a role in the protection of the insect; for example, the integument forms a highly effective barrier against the entry of many pathogens and insecticides. The integument may be quite impermeable to water, particularly that which tends to escape from the insect. Since the exoskeleton lies between the external environment and the remainder of the insect, the external portions of the sensory systems and many diverse glandular structures are of necessity intimately associated with it. Similarly, the integument is the seat of body coloration, which may be due to the presence of pigments of various sorts, physical characteristics of the integument, or both.

One of the major limitations of the insect exoskeleton is its inability to undergo extensive expansion. Increase in size of an insect usually requires the periodic shedding and renewal of the integument, the complementary processes of *molting* and *ecdysis.* Although relatively brief, the period between the shedding of the old integument, or ecdysis, and the hardening of the new leaves the insect in a highly vulnerable condition.

In the following discussion of the structure and function of the insect integument, we keep firmly in mind that this outer covering is a complex, functioning organ as well as a skeleton.

Histology of the Integument

Basic Components

The insect integument (Fig. 2–1A) can be divided into three basic parts: one cellular layer, the *epidermis,* and two noncellular layers, the *basement membrane* (entad of the epidermis), and the *cuticle* (ectad of the epidermis). The epidermis secretes the cuticle and in some cases is thought to secrete the basement membrane, although a specialized type of blood cell may be responsible for this. Since the epidermis secretes the cuticle, it should not be surprising that it shows its greatest activity during the molting process. Interspersed among the epidermal cells are *dermal glands*, which also play a part in the secretion of the cuticle. There may also be a number of other types of glandular cells associated with the epidermis which have a variety of functions, such as secretion of defensive substances, silk, and scents. The basement membrane is generally $\frac{1}{2}$ micron or less in thickness and in electron micrographs appears as a rather amorphous granular layer. It is apparently composed of a mucopolysaccharide.

Fig. 2–1

Structure of the integument (diagrammatic). A. Section of generalized integument. B. Daily growth layers and lamellar patterns. C. Generalized epicuticle.

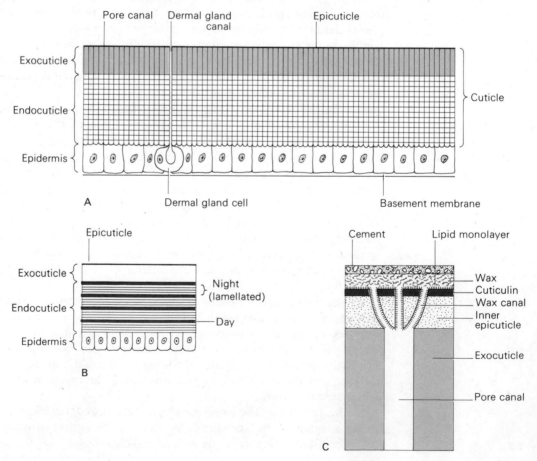

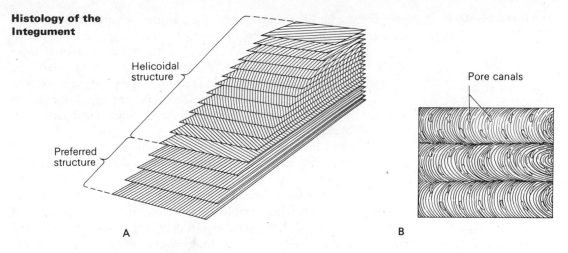

Helicoidal
structure

Preferred
structure

Pore canals

A

B

Fig. 2–2
A. Helicoidal and preferred structure of layers of endocuticle. The parallel
lines in each layer represent microfibrils. Note the parabolic effect in
transverse section. B. Transverse section of endocuticle showing parabolic
effect and appearance of pore canals as they pass through helicoidal layers.
(Redrawn with modifications from Neville, 1970.)

Cuticular Layers

Applying the techniques of light microscopy and various stains,
the insect cuticle can be seen to consist of a number of different layers.
Typically there are three major layers: a comparatively thick *endocuticle* (10 to 200 microns); an *exocuticle,* which may vary a great deal
in thickness; and a very thin *epicuticle,* somewhere in the range 0.03 to
4.0 microns thick. The endo- and exocuticles are often referred to
together as *procuticle.* The prefix of each of the cuticular layers desig-
nates the relative location of each, the endocuticle being the most
entad, followed by the exo- and epicuticles. The endocuticle may or
may not be pigmented, the exocuticle is usually pigmented, and the
epicuticle is unpigmented.

Under the light microscope, with polarizing filters, the endocuticle
can be seen to be composed of successive light and dark layers
(Fig. 2–1B), which in all insects studied thus far, except the beetles,
correspond to daily growth layers (Neville, 1970). The dark layers
are those deposited during the day and the lighter layers during the
night. The light layers are further subdivided into alternating layers of
light and dark lamellae. Also, striations running vertically through the
cuticle can be seen. The use of the electron microscope has clarified
somewhat the nature of these lamellae and striations. The lamellar
patterns are considered to be the result of the orientation of layers of
microfibrils of chitin embedded in a protein matrix. These microfibrils
are laid down in layered sheets (Fig. 2–2A). Within each sheet, the
microfibrils are parallel to one another, but in successive sheets they
are aligned at regularly changing angles. This arrangement (helicoidal)
is responsible for the parabolic patterning of endocuticle when cut
transversely and observed with the electron microscope (Fig. 2–2B).
In the dark growth layers deposited during the day, the microfibrils are
oriented in successive sheets in a "preferred" direction and hence do

not have the lamellar appearance of the lighter, nighttime layers. The vertical striations have been shown to be tiny tubes, or *pore canals* (Figs. 2–1A, C and 2–2B), which extend from the epidermal layer nearly to the external surface of the epicuticle. The pore canals are usually 1 micron or less in diameter, are ribbonlike in appearance, and apparently serve as connecting tubes between the cellular and the cuticular layers. They run a spiral course through the helicoidally arranged layers of endocuticle.

The epicuticle, despite its comparative thinness, is exceedingly complex and possesses characteristics that make it an extremely important layer of the cuticle. Although it appears as a very thin, sometimes indiscernible, line under the light microscope, the electron microscope has revealed a multilayered structure that is penetrated by wax canals 60 to 130 angstroms in diameter containing wax filaments (Locke, 1964). At least four layers, of varying thickness depending on the insect species, have been described in the epicuticle (Fig. 2–1C): an outer *cement layer,* sometimes called tectocuticle ("roof" cuticle), less than 0.1 micron thick; a *wax layer*; a *cuticulin layer;* and a *"homogeneous" inner epicuticle.* The cement layer is secreted by the dermal glands and may be similar to shellac, a substance secreted by a particular group of insects in the order Homoptera. This layer, outermost when present, probably determines the surface properties of the cuticle, that is, whether the cuticle will be water-repellent (hydrophobic) or water-attractant (hydrophilic) and also probably serves as a protective barrier for the more vulnerable layers beneath. In some insects a waxy bloom appears on the surface of the cement layer (Locke, 1964). The wax layer is thought to consist of an ordered monolayer of lipid directly associated with the cuticulin layer and a more ectad, less well-ordered lipid layer. Experimental evidence points to this layer as being responsible for the permeability characteristics of the cuticle. The cuticulin layer has been found in every insect in which the epicuticle has been investigated and covers the entire integumental surface, including the tracheoles and gland ducts. It is thought to be composed of lipoprotein. The innermost layer, the homogeneous inner epicuticle, is of unknown composition. The pore canals terminate beneath this inner layer of epicuticle; the dermal glands communicate with the surface of the epicuticle.

Chemical Composition of the Cuticle

Quantitatively speaking, the polysaccharide *chitin* and various structural proteins are the major cuticular constituents. Chitin (Fig. 2–3) is a high-molecular-weight polymer of *N*-acetyl-D-glucosamine with the empirical formula $(C_8H_{13}O_5N)_n$. It is in many ways quite similar to cellulose and may make up from one fourth to more than one half of the dry weight of the exo- and endocuticle. Chitin has not been found in the epicuticle. The cuticular proteins include *arthropodin,* a group of similar extractable proteins; *resilin,* a protein that forms a rubberlike framework and is appropriately found in skeletal articulations; and *sclerotin,* a tanned protein responsible for the hard, horny character of cuticle. It should be mentioned that chitin is not the agent responsible

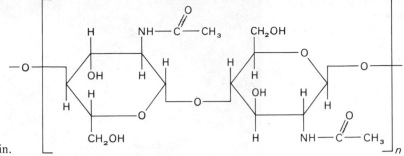

Fig. 2–3
Structural formula of chitin.

for the hardness of the cuticle, although it undoubtedly lends strength to it. In fact, highly sclerotized skeletal regions may contain considerably less chitin than softer membranous areas. Apparently the chitin chains are attached to one another by hydrogen bonds to form elongate micellae. In addition, it seems quite likely that chitin and protein are in intimate combination with one another. In a very few species calcium carbonate (lime) is responsible for the hardness of the external shell. Other constituents of the cuticle include polyhydric phenols and quinones, which play a role in the *sclerotization* (hardening) and *melanization* (darkening) processes; lipids of various sorts associated with the epicuticle; enzymes (nonstructural proteins), which catalyze the many complex biochemical reactions involved in molting and subsequent processes; and very small amounts of inorganic compounds.

Sclerotization

The hardening of insect cuticle is due to the tanning of protein, which follows the *eclosion* (emergence of an immature stage from the egg) or the ecdysis (shedding of the cuticle) and subsequent expansion of an insect. The protein is rendered hard, dark, and insoluble by the linkage of its adjacent polypeptide chains and the blocking of reactive groups by the tanning substances. The tanned protein is the sclerotin mentioned previously and is apparently the substance that fills the spaces between the chitin micellae and binds them together. The degree of hardness resulting from this process is highly variable. Segmental plates of lepidopteran larvae are nearly unsclerotized, whereas the mandibles of certain beetles are capable of biting through metals such as lead, tin, and copper.

Coloration

The coloration of the cuticle can be accounted for in two ways. It can be the result of various pigments present in the cuticle, epidermal cells, or blood and fat body, or by the physical characteristics of the cuticle. Some examples of insect pigments are the *carotenes*, variously colored (red, orange, yellow) pigments directly or indirectly derived from plant sources; *melanins* (very dark pigments); and the *pterins*,

A

B

which are the most widely distributed pigments among the insects. Physical coloration results from the light-breaking effect of minute cuticular striations or the cuticular layers acting as thin films. The brilliant metallic coloration found in such insects as the *Morpho* sp. butterflies and many species of beetles is an example of this type of

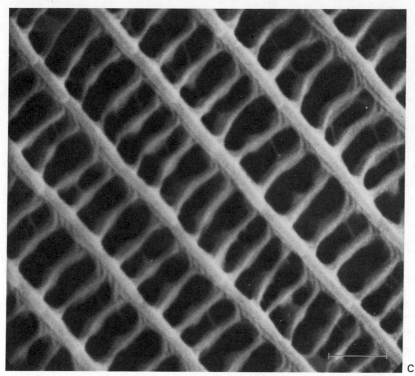

Fig. 2–4
Scanning electron micrographs of iridescent scales on a wing of the Buckeye butterfly showing the fine structure responsible for the light-breaking effect. 1, scale; 2, socket. (Scale lines: A, 20 microns; B, 2 microns; C, 1 micron.) (Courtesy of Walter J. Humphreys.)

C

coloration. Figure 2–4 shows the fine structure of iridescent scales from a wing of the Buckeye butterfly. Metallic coloration is usually due to physical characteristics; nonmetallic coloration may or may not be.

Permeability Characteristics

Insects, being essentially terrestrial animals, are continually faced with the problem of losing water, particularly those insects which live in extremely arid habitats. The generally diminutive size of insects makes this problem particularly acute, since transpiration (water loss) rate varies inversely with the ratio of surface area to volume. Thus the smaller an organism, the greater the amount of surface area per unit volume and hence the greater the tendency to lose water. During the course of their evolution this problem has lessened, partly through development of a relatively impermeable integument. Surprisingly enough, only a very small portion of the integument appears to be involved in the battle against water loss through transpiration.

Experiments such as the application of various organic solvents that are capable of dissolving a portion of the epicuticle have demonstrated that the abrasion, adsorption, or dissolution of the epicuticular layers causes an appreciable increase in the rate of transpiration, often resulting in the death of the insect. On the basis of these kinds of experiments, it is felt that the barrier to the exit of water from an insect lies in the epicuticle, particularly the wax layer. The transpiration rate in insects varies directly with temperature (Fig. 2–5).

However, this relationship does not produce a linear curve throughout its length. Above a certain temperature, depending on the species
of insect, within an approximate range between 28 and 60°C, the rate
of transpiration increases quite abruptly. The temperature at which
this occurs has been called the *critical* or *transition temperature*. The
explanation has been advanced that this abrupt change is brought
about by the disruption of the oriented lipid monolayer in the wax
layer of the epicuticle (Fig. 2–1C) at the critical temperature, which is
somewhat below the melting point of the wax. Permeability of the

Fig. 2–5
Rate of transpiration in relation
to temperature. A. From dead
insects. B. From a cockroach
nymph at constant saturation
deficiency (arrow indicates
critical temperature). +, air
temperature; •, cuticle
temperature. (A redrawn with
slight modifications from
Wigglesworth, 1945; B redrawn
with slight modifications from
Beament, 1958.)

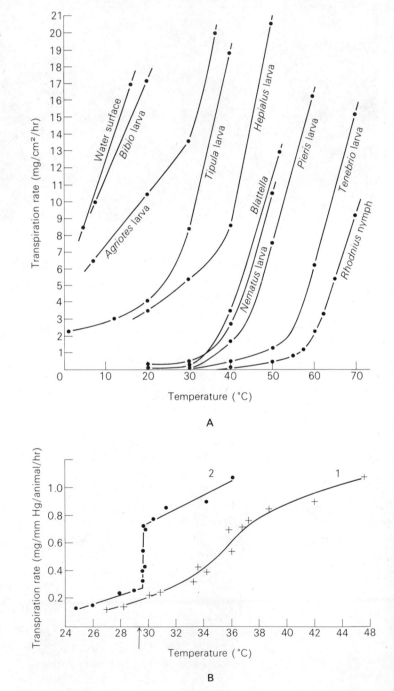

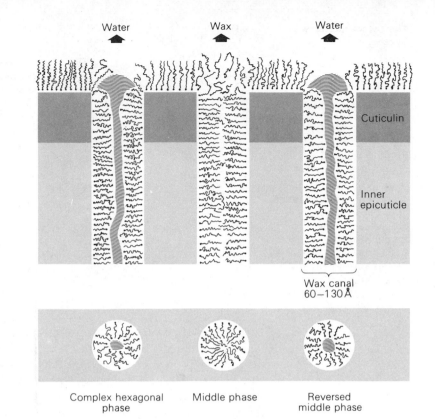

Water Wax Water

Cuticulin

Inner
epicuticle

Wax canal
60–130Å

Fig. 2–6
Lipid–water liquid
crystalline phases in wax
canals. (Redrawn from
Locke, 1965.)

Complex hexagonal
phase

Middle phase

Reversed
middle phase

insect cuticle to substances other than water may possibly be a
function of other layers of the epicuticle.

According to Locke (1965) the permeability of the integument to
water may vary at different times. He suggests that this variability is
explainable on the basis of phase changes in the polar molecules
(hydrophilic at one end) that comprise the wax filaments. These
filaments are thought to be lipid–water liquid crystals, which change
from the middle phase (Fig. 2–6), in which they impart imper-
meability, to the reversed middle phase or complex hexagonal phase,
in which they allow at least some passage of water. Phase changes
are supposed to occur in response to changes in environmental
temperature and humidity.

Molting and Ecdysis

In the introduction to the integumentary system, the periodic
shedding and resecretion of the cuticle was described as the solution
to the growth problem created by a more-or-less inflexible integument.
Molting and ecdysis have been the subject of numerous investigations,
and as a result a general description of these processes is possible
(Fig. 2–7). The cells directly involved are those of the epidermis and
the dermal glands. At the onset of molting, the epidermal cells show
much activity. They generally increase in size and may increase in
number. A rapid increase in cell number is commonly made possible
by a build-up of chromatin material, which results in a polyploid

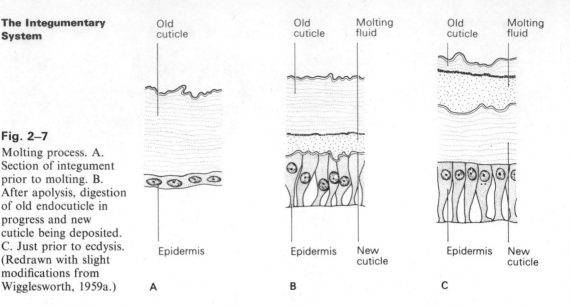

Fig. 2–7
Molting process. A. Section of integument prior to molting. B. After apolysis, digestion of old endocuticle in progress and new cuticle being deposited. C. Just prior to ecdysis. (Redrawn with slight modifications from Wigglesworth, 1959a.)

condition. At the time of increase in number, the cells may undergo rapid somatic reduction and return to the diploid condition. During the molting process the epidermal cells separate from the old cuticle and begin to secrete the new. This separation of old cuticle and epidermis is called *apolysis* (Hinton, 1966). At the same time the epidermal cells secrete a molting fluid that contains chitinase and protease capable of digesting the endocuticle, which may make up 80 to 90% of the old cuticle. In view of this, it may be said that a significant portion of the cuticle, although comparatively inert during the intermolt periods, never entirely leaves the metabolic pool of the insect.

Exo- and epicuticles are resistant to the action of the molting fluid and make up the portion of the integument that is shed at ecdysis. As the old cuticle is digested away, the new is laid down, cuticulin layer first, later accompanied by the exo- and endocuticles, the raw materials of which have come at least in part from the digestion of the old cuticle. The wax layers are laid down shortly before ecdysis, assuring the waterproofing of the newly emerged insect. The cement layer is the last secreted, forming very near the time of emergence. It is secreted

Fig. 2–8
Ecdysis (diagrammatic). A. Section of integument prior to molting. B. Section following digestion of endocuticle with line of weakness where endocuticle was previously in contact with epicuticle. C. Splitting of head capsule along preformed ecdysial line. (Redrawn from Snodgrass, 1960.)

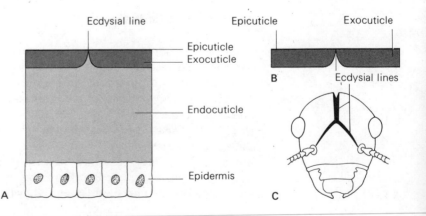

by the dermal glands. The pore canals apparently serve as routes for the secretion of the wax layer.

When the secretion of the new cuticle is complete, the insect emerges, leaving behind what remains of the old cuticle and the tracheal and gland duct linings. This process is facilitated by *ecdysial lines* beneath which only epicuticle and endocuticle are present (Fig. 2–8). Since the endocuticle is digested away during the molting process, a line of weakness develops. When ready to emerge, the insect may gulp air or water or take advantage of the hydrostatic pressure of the blood by contracting its abdomen. These actions exert an internal force on the ecdysial lines, and subsequently the old cuticle splits wherever they are located. These lines of weakness are usually located on the dorsum of the head and thorax with an anterior–posterior orientation.

External Integumentary Processes

The integument of various insects bears a great number of different external processes. Cursory examination of a number of these processes would seem to discourage attempts at trying to find an effective system of classification for them. However, Snodgrass (1935) points out that this is, in fact, a rather simple task since these processes can be classed in two groups (Fig. 2–9): noncellular and cellular. The former are composed entirely of cuticle and may take any of several forms, such as spines, ridges, or nodules. Cellular processses may be further broken down into multicellular and unicellular processes. The multicellular processes are hollow outgrowths of the integument and are lined with epidermal cells. They generally take the form of spines and are found, for example, on the hind tibiae of certain Orthoptera. The unicellular processes are all referred to as setae, although they show an extensive diversity of form. They are commonly hairlike, but may be flattened into scales (Fig. 2–4), bear branches and appear plumose, or take on other shapes. The shaft of a seta is formed by a protoplasmic outgrowth of a specialized hair-forming or *trichogen* cell. This projection is surrounded by a setal membrane and lies within a socket. The membrane and the socket are formed by a second cell, the *tormogen* or socket-forming cell. Setae have various functions,

Fig. 2–9
External
integumentary
processes.

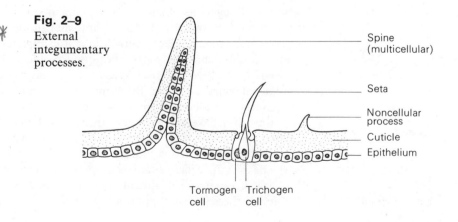

Spine (multicellular)

Seta

Noncellular process

Cuticle

Epithelium

Tormogen cell   Trichogen cell

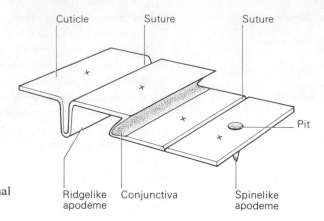

Fig. 2–10
Sclerites (denoted by ×), sutures, and internal
integumentary processes (diagrammatic).

For example, many are connected to a sensory nerve cell, some are
associated with a specialized gland that may secrete a toxic substance,
and when in the form of scales they often contain the pigment, or
possess the physical characteristics responsible for much of the
coloration and patterning.

The Insect Skeleton

The insect skeleton is composed of a series of plates, or *sclerites*,
of varying degrees of hardness (Fig. 2–10). These sclerites are separated
either by very soft, membranous areas called *conjunctivae,* or by
external grooves that indicate internal inflections or very narrow lines
of thin, flexible integument. The external grooves are referred to as
sutures or *sulci* (sing., *sulcus*). The combination of hardened plates
joined by conjunctivae allows body movement and articulation of the
appendicular joints. The internal inflections of the integument
denoted externally by sutures or pits are *apodemes* and may be
multicellular or unicellular. These inflections may be spinelike
(apophysis) or ridgelike and serve to strengthen the exoskeleton or as
points for muscle attachment.

Segmentation

The insect body, like that of other arthropods, is divided into a
series of segments. Examination of an insect will reveal that these
segments are rather different from one another in many respects.
Some bear appendages, others do not; the appendages of one segment
may be different from those of another; some segments may be quite
movable relative to the others; some segments are fused with
adjacent ones allowing little movement; and so on. One aspect
becomes immediately apparent—the obvious functional specialization
of the different segments. These specialized functions will be discussed
later, but first we want to consider briefly the hypothetical explanation
of the origin of these segments.

The primitive arthropod is thought to have been composed of a
series of essentially identical segments called *somites* or *metameres*.
Each of these somites apparently carried a pair of appendages, except

the anterior *acron,* or *prostomium,* and the posterior *telson,* or *periproct,* in which the posterior opening of the alimentary canal, the *anus,* was located. The mouth, the anterior opening to the alimentary canal, opened between the acron and the first postoral segment. During the course of insect evolution these segments fused in different ways, forming the variously divided body regions of modern arthropods, and the appendages took on specialized roles appropriate to the body region in which they were located, or disappeared altogether.

The segments in the primitive arthropod mentioned in the preceding paragraph were marked externally by constrictions of the integument (Fig. 2–11A). Internally, these constricted regions formed folds where the principal longitudinal muscle bands, the *segmental muscles,* were attached. This arrangement of body units with the intersegmental grooves forming the attachment points for longitudinal muscles has been called *primary segmentation.* This type of segmentation is found today among the soft-bodied, wormlike larvae of several insects—for example, the larvae of Lepidoptera—and in all arthropod embryos.

Another type of segmentation (Fig. 2–11B, C), *secondary segmentation,* is found in adult and many larval insects. In this type of segmentation, the membranous areas between adjacent segments do not coincide with the points of attachment of longitudinal muscles, but are slightly anterior to them. Secondary segmentation is considered to be a derivative of primary segmentation with the infolded region of a primary segment becoming sclerotized and a membranous area developing just anterior to it. Secondary segmentation is well illustrated by a pregenital abdominal segment of a generalized insect.

Fig. 2–11

Types of segmentation (tergites only). A Primary. B and C. Secondary (adjacent secondary segments overlapping in C). (Redrawn with modifications from Snodgrass, 1935.)

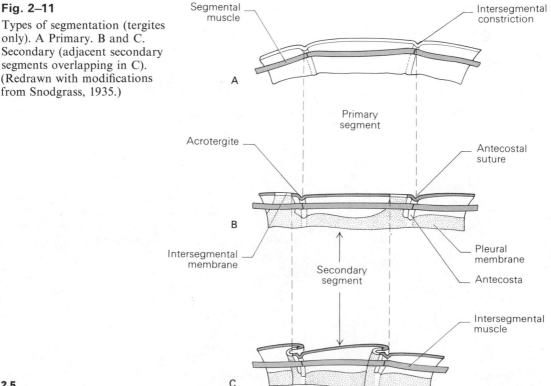

The typical abdominal segment consists of a dorsal plate, or *tergite* (Fig. 2–11B, C), and a ventral plate, or *sternite*. These plates are separated by a lateral membranous area, the *pleural membrane*. The external groove that corresponds to the infolded region forms an internal ridge, which, in some cases, may be quite pronounced. This internal ridge or apodeme is called the *antecosta*. A small sclerite is demarcated by the antecostal suture and the secondary intersegmental membrane, both in the tergite and the sternite. The sclerite associated with the tergite is the *acrotergite* and with the sternite, the *acrosternite*. Snodgrass (1935) points out the advantage of having muscles attached to hardened plates, since these plates can then have a protective function and also become involved in locomotion. Since muscle attachments in secondarily segmented insects do not coincide with the definitive segments, they are referred to as *intersegmental muscles*.

The General Insect Plan

Despite the many features that insects have in common with one another, they display a tremendous diversity in form. This being the case, it becomes necessary when considering insect structure, or morphology, to begin with a hypothetical, generalized form which subsequently will serve as a conceptual basis for interpreting variations in structure. In this section we want to discuss mainly the structure of a generalized insect.

Fig. 2–12
Generalized pterygote insect.

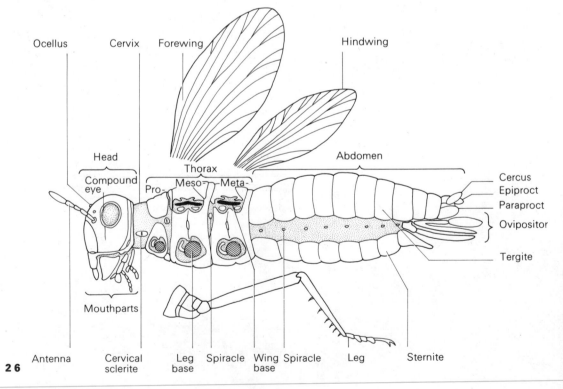

Tagmata

The segments of the insect body are divided into three well-defined regions, or *tagmata;* a head, a thorax, and an abdomen (Fig. 2–12). The head bears the organs of ingestion, or *mouthparts; compound eyes; simple eyes,* or *dorsal ocelli,* which may be lacking; and a pair of appendages called *antennae.* The thorax is composed of three basic segments named by their relative positions from anterior to posterior as follows: *prothorax, mesothorax,* and *metathorax.* All three thoracic segments bear legs in most immature and nearly all adult insects. Wings, if present, are found only on the meso- and metathorax. The class Insecta is usually divided into two subclasses on the basis of the presence or absence of wings as follows: *Apterygota,* primitively wingless insects; *Pterygota,* winged or secondarily wingless insects. The abdomen consists of a varying number of legless segments, the primitive number being considered to be 11 plus a terminal segment, the *periproct,* or *telson,* which bears the anus. The external genitalia are borne on one or more of the posterior segments.

Fig. 2–13

Generalized insect head.
A. Anterior view.
B. Posterior view.
C. Lateral view. D. Top of head capsule cut away to show tentorium. (Redrawn with slight modifications from Snodgrass, 1935.)

Head

The insect head (Fig. 2–13) is a composite structure that has evolved from the fusion of the prostomium with a number of postoral segments and modifications of the appendages of these segments into

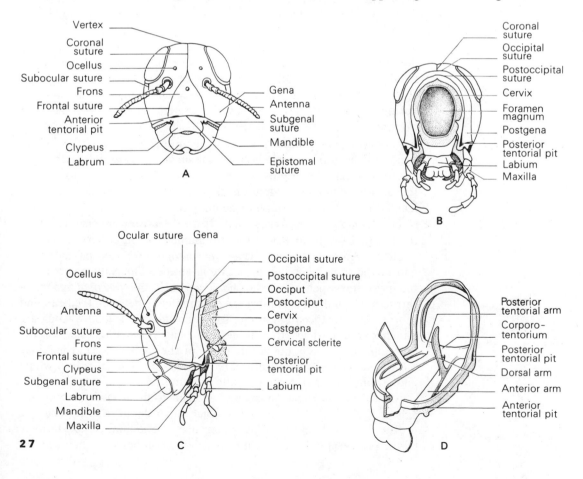

the organs of ingestion, the mouthparts. There is no general agreement as to the number of primitive segments that contributed to the evolution of the head. Opinions vary from four to six segments plus the acron. Basically, the head is composed of a hardened capsule, the *cranium,* which bears the antennae, eyes, and mouthparts. There is little evidence of segmentation in the mature insect head, and some of the lines or sutures that give it a segmented appearance are probably of functional significance rather than relating to any primitive segmentation. However, one of the sutures, the postoccipital suture of the cranium, is considered to have persisted, separating the maxillary and labial segments. The head is attached to the thorax by means of a membranous region, the *neck,* or *cervix.* The cervical membrane is quite flexible and allows movement of the head. The neck typically bears two pairs of lateral plates, the *cervical sclerites.*

Cranial Structure. The cranium is divided into various regions by a series of sutures. These sutures may be quite apparent in some insects and completely lacking in others. Therefore, this description will pertain to a generalized insect cranium and will mention sutures and regions that may or may not be readily discernable in a given insect. The first suture we want to consider is the *epicranial.* It is shaped like an inverted Y, the stem forming the dorsal midline of the cranium and the arms diverging ventrally across the anterior portion of the head. The stem of the Y is called the *coronal suture;* the arms form the *frontal sutures.* The frontal sutures are commonly obscure and often lacking altogether.

The region delimited by the arms is called the *frons* and dorsal portion of the cranium bissected by the coronal suture is the *vertex.* Snodgrass (1963a) explains that these three lines are really not sutures at all but lines along which the shed cuticle of the cranium splits during ecdysis. For this reason he prefers to call them *ecdysial cleavage lines.* In addition, since the ecdysial cleavage lines are poorly developed in many insects, a better definition of frons is the frontal region of the cranium, which bears the *median ocellus,* if present, and internally bears the origins of the muscles of the anterior mouthpart, the *labrum.* Using this definition, the ecdysial cleavage lines would be located on the frons instead of defining it.

The *occipital suture* forms a line from the posterior termination of the coronal suture to just above the mandibles on either side of the cranium. Another suture, the *postoccipital suture,* lies posterior to, and in the same plane as the occipital suture. This suture surrounds the posterior opening of the head capsule, the *foramen magnum,* through which the internal organs communicate between the head and thorax. The postoccipital suture forms an internal ridge to which are attached the muscles that move the head. On either side of the cranium immediately above the bases of the mandibles and maxillae are the *subgenal sutures.* Internally, these sutures form ridges which add strength to that portion of the cranium.

Above each subgenal suture and anterior to the occipital suture is a "cheek," or *gena.* A *postgena* lies adjacent to the gena, but posterior to the occipital suture. The postgenae are delimited posteriorly by the postoccipital suture. Dorsally, the region between the occipital and postoccipital sutures is called the *occiput*. The plate posterior to the postoccipital suture, which surrounds the better part of the foramen magnum, is the *postocciput*. This structure generally bears a bilateral pair of processes upon which the cervical sclerites articulate. The subgenal sutures may be connected across the front of the cranium, just beneath the frontal sutures, by the *epistomal suture*. In a fashion similar to the postoccipital and subgenal sutures, the epistomal suture forms an inflected ridge.

A lobelike structure, the *clypeus,* lies immediately ventral to the epistomal suture. In many insects, the clypeus is hinged with the foremost mouthpart, the labrum. The compound eyes are commonly surrounded by the *ocular sutures*. Similarly, an *antennal suture* surrounds the base of each antenna. In addition, *subocular sutures* may be present, running vertically beneath the compound eyes.

Tentorium.   At four points on the cranium, the cuticle forms an armlike inflection. Internally these inflections join, forming a framework. This internal framework is the *tentorium* (Fig. 2–13D) and affords many points for muscle attachment and contributes appreciably to the rigidity of the head capsule. Externally the points of inflection form the anterior and posterior *tentorial pits*. The anterior pits are generally contained in the epistomal suture, although in some insects they may be more closely associated with the subgenal sutures. The posterior tentorial pits open bilaterally in the postoccipital suture immediately adjacent to the subgenal sutures. Internally the tentorium is composed of a pair of anterior arms, which arise from the anterior pits and fuse with the posterior arms, which correspondingly arise from the posterior pits. At the point of fusion, a transverse bar, the *tentorial bridge,* or *corporotentorium,* is formed. In addition, a third pair of arms may arise dorsally from the anterior arms.

Compound Eyes and Ocelli.   On either side of the head capsule are located the compound eyes, which are composed of a number of usually hexagonal facets or *corneal lenses*. These are the cuticular parts of the eye units, or *ommatidia*. The dorsal ocelli, or simple eyes, are usually three in number and are located on the anterior portion of the cranium, one on either side of the coronal suture and the third between the frontal sutures.

Antennae.   The antennae are paired appendages that articulate with the head capsule and are located on the anterior portion near the compound eyes. Typically insect antennae are composed of a series of segments, and the generalized form (Fig. 2–14) would be filament-like in appearance. There are three basic parts of an antenna in most insects, a basal *scape*, a *pedicel*, and a distally located *flagellum*, which is usually long and composed of a number of subsegments. The pedicel in most insects contains *Johnston's organ*, a special sensory structure that will be discussed later. The scape articulates in an antennal socket, the integument between the antenna and head capsule being membranous and flexible. The rim of the antennal socket associated with the cranium characteristically contains an articular point, the *antennifer*.

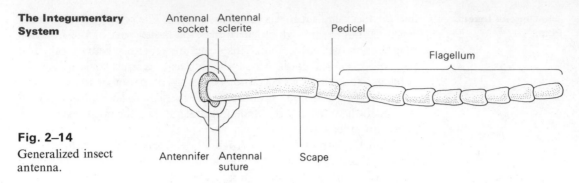

Fig. 2–14
Generalized insect antenna.

The antennae are moved about by muscles inserted on the base of the scape.

Mouthparts. The paired mouthparts apparently arose from the appendages of three of the segments posterior to the acron of the primitive arthropod ancestor. The mouthparts of the generalized mandibulate type (Fig. 2–15) are considered to be the primitive form. They typically consist of an anterior "upper lip", or *labrum*; the *hypopharynx;* a pair of *mandibles;* a pair of *maxillae;* and a posterior "lower lip," or *labium*. The labrum is suspended from and articulates with the clypeus by a narrow membrane, which allows considerable movement. The integument of the labrum forms a lining that runs dorsally across the inside of the clypeus and terminates at the true

Fig. 2–15
Generalized mandibulate mouthparts as illustrated by a cockroach. (Redrawn from James and Harwood, 1969.)

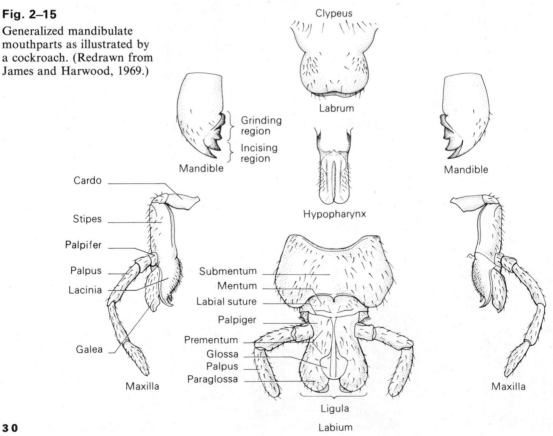

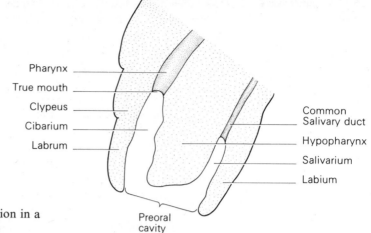

Fig. 2–16

Sagittal section of mouthparts region in a
generalized insect (diagrammatic).

mouth. This epipharyngeal wall or palate forms the dorsal lining of
the *preoral* or *intergnathal cavity* formed by the mouthparts and is
continuous with the lining of the pharynx. Commonly the epipharyngeal wall bears a lobe referred to as the *epipharynx*. The *hypopharynx*
lies in the preoral cavity somewhat in the way a tongue does (Fig.
2–16). Generally the duct from the salivary glands opens between the
hypopharynx and the epipharyngeal wall. The portion of the preoral
cavity between the hypopharynx and labrum is the *cibarium*. The
cavity between the hypopharynx and labium forms the *salivarium*.

The mandibles (Fig. 2–15) are a pair of highly sclerotized, unsegmented jaws each of which articulates with the cranium at two
points, an anterior secondary articulation near the tentorial pit and a
posterior articulation with the postgena. Each mandible has a proximal molar or grinding region and a distal incisor or cutting region.
The maxillae are somewhat more complex than the mandibles, being
paired, segmented structures. Each is composed of a proximal segment, the *cardo*, which bears the *stipes*, which in turn bears two
distal lobes, the *galea* and *lacinia*. The galea, a comparatively unsclerotized lobe, is located lateral to the mesal lacinia, which is more
sclerotized and contains teeth on its inner edge. In addition, the
stipes also bears a lateral sclerite, the *palpifer* to which is attached
a five-segmented maxillary palpus.

The labium is a composite structure formed from the fusion of two
primitive segmental appendages, and its parts can be seen to correspond to those of the maxillae. It consists of a basal *postlabium*
attached to the cervix ventral to the foramen magnum and is quite
closely associated with the postoccipital region near the posterior
tentorial pits. The postlabium is commonly divided transversely into
two portions, a proximal *submentum* and a distal *mentum*. The *prelabium* is hinged to the postlabium by the labial suture. It is composed of a basal *prementum*, which bears laterally a pair of segmented
labial palpi and distally four lobes, two inner lobes, the *glossae*,
located between two outer lobes, the *paraglossae*. The labial palpi
are attached to lateral sclerites on the prelabium, the *palpigers*.

The muscles responsible for the movement of the mouthparts are
attached at various points on the head capsule and tentorium and on

the appendages described in the preceding paragraph. One needs only to observe the ingestatory activities of a grasshopper or a cockroach to appreciate the intricate, highly coordinated movements of which the mouthparts are capable.

Thorax

As mentioned previously, the insect thorax is composed of three segments: an anterior *prothorax*, a *mesothorax*, and a posterior *metathorax*. Functionally, the thorax is the locomotive tagma since it bears the legs and, if present, the wings. Wings are borne on either or both the meso- and metathoracic segments. These two segments are thus often referred to collectively as the *pterothorax* (ptero = wing). In most winged insects the prothorax is usually separate from the remaining segments and somewhat less developed. In many insects at least part of the first abdominal segment has become intimately associated with the thorax, and in many of the Hymenoptera it has literally become a part of the thorax, being separated from the rest of the abdomen by a constriction.

Each thoracic segment typically can be divided into four distinct regions (Fig. 2–17A): a dorsal tergum, or *notum;* a pair of bilateral *pleura* (sing., *pleuron*); and a ventral *sternum*. Each of these regions is commonly subdivided into two or more sclerites. The legs arise on the pleura; the wings articulate between the notal and pleural regions. Spiracles (Fig. 2–12), the external openings of the ventilatory system, are usually found, one on each side as follows: the mesothoracic spiracles in the pleural regions between the pro- and mesothorax, and the metathoracic spiracles between the meso- and metathorax. Prothoracic spiracles are atypical.

Fig. 2–17
A. Cross section of a generalized thoracic segment. B. Sternal apophyses in the form of furcae (diagrammatic).

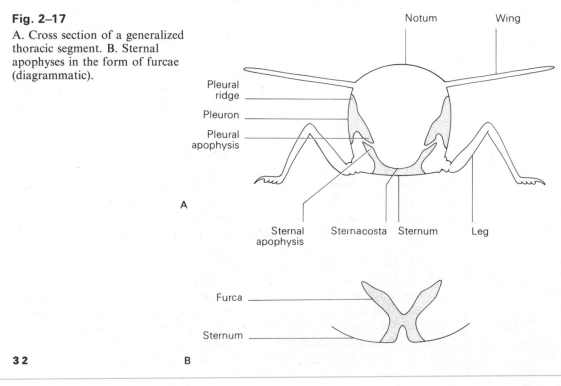

Matsuda (1970) should be consulted for an up-to-date, rigorous treatment of the morphology of the insect thorax.

Thoracic Terga.　The thoracic terga (nota) of apterygote and some immature pterygote insects are rather simple when compared to the more modified terga of the adult winged forms. Each is merely a single plate associated with the remaining two by secondary segmentation. The dorsal longitudinal muscles of the thorax are attached to the antecostae as described in the discussion of secondary segmentation in an abdominal segment. The terga of most pterygote insects are more complex, being divided into smaller sclerites by various sutures. These divisions of the terga, as well as the other modifications of the

Fig. 2–18

Generalized pterygote insect thorax. A. Notum. B. Pleuron. C. Sternum. (A and B redrawn from Snodgrass, 1935, A with modifications.)

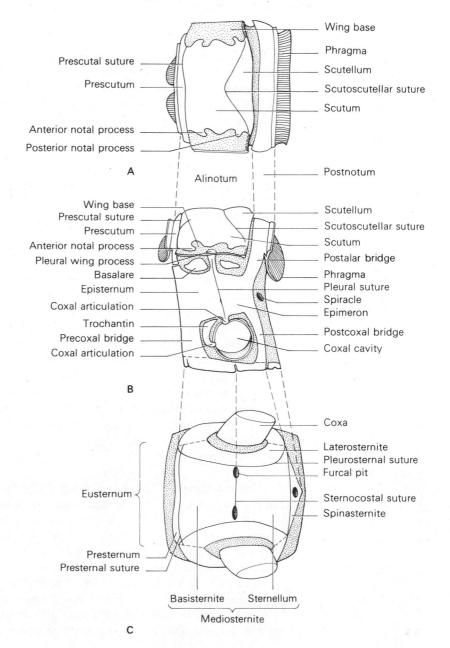

A

Prescutal suture
Prescutum
Anterior notal process
Posterior notal process
Wing base
Phragma
Scutellum
Scutoscutellar suture
Scutum
Alinotum
Postnotum

B

Wing base
Prescutal suture
Prescutum
Anterior notal process
Pleural wing process
Basalare
Episternum
Coxal articulation
Trochantin
Precoxal bridge
Coxal articulation
Scutellum
Scutoscutellar suture
Scutum
Postalar bridge
Phragma
Pleural suture
Spiracle
Epimeron
Postcoxal bridge
Coxal cavity

C

Coxa
Laterosternite
Pleurosternal suture
Furcal pit
Sternocostal suture
Spinasternite
Eusternum
Presternum
Presternal suture
Basisternite
Sternellum
Mediosternite

33

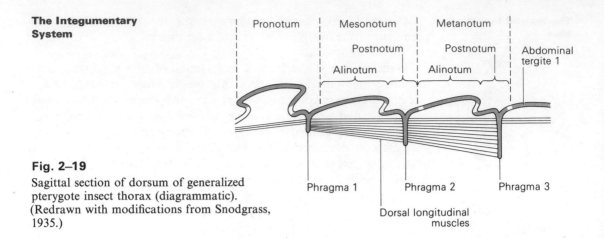

Fig. 2–19
Sagittal section of dorsum of generalized
pterygote insect thorax (diagrammatic).
(Redrawn with modifications from Snodgrass,
1935.)

generalized thoracic segment, arose as a result of the evolution of
wings and flight. Typically the tergum of a wing-bearing segment is
composed of two parts, an *alinotum* and a *postnotum* (Figs. 2–18A and
2–19). The alinotum bears the wings; the postnotum bears the
internally inflected *phragma,* which is the modified antecosta of the
next segment posterior. In the case of the metathorax this would be
the antecosta of the first abdominal segment. The phragmata are more
platelike than ridgelike and afford a comparatively large surface area
for the attachment of the dorsal longitudinal wing muscles, which are
extremely important in the flight of insects.

According to Snodgrass (1935) the intersegmental membrane
present in the more primitive apterygote insects is reduced or lacking
and the acrotergite of the next segment posterior has become the
postnotum. Correlated with the presence of wings is a strengthening
of the alinotum afforded by various internally inflected ridges.
Although there are different sutures associated with these internal
ridges in various insects, one, the *scutoscutellar suture* (Fig.2–18A) is
present in nearly all winged forms. It lies in the posterior portion of
the alinotum and it is shaped somewhat like a V, with the bottom part
directed anteriorly. This suture divides the alinotum into an anterior
scutum and a posterior *scutellum.* Many insects also have a transverse or *prescutal suture,* which lies on the anterior part of the alinotum
dividing off a small anterior plate, the *prescutum.* Processes that serve
as articular points for the wings have developed on the lateral margins
of the scutum. These are the *anterior notal wing process* and the
posterior notal wing process.

Thoracic Pleura. The pleural region in the apterygotes is composed
of a group of subcoxal sclerites associated with the *coxa,* the basal
segment of the leg. On the basis of comparative morphological
studies it has been theorized that the primitive subcoxal sclerites were
three in number (Fig. 2–20), a dorsally located *anapleurite,* forming a
rather crude semicircle around the coxa; a smaller *coxopleurite,*
concentric with and ventral to the anapleurite; and a ventral *sterno-*
pleurite. Presumably the coxopleurite and the sternopleurite each
carried a process with which the coxa articulated. With the advent of
wings, the anapleurite and coxopleurite apparently fused, forming a

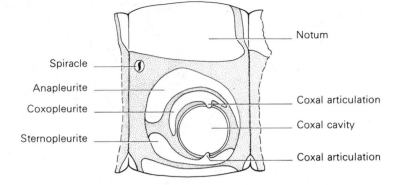

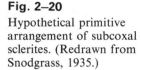

Fig. 2–20

Hypothetical primitive
arrangement of subcoxal
sclerites. (Redrawn from
Snodgrass, 1935.)

Notum

Spiracle

Anapleurite

Coxopleurite

Coxal articulation

Coxal cavity

Sternopleurite

Coxal articulation

sclerotic pleuron, which assumed the role of supporting the ventral
part of the wing by means of the *pleural wing process*. The ventral
sternopleurite is thought to have fused with the sternum, forming a
continuous sclerotic body beneath the notum and the wings.

A vestige of the coxopleurite, the *trochantin* (Fig. 2–18B), is present
in many of the more generalized wing-bearing insects. When present
it usually bears one of the points of articulation of the coxa. A second
articular process is generally located at the ventral extremity of the
pleural suture. In pterygote insects an internal inflection, denoted
externally by the pleural suture, divides the pleuron of each thoracic
segment into two parts, an anterior *episternum* and a posterior
epimeron (Fig. 2–18B). Internally this inflection forms the pleural ridge
(Fig. 2–17A), which gives additional strength to the pleuron and which
bears, in pterothoracic segments, the *pleural apophysis,* an armlike
projection that is directed ventrally and is usually associated with a
sternal apophysis (Fig. 2–17A). The prefixes pro-, meso-, and meta-
are commonly used in combination with epimeron and episternum.
Thus the epimeron of the mesothorax becomes the mesepimeron, and
so on. The postnotum of a wing-bearing segment is usually united
with the epimeron, forming the *postalar bridge*. The pleuron is usually
supported ventrally by the *precoxal* and *postcoxal bridges*, which are
fused with the sternum.

Thoracic Sterna. The morphology of the thoracic sternum of
both apterygote and pterygote insects has not as yet been completely
clarified, particularly in the more advanced forms. Supposedly in
pterygote insects the sternum of a typical wing-bearing segment has
been formed from the fusion of four distinct sclerites (Fig. 2–18C):
(1) a large segmental ventral plate, the *mediosternite*; (2) the *coxo-
sternites*, ventral to the coxae; (3) the *laterosternites,* or *pleurosternites*,
which are separated by the *pleurosternal suture*; and (4) a small
intersegmental sclerite which bears an internal median process, the
spina, and is hence referred to as the *spinasternite*. The pleurosternal
suture may be obscure or lacking. The mediosternite and pleurosternites together form the *eusternum*. A pair of internal projections,
the *sternal apophyses,* arise from the eusternum (Fig. 2–17). Externally, these apophyses are indicated by the presence of two *furcal
pits* (Fig. 2–18C).

A ridge, the *sternacosta* (Fig. 2–17A), is usually present between the
anterior edges of the bases of the sternal apophyses and extends

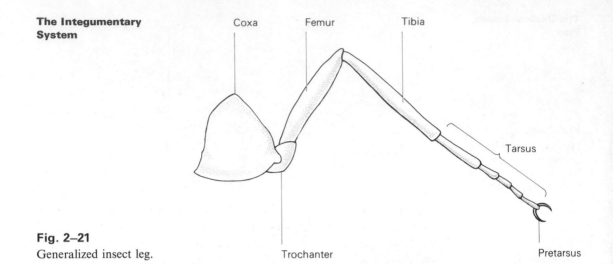

Fig. 2–21
Generalized insect leg.

laterally across the pleurosternites. This ridge is indicated externally by an inflection line, the *sternocostal suture*. This suture then divides the mediosternite into an anterior *basisternite* and a posterior *sternellum* or *furcasternite*. The sternal apophyses are often in the form of a V diverging from the sternacosta (Fig. 2–17B). In this case they are termed *furcae* (sing., *furca*). Furcalike structures also arise from the sternites of some apterygotes. The various internal inflections of the sternites, of course, lend both strength and areas for muscle attachment in the thorax. There may be a presternal suture delineating an anterior sclerite of the eusternum, the *presternum*. In the higher insects, the sterna, as is the case with the other thoracic sclerites, undergo extensive modification, involving, for example, the apparent fusion of sclerites recognized in the more generalized forms and formation of secondary sutures.

Legs

The generalized insect leg consists of six segments as follows (Fig. 2–21): a basal *coxa*, which articulates with the thorax in the pleural region; a small *trochanter*; a *femur*; a *tibia*; a segmented *tarsus*; and a *pretarsus*, which usually bears a pair of movable claws. The legs are usually looked upon as representing the principal organs of terrestrial locomotion, although, as will be seen, they have undergone many modifications and have been adapted to a wide variety of functions, including swimming, prey capture, and digging.

Wings

The wings arise as outgrowths of the integument between the tergal and pleural sclerites (Fig. 2–17A). They are thus composed of two layers of integument (Fig. 2–22). A series of tracheae grow between these integumentary layers, and when a wing is fully developed, they run within the longitudinal and transverse supportive framework, the *wing veins*. The cuticle is often thicker in the region of these veins, lending further rigidity. Since the wings are outgrowths of the integu-

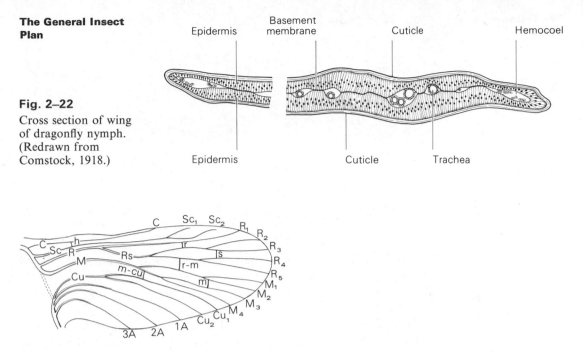

Fig. 2–22
Cross section of wing of dragonfly nymph. (Redrawn from Comstock, 1918.)

Fig. 2–23
Hypothetical primitive pattern of wing venation. Longitudinal veins: A, anal; C, costa; Cu, cubitus; M, media; R, radius; Rs, radial sector; Sc, subcosta. Cross veins: h, humeral; m, medial; m-cu, mediocubital; r, radial; r-m, radiomedial; s, sectorial. (Redrawn from Comstock, 1918.)

ment, the space between the epidermal layers is continuous with the body cavity, or *hemocoel,* of the insect. This space is usually evident only around the veins, since in the other parts of the wing, the cells, for example, the two layers of integument become closely appressed to one another. In many instances, blood cells, or *hemocytes*, can be seen circulating in the wing immediately on either side of a vein. Complete circulation of hemolymph in the wing is made possible by the presence of numerous cross veins in many insects. The wings are, of course, the organs of aerial locomotion in most cases, and like the legs have undergone extensive adaptive modification. Comparative studies of wing venation in many species have led to the development of various hypothetical generalized patterns. One such pattern is given in Fig. 2–23. The patterns of wing venation in different insects, interpreted on the basis of a generalized pattern, are extremely useful in the identification of many insects.

Abdomen

The segmentation of the insect abdomen has already been discussed. As previously mentioned, the primitive number of abdominal segments is considered to have been 11 true metameres plus a terminal segment that contained the anus, the periproct or telson. The tendency in insectan evolution has been toward a reduction in the number of segments, and in the generalized insect abdomen (Fig. 2–24) there are 11, the eleventh being reduced and divided into lobes that surround the anus. This terminal segment may bear a pair of appendages, the *cerci*. These are considered to be serially homologous with the legs and mouthparts. The plates of the eleventh segment are generally three in number, one being located dorsal to the anus, the *epiproct*, and one on either side of the anus, the *paraprocts*. The abdominal segments are

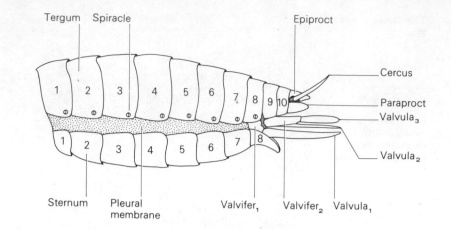

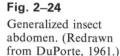

Fig. 2–24

Generalized insect abdomen. (Redrawn from DuPorte, 1961.)

usually numbered from anterior to posterior, number one being immediately posterior to the metathorax. In the generalized female pterygote insect modified appendages of the eighth and ninth abdominal segments form the *ovipositor*, or egg-laying apparatus, which is composed of two pairs of basal *valvifers*, which in turn bear the *valvulae*, one pair on the eighth and two pairs on the ninth. The male external copulatory apparatus is usually borne on the ninth abdominal segment. A pair of spiracles, the external openings of the ventilatory system, is typically found one on either side of the first eight abdominal segments.

Variations on the General Insect Plan

One of the major reasons for the tremendous success of insects has undoubtedly been the seemingly endless potential of the basic plan to undergo evolutionary modification. It appears as though no structure has escaped modification in some way or another. Because of this fact, insects are excellent organisms with which to demonstrate the phenomenon of adaptation. When considering modifications of anything, one needs a base from which to interpret a given modification. The external structure of a hypothetical, generalized insect has been presented to subserve this purpose. The remainder of this chapter will be devoted to the presentation of some of the more obvious variations in the insect skeleton.

Modifications of the Head

Antennal Variations. Although insect antennae vary greatly in length, overall size, size of the individual segments, segmentation, setation, and other aspects, they can usually be described as being of a particular type or combination of types (e.g., capitate–lamellate). Antennal type is commonly of value in the identification of a given insect to family and may, in certain instances, serve as a basis for differentiating the sexes, as in most male and female mosquitoes. Male mosquitoes bear distinctly plumose antennae; the female's antennae are less feathery in appearance and bear comparatively few whorls of hairs. This is an example of *sexual dimorphism*, a structural difference between the two sexes. Most modifications of the antennae occur in the flagellum. Commonly occurring antennal types with examples of insects possessing them are presented in Figure 2–25.

Compound Eyes and Ocelli. Compound eyes are commonly lacking altogether. For example, many larval forms, including certain dipteran larvae (maggots), the soldier castes of some termite species,

Fig. 2–25

Antennal types. A. Filiform, grasshopper. B. Moniliform, wrinkled bark beetle. C. Capitate, skin beetle. D. Clavate, carrion beetle. E. Setaceous, dragonfly. F. Serrate, click beetle. G. Pectinate, fire-colored beetle. H. Plumose, male mosquito. I. Aristate, flesh fly. J. Stylate, horse fly. K. Lamellate, scarab beetle. L. Flabellate, cedar beetle. M. Geniculate, honey bee.

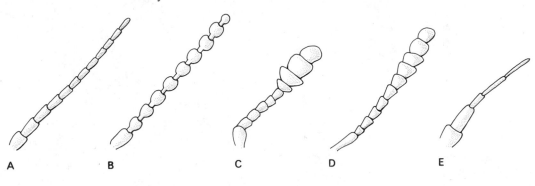

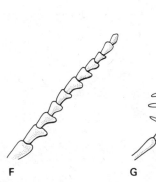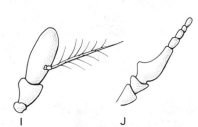

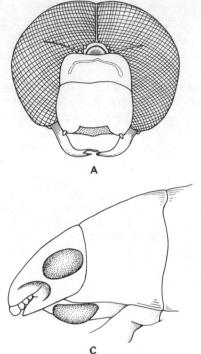

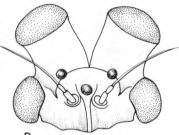

Fig. 2–26

Variations in compound
eyes. A. Dragonfly.
B. Thrips. C. Whirligig
beetle. D. Mayfly,
Cloeon sp. E. March
fly, *Bibio* sp. F. Blow
fly, *Phormia* sp., male.
G. Blow fly, female.
(A redrawn from
Snodgrass, 1954;
B redrawn from Essig,
1958; C redrawn from
CCM General
Biological, Inc., key
card; F and G
redrawn from Folsom
and Wardle, 1934.)

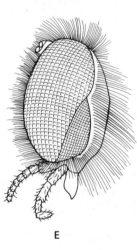

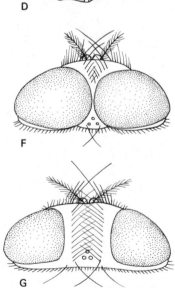

some species of fleas, and certain species of springtails all lack compound eyes. When present (Fig. 2–26), compound eyes occur in many diverse, sometimes bizarre forms. Some are exceedingly large and contain many facets, as in adult dragonflies (Fig. 2–26A); others are quite small and have few facets, as in many hemipterous insects (Fig. 2–26B). In some insects, the compound eyes are actually divided, appearing as two pairs. Examples are found among the beetles in the family Gyrinidae (Fig. 2–26C) and certain members of the family Cerambycidae, and among mayflies in the genus *Cloeon* (Fig. 2–26D). In the latter the anterior division is borne upon a stalklike outgrowth of the head capsule. The compound eyes of males of the flies in the

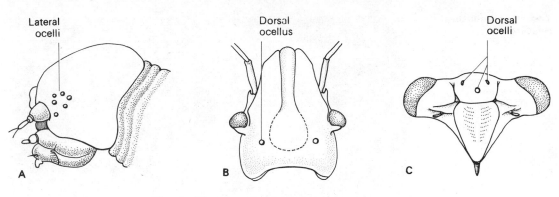

Fig. 2–27

Ocelli. A. Lateral ocelli (stemmata) of a caterpillar. B. Two dorsal ocelli, stink bug. C. Three dorsal ocelli, cicada. (A redrawn from Snodgrass, 1961; B redrawn with modifications from Snodgrass, 1935; C redrawn from CCM General Biological, Inc., key card.)

genera *Bibio* (Fig. 2–26E) and *Simulium* have two rather distinct areas of different-sized facets, giving the appearance of a compound compound eye. The compound eyes, like the antennae, may be involved in sexual dimorphism, as is the case in the fly *Phormia regina*. In the male fly the compound eyes are large and meet at the dorsal midline of the head; in the female they are smaller and do not meet at the dorsal midline of the head. (Fig. 2–26F, G).

In many larval and adult insects, compound eyes are lacking and in their place are the *stemmata* or *lateral ocelli* (Fig. 2–27A). In the larvae of insects with complete metamorphosis, the lateral ocelli are the precursors of the compound eyes of the adult. In other insects (e.g., springtails, silverfish, fleas) lateral ocelli are the only "eyes" of the imaginal stage; compound eyes never appear. The number of ocelli is quite variable, some insects having as few as one on each side of the head (e.g., many fleas) and others having many (e.g., 12 in *Lepisma* and 50 in some adult Strepsiptera).

Many insects possess simple eyes in addition to the compound eyes, the *dorsal ocelli*. When present (Fig. 2–27B, C), they vary in number from one to three: one is very uncommon; two are found in most Heteroptera, in many hymenopterous larvae, and in several other insects; and three occur in many members of the suborder Homoptera and others. Ocelli may also vary in position. For example, the most common position for three ocelli is one in each of the angles formed by the ecdysial cleavage lines. In members of the genus *Perla* in the order Plecoptera all three lie within the angle formed by the arms of the Y formed by the ecdysial cleavage line.

Mouthpart Structure. Upon examination of the mouthparts of various insects, it seems as though there are nearly as many variations in mouthpart structure and function as there are different feeding situations. The mouthparts discussed previously are considered to represent the primitive condition since they are found in the more generalized insectan forms and in most cases show greater resemblance to primitive locomotor appendages from which, based on comparative morphological and ontogenetic data, they were derived. Specialized

A

B

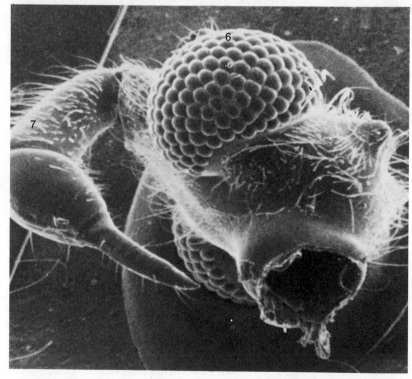

Fig. 2–28
Scanning electron micrographs of mandibulate and haustellate mouthparts. A. Head of a termite, mandibulate type. B. Head of the spider bug, *Stenolemus* sp., haustellate type. 1, Labrum; 2, mandible; 3, maxillary palp; 4, labial palp; 5, antenna; 6, compound eye; 7, beak (contains stylets). Highly magnified. (Courtesy of Walter J. Humphreys.)

mouthparts can, in most instances, be homologized with the generalized type.

Mouthparts can be very broadly classified into two groups, *mandibulate* and *haustellate* (Fig. 2–28). The mouthparts already discussed represent the mandibulate type. An outstanding characteristic of this type is the presence of a pair of well-developed, usually highly sclerotized, mandibles, which articulate at two points with the

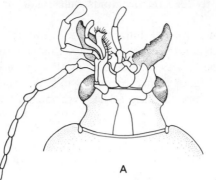

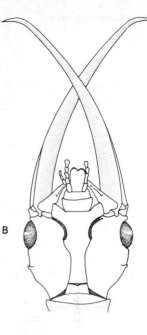

Fig. 2–29

Variations in form of mandibles. A. Ground beetle, *Calosoma* sp. B. Male dobsonfly, *Corydalus cornutus*. C. Antlion larva, *Myrmeleon* sp. (A redrawn from Essig, 1958; B redrawn from Packard, 1898; C redrawn from Peterson, 1951.)

head capsule and are capable of lateral movement (Fig. 2–28A). Mandibulate mouthparts are generally adapted to chewing activities, the mandibles acting as cutting and grinding structures. However, there are many exceptions to this (Fig. 2–29). For example, in many predaceous beetles and ants, they are elongate, grasping structures, well adapted for catching and holding prey (Fig. 2–29A). Similarly developed mandibles in the male dobsonfly hold the female during copulatory activities (Fig. 2–29B). In some insects, such as antlion larvae (Fig. 2–29C), the maxillae and mandibles are elongate and grasping and together form a food channel through which the body fluids of prey are sucked. Although these particular mouthparts are functionally sucking, they are obvious modifications of the chewing, mandibulate type. In pollen-feeding and dung-feeding beetles, the mandibles are more or less flattened and serve to mold dung or pollen into small pellets or balls.

Haustellate mouthparts (Fig. 2–28B) are generally adapted for sucking activities of various sorts. Many are characterized by the presence of *stylets* (Fig. 2–30), which are swordlike or needlelike modifications of one or more of the generalized mouthpart structures. Stylets may be formed from a combination of one or more of the mouthparts and the hypopharynx. Stylets enable the insects that possess them to pierce or at least abrade plant or animal tissues and subsequently feed on the fluids that exude or are pumped from the host.

However, not all haustellate mouthparts have piercing stylets. Three outstanding exceptions to this are the mouthparts found in most butterflies and moths, in the nonbiting muscoid flies, and in many higher hymenopterous insects. The mouthparts of each of these groups, lacking stylets, are incapable of the penetration of

Fig. 2–30

Examples of stylate–haustellate mouthparts. A. Sagittal section of head of a sucking louse. B, C, and D. Mouthparts spread out to show details of structure: B, mosquito; C, cicada; D, flea. (A after Vogel, 1921; B redrawn from Snodgrass, 1959; C redrawn from Snodgrass, 1935; D redrawn with slight modifications from James and Harwood, 1969.)

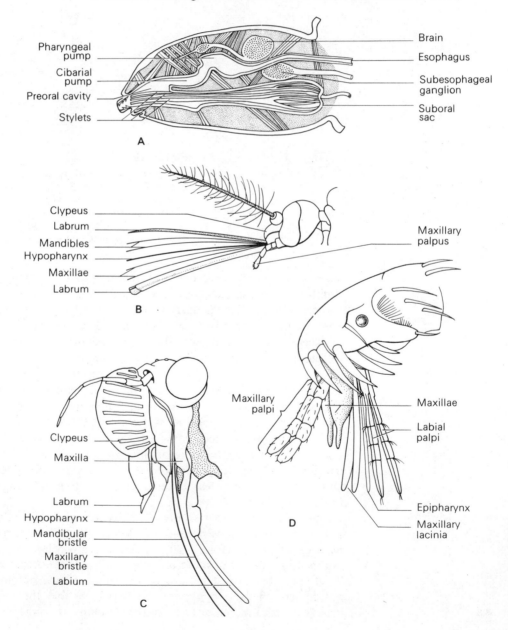

tissues. These insects are then obliged to feed on exposed fluids or soluble solids of various sorts, such as sugar. In the vast majority of butterflies and moths, an elongate sucking tube or proboscis is formed from the galeae of the maxillae (Fig. 2–31A). The remaining mouthparts are reduced or absent. These insects feed mainly on flower nectar. When inactive, the proboscis is coiled beneath the head. This type of mouthpart structure and method of feeding is commonly referred to as *siphoning*. The nonbiting muscoid flies have a rather peculiar method of feeding, often referred to as *sponging* (Fig. 2–31B). A basal segment, the *rostrum*, which is made up of a part of the clypeus and basal portions of the maxillae, bears distally a fleshy, retractile proboscis that represents the labium. The apical portion of this proboscis bears the spongelike *labella*, in which are located many tiny channels, which ultimately converge into the food channel formed by the labrum–epipharynx and hypopharynx. The labella are capable of

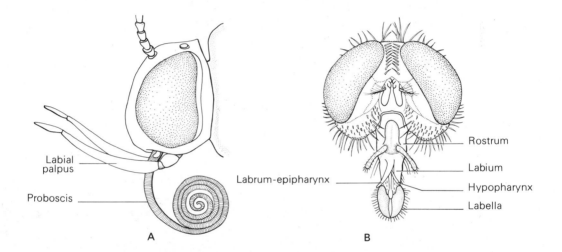

Labial palpus

Proboscis

Labrum-epipharynx

Rostrum

Labium

Hypopharynx

Labella

A

B

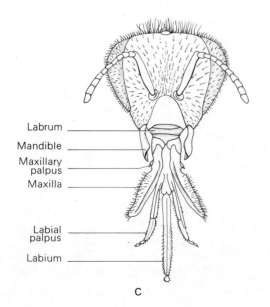

Labrum

Mandible

Maxillary palpus

Maxilla

Labial palpus

Labium

C

Fig. 2–31
Examples of nonstylate–haustellate mouthparts. A. Moth. B. House fly, *Musca domestica*. C. Honey bee, *Apis mellifera*. (A redrawn from Snodgrass, 1961; B redrawn from James and Harwood, 1969; C redrawn from Herms and James, 1961.)

taking up exposed liquids. These insects egest salivary secretions onto solid foods, so they are quite able to feed on such materials as solid sugar.

The more advanced hymenopterous insects have an altogether different "sucking" arrangement (Fig. 2–31C). The labrum and mandibles usually resemble those found in typically chewing insects. For this reason these mouthparts could as easily be included with the mandibulate group as with the haustellate group. However, the

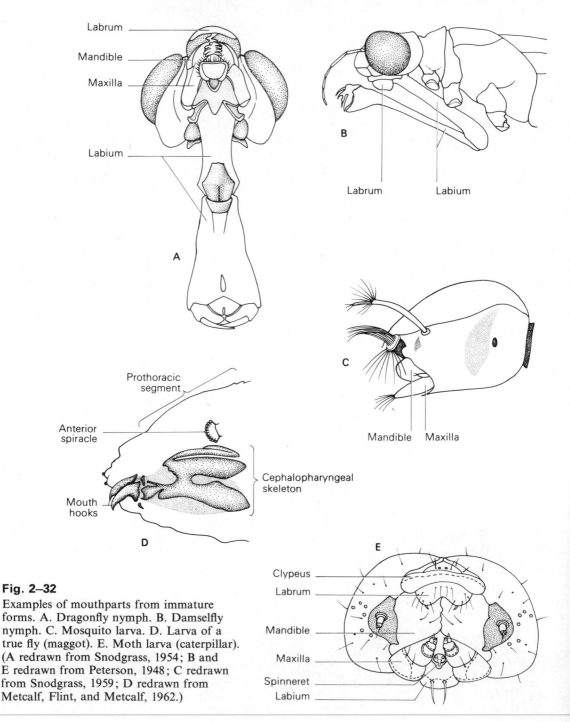

Fig. 2–32

Examples of mouthparts from immature forms. A. Dragonfly nymph. B. Damselfly nymph. C. Mosquito larva. D. Larva of a true fly (maggot). E. Moth larva (caterpillar). (A redrawn from Snodgrass, 1954; B and E redrawn from Peterson, 1948; C redrawn from Snodgrass, 1959; D redrawn from Metcalf, Flint, and Metcalf, 1962.)

maxillae and labium are quite different from the labrum and mandibles, having become united, as in the honey bee, forming a sort of lapping structure. The elongate glossae of the labium form the tubular part of this structure, which is well adapted for thrusting into flower nectaries. This type of mouthpart structure has been functionally classified as chewing–lapping.

In most sucking insects, the cibarial region has become modified and functions as a pump. The pharyngeal region also commonly forms a pump. An example of both types of food pumps is found in adult mosquitoes (see Fig. 3–2).

Mouthparts of Immature Forms.   In many instances the mouthparts of immature forms are essentially identical to those of the adult. However, in some, the differences are rather profound. For example, the nymphs (aquatic immatures) of dragonflies and damselflies (Fig. 2–32A, B) present an unusual modification of the labium. This structure is quite enlarged and an "elbow" is found at the junction of the prementum and postmentum. The apical portion of the prementum bears a pair of grasping jaws. When the nymph is at rest or stalking prey, the elbow of the labium is flexed and part of the prementum with its jaws forms a "mask" over the other mouthparts. When prey is within striking range, the labium is thrust out extremely fast, the jaws grasping the prey until it can be brought within reach of the maxillae and mandibles. In the lower Diptera, the mouthparts, although fundamentally chewing, are adapted for filter feeding, as in larval mosquitoes (Fig. 2–32C). In the larvae of higher Diptera the mouthparts are extremely reduced and have been invaginated into the head, forming the *cephalopharyngeal skeleton*, which anteriorly bears the vertically moving *mouth hooks* (Fig. 2–32D). Lepidopterous larvae (Fig. 2–32E) and others, such as larval trichopterans and hymenopterans, have a rather elaborate apparatus for spinning silk into cocoons. This spinning apparatus, the *spinneret*, is composed of the maxillae, hypopharynx, and labium. Neuropteran larvae, such as antlions and aphidlions, have a grasping–sucking modification of the mandibles and maxillae similar to that in the predaceous diving beetle described earlier.

Position of the Mouthparts.   There are basically three positions of the mouthparts relative to the head capsule: *hypognathous, prognathous,* and *opisthognathous* (Fig. 2–33A–C). In the hypognathous condition, the mouthparts hang ventrally from the head capsule. This is considered to be the most primitive condition of the three since the mouthparts are apparently modified locomotor appendages and have retained a similar position relative to the insect body. The prognathous condition is characterized by the anteriorly directed position of the mouthparts. Correlated with this modification is an elongation of the genal and postgenal regions and a more or less flattening of the head capsule. Opisthognathous insects are those in which the mouthparts are directed ventroposteriorly relative to the head capsule. This condition is found mainly among the Hemiptera and Homoptera and enables them to place the sucking beak between the legs and out of the way when not feeding.

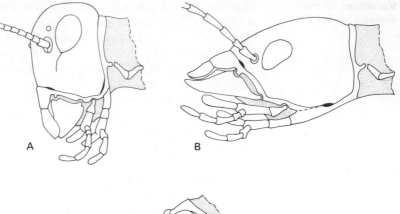

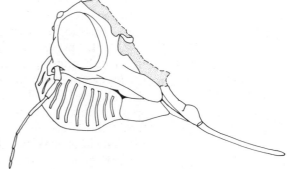

Fig. 2–33

Positions of mouthparts relative to the
head capsule. A. Hypognathous.
B. Prognathous. C. Opisthognathous.
(Redrawn from Snodgrass, 1960.)

Head Capsule. The overall shape and structure of the head varies
with the position of the mouthparts relative to the head capsule and
the demands placed upon it for rigidity and muscle attachment from
the mouthparts. The various sutures and cleavage lines described
earlier for the generalized insect may or may not be present, and if
present may be highly modified or reduced. As a result, the various
regions, which are largely defined by the sutures, also show a great
amount of variation. The internal framework, the tentorium, also
varies considerably, again correlated with the demands for muscle
attachment and strengthening support.

In many insects with the hypognathous condition, the occipital
foramen (foramen magnum) is closed ventrally, separating the postmentum of the labium from the cervical membrane. This situation is
due to the fusion of lobes of the posterior or hypostomal region of the
subgenae and is referred to as a *hypostomal bridge* (Fig. 2–34A, B). In
other insects, lobes of the postgenae converge mesially on the hypostomal bridge, and in some cases may fuse, forming a *postgenal bridge*
(Fig. 2–34C, D).

In some prognathous insects, the parts on the ventral side of the
head are essentially equivalent to the posterior parts in the hypognathous forms. However, in some, for example several species of
beetles and neuropterous insects, the hypostomal regions of the postgenae have fused with one another in a fashion similar to the formation
of the hypostomal bridge in hypognathous insects, forming a structure
called the *gula* (Fig. 2–34E). In many instances, the gula fuses with the
submentum, forming a composite structure, the *gulamentum*.

Modifications of the frontoclypeal region of the head are often

correlated with the size of the cibarial region and the pharynx or with
the development of a cibarial pump. The clypeus may be quite pronounced, as in many Hemiptera and Homoptera (Fig. 2–34F), where
it presents a broad surface area for the attachment of the muscles of
the sucking pump.

Fig. 2–34
Modifications of the head capsule. (See text for explanation.)
A, B, and E. Diagrammatic. C. Honey bee. D. Vespid wasp. F. Cicada.
(A–D redrawn from Snodgrass, 1960; E redrawn from Snodgrass, 1959;
F redrawn from Snodgrass, 1935.)

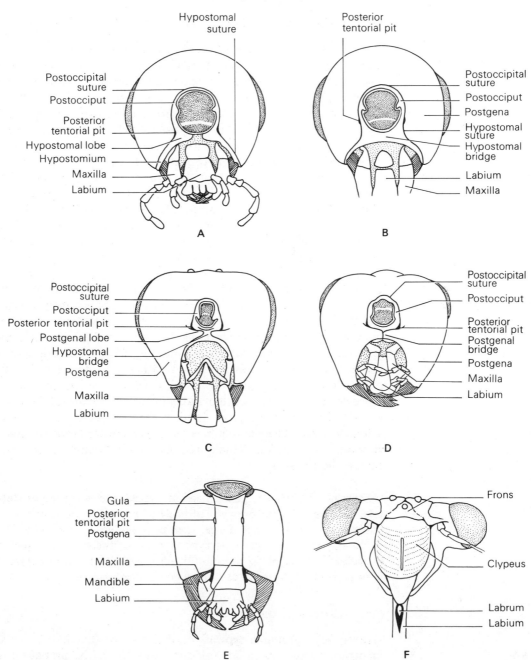

The tentorial structure already described was that of a generalized pterygote insect and would be found, for example, in members of the order Orthoptera. In the entognathous apterygote insects (those in which the mouthparts are not externally apparent), the tentorium is absent. In the ectognathous apterygotes (mouthparts externally evident), the thysanurans, anterior arms arise from the anterior tentorial pits. However, these arms and the posterior bridge do not connect. In some of the higher orders there has been a tendency toward a reduction of the tentorium, and in some, as in the Thysanura, the anterior and posterior arms do not connect.

Modifications of the Thorax

Nota. In wing-bearing segments, the terga or nota vary as to the number, nature, and location of internally inflected ridges they possess. These ridges are marked externally by sutures and serve to lend strength and rigidity to the tergum. The terga of the prothorax of pterygote insects are quite different from those of the wing-bearing segments since they are not directly associated with the wing mechanism. Consequently, any sutures associated with internal ridges on the pronotum cannot be considered equivalent to those in the pterothorax. The pronotum is commonly quite small and undeveloped compared to those of the wing-bearing segments. However, in many insects it is quite pronounced, forming a pronotal shield (Fig. 2–35A, B). In some insect species it has taken on rather bizarre shapes, often mimicking an environmental characteristic such as a thorn (Fig. 2–35C) or an elongate projection over the head (Fig. 2–35D).

Pleura. There is generally more similarity between the pleuron of a winged segment and that of the prothorax. However, the propleuron is usually less developed and there may be secondary modification in the pterothoracic pleuron since, like the nota in these segments, they are intimately involved with the wing mechanism. For example, the episternum or epimeron may be subdivided into dorsal and ventral plates. The precoxal and postcoxal areas of the pleuron may be separated from the episternum and epimeron. The trochantin is fairly well developed in more generalized pterygotes and somewhat less so in higher orders, being absent in some. In certain dipterous species, the meron of the coxa has actually become quite pronounced, forming a part of the pleuron.

Sterna. In the more generalized pterygote insects the major sternal sclerites are usually present, although they may be considerably reduced, leaving rather large membranous areas between them. However, in the higher pterygote orders, there is considerable modification and it is commonly very difficult to homologize the various parts to those of the more generalized insects.

Modifications of the Legs

Insect legs, although typically ambulatory in function, have been modified extensively in several directions. Both the immature and

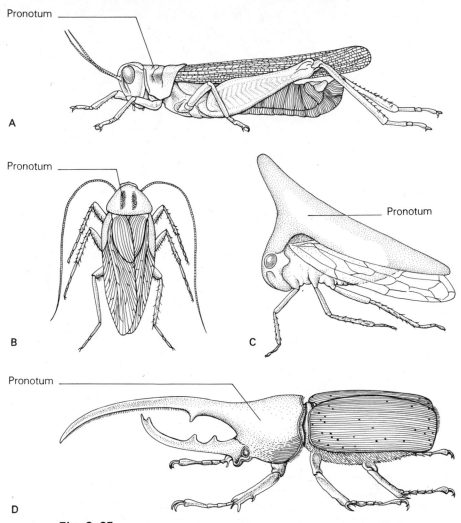

Pronotum

Pronotum

Pronotum

Pronotum

Fig. 2–35
Modifications of the pronotum. A. Grasshopper, *Melanoplus bilituratus*.
B. German cockroach, *Blattella germanica*. C. Treehopper, *Thelia bimaculata*. D. Hercules beetle, *Dynastes hercules*. (A redrawn from Ross, 1962a; B courtesy of U.S. Public Health Service (redrawn);
C redrawn from Borror and Delong, 1964; D redrawn from Essig, 1942)

adult stages of most insects have thoracic legs. However, there are many examples of apodous larvae (those lacking legs) (e.g., fly maggots) and even of apodous adults (e.g., female scale insects). Typically developed insect legs are *cursorial,* meaning that they are adapted for walking and running. The cockroach is a good example of an insect with cursorial legs (Fig. 2–36A). In some insects, such as the mole cricket and the nymphs of the periodical cicada, the forelegs are highly modified, bearing heavily sclerotized digging claws (Fig. 2–36B). *Fossorial* is the term commonly used to denote this adaptation. The forelegs of other insects [e.g., the praying mantis (Fig. 2–36C)] are *raptorial* or modified for grabbing and holding prey. The forelegs have not been the only ones to undergo modification. For example, the femora of the hindlegs of grasshoppers and katydids

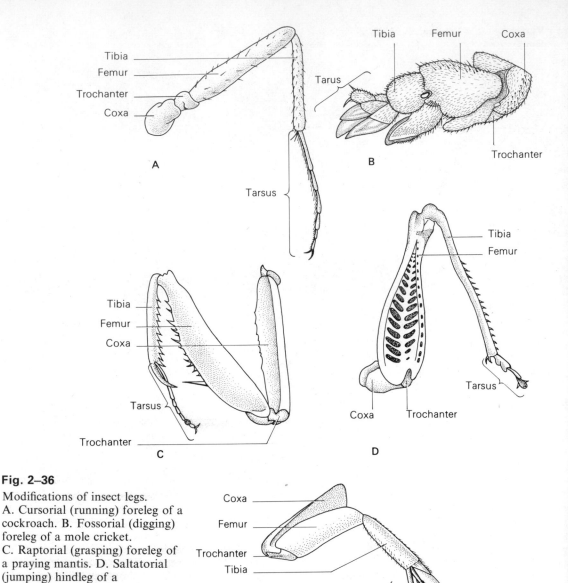

Fig. 2–36

Modifications of insect legs.
A. Cursorial (running) foreleg of a cockroach. B. Fossorial (digging) foreleg of a mole cricket.
C. Raptorial (grasping) foreleg of a praying mantis. D. Saltatorial (jumping) hindleg of a grasshopper. E. Natatorial (swimming) leg of a water beetle.
(A–D redrawn from CCM General Biological, Inc., key card; E redrawn from Folsom and Wardle, 1934.)

are enlarged, accommodating the muscles used in jumping (Fig. 2–36D). Legs adapted for this kind of activity are commonly referred to as *saltatorial*.

The legs of several aquatic insects are modified in such a way that they facilitate swimming, as is the case with adult dytiscid beetles (Fig. 2–36E), which bear two rows of "swimming hairs" on the edges of the flattened tibiae and tarsi of the middle and hindlegs. These hairs are attached to the legs by movable joints. When the legs are thrust posteriorly during the act of swimming, the distal ends of these hairs move out from the legs, greatly expanding the surface area being applied against the water in the paddling action. As the legs are brought anteriorly in the recovery stroke, the hairs become pressed very close to the legs, reducing the surface area applied against the water, much like the feathering of a paddle when paddling a boat in the wind. The term *natatorial* applies to swimming legs. Although their gross morphology is similar to the middle and hindlegs, the forelegs of the tiny members of the order Protura are carried in an elevated position anterior to the body. It has been said that these are principally sensory in function and that they suggest a step in the evolution of antennae.

The legs of many insects bear various specialized structures. For example, the legs of honey bees bear structures that are used during their pollen-collecting activities. One of these structures is the *corbiculum* (Fig. 2–37A), or *pollen basket*, composed of two rows of hairs on the outer surface of the hind tibia, where the pollen collected by a

Fig. 2–37
Specialized structures borne on the legs. A. Corbiculum or pollen basket on the hind tibia of the honey bee. B. Suction discs on the fore tarsus of a diving beetle. C. Tympanic organs on the fore tibia of a long-horned grasshopper. (A redrawn from Snodgrass, 1956; B redrawn from Folsom and Wardle, 1934; C redrawn from Packard, 1898.)

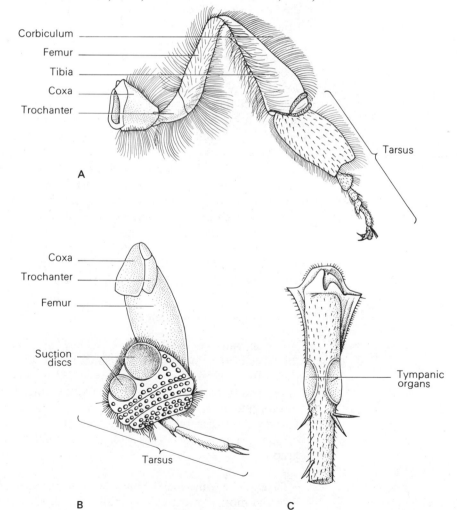

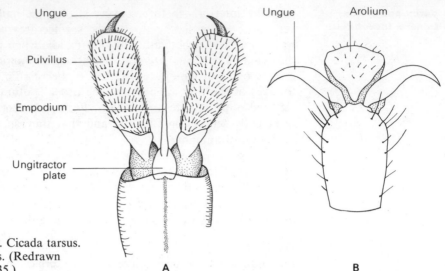

Fig. 2–38
Tarsal structures. A. Cicada tarsus. B. Cockroach tarsus. (Redrawn from Snodgrass, 1935.)

foraging worker is stored for transport back to the hive. The forelegs of males of some species of diving beetles bear large suction discs on the tarsi (Fig. 2–37B). They are used to hold the female during copulation.

The hind femora of certain species of short-horned grasshoppers have short peglike structures with which they rub the forewings and produce a sound. The legs of different insects may bear sensory structures of various types (see Chapter 5). Several species of flies (e.g., blow flies and house flies) "taste" by means of sensilli on the tarsi of their forelegs. The long-horned grasshoppers and crickets possess oval auditory organs, or *tympana,* at the base of each front tibia (Fig. 2–37C).

The tarsal and pretarsal segments are also variously modified. Padlike *pulvilli* may be found on the lower surface of each tarsal segment as in several members of the order Orthoptera or in association with each *ungue* or *pretarsal claw* as in the flies (Fig. 2–38A). A bulbous, lobelike structure, the *arolium* (Fig. 2–38B), may be present between the claws. Some insects have a spinelike or lobelike structure, the *empodium* (Fig. 2–38A), which arises from the distal part of the *ungitractor plate* and also is located between the claws.

Modifications of the Wings

Insects may bear a single pair of wings, two pairs, or none at all. On the basis of their morphological similarity to the wing-bearing pterygotes, their apterous condition is considered to be secondary, having developed from a winged ancestor. On the other hand, the apterygote insects and their ancestors never had wings. There is considerable variation between the wings of insects that do have them. Examples of some of these variations will be presented in the following paragraphs.

Size.　The wings may be quite large, as in many of the larger butterflies and moths, or extremely small, as in many of the wasps and flies.

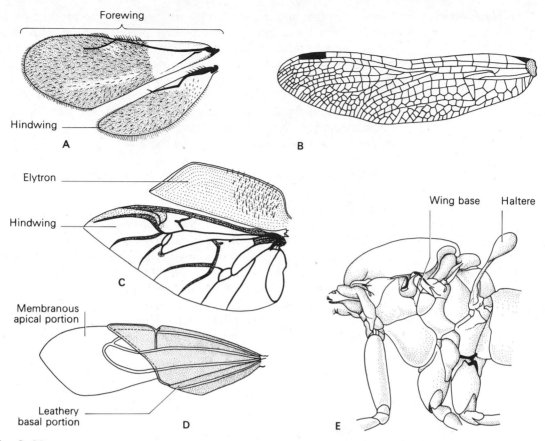

Forewing

Hindwing

A

B

Elytron

Hindwing

C

Membranous
apical portion

Leathery
basal portion

D

Wing base   Haltere

E

Fig. 2–39

Variations in wing
structure. A. Wings of
a wasp with reduced
venation. B. Wings of a
dragonfly with an
elaborate network of
veins. C. Ladybird
beetle with left elytron
(forewing) and hindwing
extended.
D. Hemelytron
(forewing) of a true bug.
E. Lateral view of the
thorax of a true fly.
(A, C, and D redrawn
from Essig, 1958;
B redrawn from CCM
General Biological,
Inc., key card;
E. redrawn from
Smart, 1959.)

The Atlas moth of Australia has been said to reach a wingspan of
14 inches! In many insects, there is a tendency for the hindwings to be
smaller than the forewings (Fig. 2–39A).

Venation.   Because of the tremendous variation in wing venation,
it has been used as a source of taxonomic characters. Venation
ranges from the extensively reduced and simplified system found in
many of the wasps (Fig. 2–39A) to the highly complex network in the
wings of dragonflies and damselflies (Fig. 2–39B). Veins also vary in
thickness; for example, those of the periodical cicada are quite thick,
whereas those in the scorpionfly are very thin and delicate.

Function.   The most obvious function of wings is, of course, flight.
However, in several instances the wings have been modified or are at
least used for different purposes. In the beetles, the hindwings are
membranous and fold beneath the forewings, which are usually quite
hard and form a protective armor for the membranous hindwings
when not in use. These modified forewings (Fig. 2–39C) are called
elytra (sing., *elytron*). A similar situation exists among the Hemiptera,
although the forewings of these insects are only partly hardened, the
distal portions remaining membranous and containing veins. These
structures (Fig. 2–39D) are appropriately named *hemelytra* (sing.,
hemelytron), or "half"-elytra. The forewings of orthopterans are
parchmentlike and probably afford similar protection to the hind-

wings. In other insects the wings are used for the production of sound. Several examples of insects that produce sound with their wings are to be found among the Orthoptera. The field cricket is well known for this activity. The true flies possess a single pair of well-developed forewings and a pair of highly modified hindwings, the *halteres* (Fig. 2–39E). These club-shaped structures are important in the stability of flight of these insects. In the very hot days of mid-summer, honey bees fan their wings in a community effort and thereby reduce the temperature within the hive.

Relationship to One Another. Among the insects with two pairs of wings, the wings may work separately as in the dragonflies and damselflies, mayflies, and Neuroptera. However, in many of the higher pterygote insects, the fore- and hindwings are coupled to one another in various ways, resulting in each pair of wings acting together as a unit. Examples of wing-coupling mechanisms include tiny hooks or *hamuli* found among the Hymenoptera (Fig. 2–40A) and the spinelike *frenulum* (Fig. 2–40B) and lobelike *jugum* (Fig. 2–40C) characteristic of several Lepidoptera. In insects that have wing-coupling mechanisms, the hindwings are usually somewhat smaller than the forewings. The tendency toward a reduction of the hindwings has, of course, reached its maximum in the true flies, which have lost

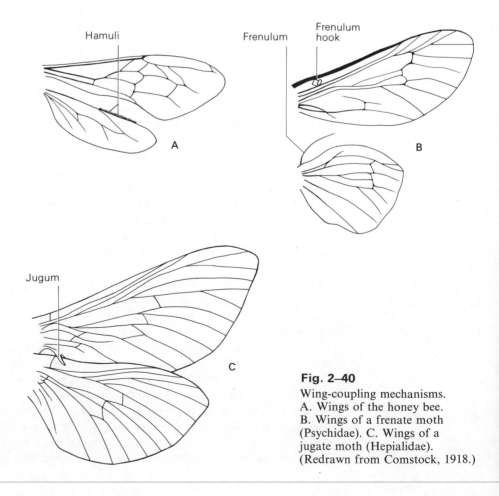

Fig. 2–40
Wing-coupling mechanisms.
A. Wings of the honey bee.
B. Wings of a frenate moth
(Psychidae). C. Wings of a
jugate moth (Hepialidae).
(Redrawn from Comstock, 1918.)

the hindwings as such, the remnants being the previously mentioned haltères.

Resting Position. When not being utilized for flight, the wings are held in various positions relative to the body. Members of the orders Ephemeroptera and Odonata are unable to flex the wings over the abdomen and hence when at rest hold them vertically over the dorsum (mayflies and damselflies) or horizontally (dragonflies). Other insects, bees, wasps, and so on, are able to flex the wings over the abdomen at rest. Homopterous insects typically hold the wings rooflike over the abdomen.

Coloration. Although many insects have *hyaline* or opaque, unpigmented wings, there are groups in which wing coloration is especially well developed. The coloration may be either due to pigmentation within the integument itself or may be the result of a covering of minute scales, which are pigmented or mechanically resemble thin layers or diffraction gratings in their effects upon impinging light. The bright and decorative colorations on the wings of some dragonflies are a good example of the first kind of coloration. The butterflies and moths as a group possess the second type of coloration, colored scales. Variety and complexity of coloration abounds in the Lepidoptera, and many behavioral patterns are intimately tied up with it. In several instances the coloration mimics the environmental background and in this way affords a degree of protection from predators.

Texture. The wings of many insects are quite smooth and membranous. However, in some they are quite hard (e.g., beetle elytra) and may be sculptured in various ways. Other wings are leathery or parchmentlike, for example the forewings of orthopterans.

Presence of Hairs and Scales. The scales of butterflies and moths have already been mentioned. These are considered to be modified setae. The wings of the caddisflies are covered with tiny hairs, which are also modified setae. Wings of other insects bear various types of large and small hairs, called *macro-* and *microtrichia,* respectively.

Modifications of the Abdomen

The abdominal segments, other than the first, of adult pterygote insects anterior to those which bear the external genetalia are usually quite simple and uniform, each consisting of a tergum and sternum separated by a pleural membrane and never bearing appendages. As explained earlier, the first abdominal segment in pterygote insects is associated more with the thorax than the abdomen, since the antecostal portion of the tergum furnishes the third phragma, to which the dorsal longitudinal wing muscles are attached. In many of the Hymenoptera the first abdominal segment, the *propodeum* (Fig. 2–41A), is completely associated with the thorax and is separated from the remaining abdominal segments by a constriction, the *petiole.*

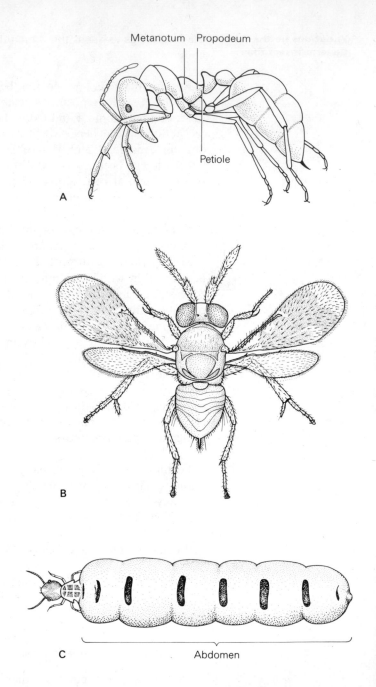

Fig. 2–41

Modifications of the abdomen.
A. Fire ant. B. Parasitic wasp,
Eusemion sp. C. Gravid termite
queen. (A courtesy of U.S. Public
Health Service (redrawn);
B redrawn from Essig, 1958;
C redrawn from Skaife, 1961.)

The abdomen as a whole varies considerably, both in size and in the number of segments. As pointed out earlier, the primitive number of segments is considered to have been 12. This condition occurs, however, only among adult Protura and in the embryos of some higher insects. The tendency is toward a reduction in the number of abdominal segments. The springtails have 6 segments in their abdomen, the more generalized pterygotes have 11, and the higher pterygotes usually have 10 or less. The size of the abdomen relative to the remainder of the body ranges from the tiny abdomen characteristic of several parasitic wasps (Fig. 2–41B) to the extremely large abdomen of a gravid termite queen (Fig. 2–41C).

Nongenital Abdominal Appendages. In contrast with the adult pterygotes, the pregenital and genital segments of many larval pterygotes and many apterygote insects do bear appendages of various sorts. For example, the first three abdominal segments of adult proturans bear rather simple, bilateral appendages, the *styli*. Similarly, styli are usually borne on several of the abdominal segments of adult thysanurans (Fig. 2–42A). Members of the order Collembola are in many ways quite different from other insects. One of the outstanding differences is the presence of three rather unique abdominal structures (Fig. 2–42B). The most anterior structure, the *collophore*, is located on

Fig. 2–42

Nongenital abdominal appendages. A Venter of posterior portion of the abdomen of a silverfish. B. Springtail. C. Caterpillar. D. Mayfly nymph. E. Aphid. (A redrawn with modifications from Essig, 1942, after Oudemans; B courtesy of U.S. Public Health Service (redrawn); C redrawn from Snodgrass, 1961; D and E redrawn from CCM General Biological, Inc., key card.)

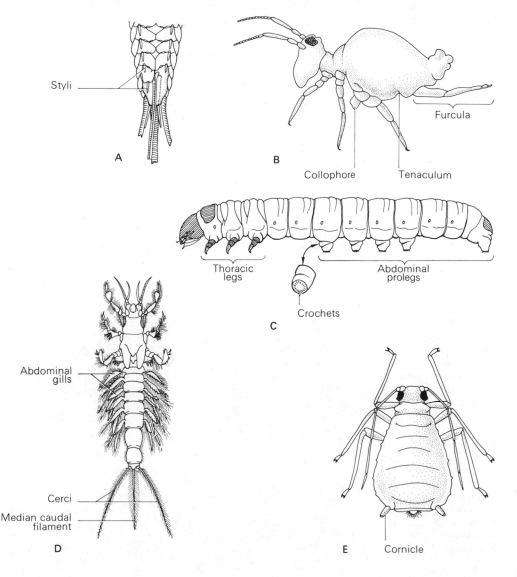

the venter of the first abdominal segment. This structure is roughly cylindrical in appearance when it is protruded by the hydrostatic pressure of the hemolymph. The function of the collophore is not clear. Its name means literally "that which bears glue," based on an early suggestion that it might serve as an organ of adhesion. The two posterior structures consist of a *tenaculum* on the venter of the third segment and the *furcula* on the venter of the fifth segment. The furcula is capable of being moved anteriorly, engaged by the tenaculum, and subsequently released, exerting a force against the substrate and propelling the insect through the air. Hence the name "springtail."

Abdominal appendages of pterygote larvae usually serve either a walking or a ventilatory function. An example of the first is found among the larvae in the order Lepidoptera (Fig. 2–42C), which generally bear four or five pairs of bilateral outgrowths of the first four, and commonly the tenth, abdominal segments. These are the *prolegs* and complement the three pairs of thoracic appendages in the locomotion of the insect. Each proleg bears a series of minute hooks or *crochets*. Bilateral abdominal appendages on the abdominal segments of mayfly nymphs serve as gills (Fig. 2–42D), facilitating the absorption of oxygen from and release of carbon dioxide into the surrounding water. A pair of lobelike projections, *cornicles,* on the posterior dorsum of the abdomen is characteristic of Aphids (order Hemiptera; suborder Homoptera).

The structure of the terminal abdominal segments (e.g., the cerci and the plates surrounding the anus, the epi- and paraprocts) is variously modified. For example, the cerci may be forceps—or clasperlike (Fig. 2–43A), feelerlike (Fig. 2–43B), reduced, or absent. Likewise, the epiprocts and paraprocts may be long and feelerlike, may bear anal gills (Fig. 2–43C), or may be reduced or quite inconspicuous.

External Genitalia. In insects that have an ovipositor, this structure may show considerable variation, depending upon the situation into which the eggs must be placed. For example, cicadas lay their

Fig. 2–43
Modifications of the terminal abdominal segments.
A. Posterior portion of earwig abdomen with forcepslike cerci.
B. Stonefly nymph with feelerlike cerci.
C. Posterior portion of damselfly abdomen with gill-bearing epi- and paraprocts.
(A redrawn from Hebard, 1934; B redrawn from Ross, 1962a; C redrawn from Snodgrass, 1954.)

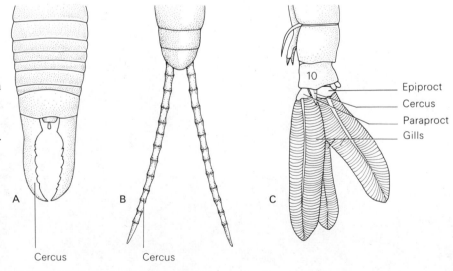

Cercus

Cercus

10

Epiproct
Cercus
Paraproct
Gills

eggs beneath the bark of twigs of trees. Their ovipositor (Fig. 2–44A) is well adapted for this function, being composed of three rather sharp and rigid blades. Other ovipositors [e.g., those of the katydid (Fig. 2–44B)] are constructed such that they enable the insect to deposit eggs beneath the surface of the soil. Parasitic ichneumon wasps have extremely long ovipositors (Fig. 2–44C), which enable them in some instances to penetrate the bark of a tree and deposit an egg in a wood-boring larva.

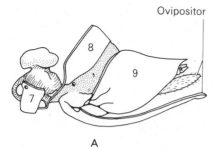

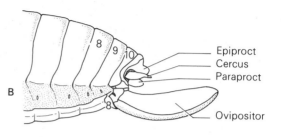

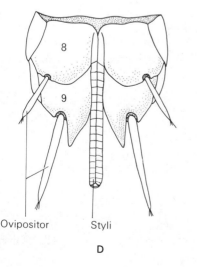

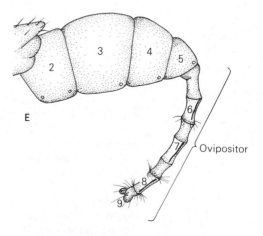

Fig. 2–44

Modifications of the ovipositor. A. Cicada, *Magicicada septendecim.* B. Katydid, *Scudderia* sp. C. Ichneumon wasp, *Megarhyssa lunator.* D. Firebrat, *Thermobia domestica.* E. Telescoping terminal abdominal segments serving as an ovipositor in the house fly, *Musca domestica.* Numerals indicate abdominal segments. (A, B, and D redrawn from Snodgrass, 1935; C redrawn from Riley, 1888; E. redrawn from West, 1951.)

The adult female members of the apterygote order Thysanura possess a very primitive ovipositor composed of paired appendages borne on the venter of the eighth and ninth abdominal segments (Fig. 2–44D). Each appendage or *gonopod* is borne on a basal *coxopodite*, which may or may not bear a stylus. Many similarities are apparent between this primitive ovipositor and that found among the pterygotes. Many insects lack an ovipositor altogether and have devised other means of egg deposition. For example, certain members of the orders Thysanoptera, Mecoptera, Lepidoptera, Coleoptera, and Diptera use their abdomen as an ovipositor (Fig. 2–44E). Some are capable of telescoping the abdomen to great lengths, thus being able to place eggs in very small crevices or similar tight spots.

The external genitalia of male insects are commonly extremely complex structures and probably show more inherent variation than any other single insectan structure. It is for this reason that taxonomists have made extensive use of these structures for their work. Since taxonomists are usually specialists in a single group, there has been a proliferation of terminologies applied to male genitalia. This obviously creates a problem for the entomology student and he is best referred to the work of Tuxen (1970). The most fruitful attempts to homologize the male genitalia in the diverse insectan groups have been by Snodgrass (1957) and Smith (1969). Because of the rather complex and specialized nature of this subject, the reader is referred to either or both of these works and the extensive bibliographies contained in each.

Modifications of the General Body Form

The shape of many insects is an obvious adaptation to a given environmental situation. An outstanding example of this is found in the fleas. These insects are bilaterally flattened, a characteristic that enables them to move easily between the feathers or hairs of their hosts. One does not realize the efficiency of this adaptation until he attempts to remove one of these insects from a pet cat or dog. An example of an insect flattened dorsoventrally is the bed bug. This body shape enables the bed bug to hide in tiny cracks and crevasses between feedings. Many insects possess rather bizarre shapes and in many instances actually mimic an environmental characteristic, as in the case of protective coloration, which provides a certain amount of protection from potential predators. Some insects (e.g., walkingsticks) are quite elongate and tubular in shape.

Selected References

PHYSIOLOGICAL ASPECTS
Ebeling (1964); Hackman (1964); Locke (1964, 1965); Richards (1951); Wigglesworth (1965).

MORPHOLOGICAL ASPECTS
Butt (1960); DuPorte (1961); Matsuda (1970); Scudder (1971); Smith (1969); Snodgrass (1935, 1950, 1952, 1957, 1958, 1960, 1963a); Tuxen (1970).

Alimentary, Circulatory, Ventilatory, and Excretory Systems

Every cell in the insect body, regardless of its function, requires a source of energy, a source of oxygen, and the raw materials with which to carry out its own maintenance and synthesizing activities and produces, as a result of carrying out these processes, carbon dioxide and other waste products of various sorts. In this chapter we shall consider those systems directly involved in the transport of nutrients and oxygen to the individual cells and the removal of accumulated waste materials and carbon dioxide from the insect.

The Alimentary System

The alimentary system is concerned with the initial steps involved with the transport of nutrients to the individual cells. The processes of ingestion, triturition (chewing), digestion, absorption into the hemolymph, and egestion are all associated with this system. Insects are basically *holotrophic* organisms; that is, they possess a tube (often somewhat coiled), the alimentary canal, which extends from an anterior oral opening, the *mouth*, to a posterior *anus*.

The insect alimentary canal (Fig. 3–1) can be divided into three usually quite distinct regions, an anterior *foregut*, or *stomodaeum*; a *midgut,* or *mesenteron*; and a posterior *hindgut*, or *proctodaeum*. The fore- and hindguts are lined with a chitinous *intima*, which is continuous with the cuticle of the integument. The presence of this intima is not surprising when one realizes that the fore- and hindguts arise as invaginations of the ectodermal tissue, which also gives rise to the integument. The midgut is generally believed to be of endodermal origin. Longitudinal and circular muscles are usually associated with each of the three regions and by means of rhythmical peristaltic contractions move the food along the alimentary canal. These muscles are innervated by the *stomodael system* in the anterior portion of the gut and by nerves from the posterior ganglion of the ventral nerve cord in the hindgut.

Foregut

Although there is considerable variation in the foregut, it can usually be divided into several fairly distinct morphological regions. The true

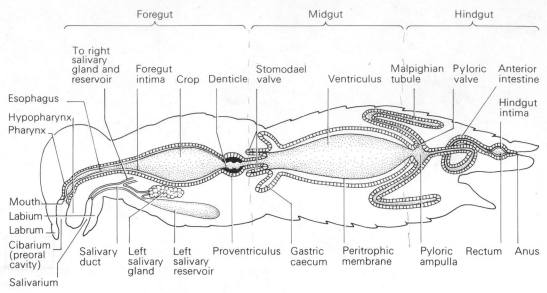

Fig. 3–1

Generalized insect alimentary system.

mouth lies at the base of the hypopharynx within the preoral cavity formed by the mouthparts, the *cibarium*. The true mouth communicates directly with the *pharynx*, a structure that varies considerably among different insects. The cibarium or pharynx or both may be highly modified, forming well-musculated pumps (Fig. 3–2). The pharynx is followed by the *esophagus,* which is commonly enlarged posteriorly, forming the *crop* (Fig. 3–1). In some insects the posterior enlargement may be in the form of one or more blind sacs, or *diverticula* (Fig. 3–2). Immediately posterior to the crop region is the *proventriculus* (Fig. 3–1). The luminal side of this structure often bears sclerotized denticles (teeth) or spines. The proventriculus usually communicates with the midgut by means of the *esophageal intussusception,* which consists of both foregut and anterior midgut tissue. The esophageal intussusception has been variously referred to as the *stomodael, esophageal,* or *cardiac valve.* (It will be called the stomodael valve in this text.) The midgut tissue *(cardia)* lies ectad of and surrounds the foregut portion of the intussusception. The

Fig. 3–2

Alimentary canal of a mosquito.

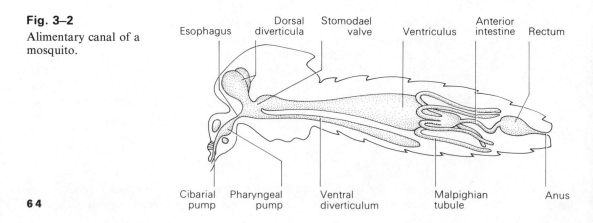

esophageal intussusception is commonly well developed and in the larvae and adults of true flies (order Diptera), which lack a proventriculus in the sense described above, lies between the esophagus or diverticulum(a), if present, and the midgut posterior to the cardia. It is commonly referred to by investigators studying the members of this order as a proventriculus.

The foregut with its various morphological divisions serves mainly as a conducting tube, carrying food from the cibarial cavity to the midgut. The enlarged crop and diverticulum(a), when present, function, at least in part, as temporary food-storage organs. An interesting phenomenon occurs in adult mosquitoes relative to the diverticula (Fig. 3–2). When the female takes a blood meal, it is directed to the midgut. However, when she feeds on a sugar solution, such as flower nectar, it is shunted to the diverticula and apparently released gradually into the midgut. This phenomenon has been appropriately termed the *switch mechanism*.

The proventriculus, when armed with denticles, may serve as a grinding or triturating structure in addition to the mouthparts. Spines in the proventricular region probably act as a food sieve or filter, as in the honey bees, in which the proventriculus allows the movement of pollen into the midgut without admitting ingested flower nectar. The stomodael valve is developed to varying degrees in different insects. Whether or not this structure actually acts as a valve in most insects has been seriously questioned. Wigglesworth (1930) explains one of its major functions in many insects to be the secretion and formation of the peritrophic membrane. Although the foregut is not the major digestive region of the alimentary canal, there is probably some digestion in the crop brought about by the presence of salivary enzymes and possibly enzymes regurgitated from the midgut.

Midgut

The midgut (Fig. 3–1) begins with the cardial epithelial layer associated with the stomodael valve in most insects. Immediately posterior to the cardia there is commonly a group of diverticula, the *gastric caeca*. The number of these caeca varies in different insectan species. Similar pouches may be present on other sections of the midgut in some insects. The remainder of the midgut is usually a rather enlarged sac and serves as the insect's stomach.

At the bases of the midgut epithelial cells are small regenerative or replacement cells, which replace the actively functioning gut cells when they degenerate (Fig. 3–3A). Regenerative cells may be dispersed individually among the epithelial cells (e.g., caterpillars and true flies) or be concentrated in discrete groups in the form of *nidi* (e.g., grasshoppers and relatives, dragonflies, and damselflies) or *crypts* (e.g., many beetles). Nidi ("nests") are groups of regenerative cells located at the bases of the active epithelial cells but are within the confines of the muscle layers (Fig. 3–3B). Crypts are packets of cells that project through the muscle layers of the gut (Fig. 3–3C).

The midgut is the principal site of digestion and absorption. It may be divided into different regions, which absorb different components of the food materials. A chitinous *peritrophic membrane* (Fig.

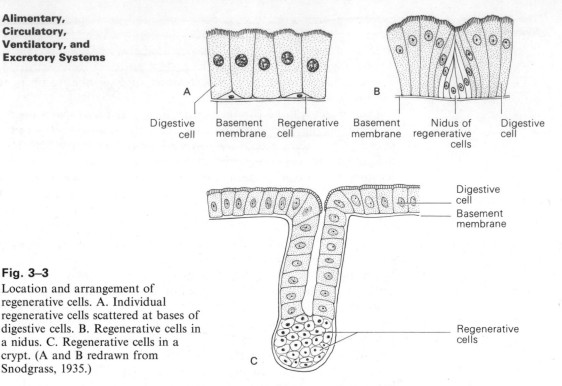

Digestive cell   Basement membrane   Regenerative cell   Basement membrane   Nidus of regenerative cells   Digestive cell

Digestive cell

Basement membrane

Regenerative cells

Fig. 3–3
Location and arrangement of
regenerative cells. A. Individual
regenerative cells scattered at bases of
digestive cells. B. Regenerative cells in
a nidus. C. Regenerative cells in a
crypt. (A and B redrawn from
Snodgrass, 1935.)

3–4) is secreted by either the cardial epithelium, the general ventricular epithelium, or both. This membrane, as indicated by its name, peri(around)-trophic(food), surrounds the food bolus and has been suggested to function as a mechanical barrier between the food bolus and epithelial cells, protecting these cells from possible abrasion. That the peritrophic membrane acts in other ways is suggested by its presence in insects that do not ingest solid, potentially abrasive food particles (i.e., fluid feeders of various types). A recent paper suggests that it may function as a "semipermeable" membrane.

Hindgut

The hindgut (Fig. 3–1) commences with the *pylorus*, which is associated with a variable number of usually rather slender, elongate excretory structures, the *Malpighian tubules*, and which usually contains a valvular structure, the *pyloric valve*. The Malpighian tubules are generally used as landmarks signaling the beginning of the hindgut. The hindgut is often divisible into a tubular *anterior intestine*, just posterior to the Malpighian tubules, and a highly musculated, enlarged *rectum*, which terminates with the proctodael opening, the *anus*. The anterior intestine may be further differentiated into an anterior *ileum* and posterior *colon*. The hindgut usually does not carry on many digestive activities and serves to carry undigested food material away from the midgut, ultimately egesting it from the insect. However, before egestion, the hindgut absorbs, to varying degrees, water, salts, and amino acids previously removed from the hemolymph by the Malpighian tubules. Thus it plays a major role in the water and salt balance of an insect.

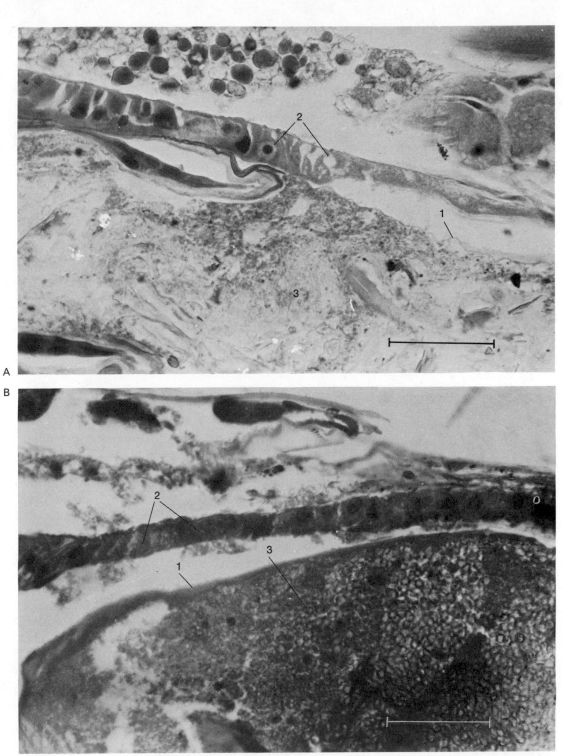

A

B

Fig. 3–4

Photomicrographs of the peritrophic membrane in the mosquito *Aedes triseriatus*. A. Sagittal section of anterior portion of midgut of fourth instar larva. B. Sagittal section of a portion of the midgut of an adult female. The peritrophic membrane in the adult has formed around a recent blood meal. 1, Peritrophic membrane; 2, midgut epithelium; 3, food in lumen of gut; red blood cells in B. (Scale lines = 50 microns.)

Digestion

Although the initial stages of the breakdown of ingested food may occur in the foregut, the insect midgut enzymes are the major entities involved in this process. Several proteases, lipases, and carbohydrases have been found, some of which are rather unusual. For example, cellulase has been found in several wood-boring insects. In some, microbes may furnish this enzyme, but others such as *Ctenolepisma lineata*, a thysanuran, are able to digest cellulose in the absence of intestinal microflora. Other enzymes in the unusual category include chitinase, lichenase, lignocellulase, hemicellulase, and hyaluronidase. Certain insects are able to digest substances that are ordinarily highly stable. For example, chewing lice and a few other insects are able to break down a very stable protein, keratin, which occurs in wool, hair, and feathers. Larvae of the wax moth, *Galleria mellonella*, possess the necessary enzymes to digest beeswax, which makes them the bane of the beekeepers. In any case, digestion occurs by a series of progressive steps, each producing a simpler substance until molecules of absorbable size or nature are produced. For example, polysaccharides are broken down into small chains, disaccharides, and finally into simple absorbable hexoses or proteins, to peptones, polypeptides, dipeptides, and finally amino acids, which are absorbable.

Control of Digestion

Three possible mechanisms for the control of enzyme secretion in the insect gut have been suggested:

1. Secretogogue; this is a substance in the ingested material that stimulates enzyme secretion.
2. Neural; the stimuli that induce enzyme secretion are mediated via nervous impulses initiated at some point along the gut or as a result of distension of the abdomen.
3. Hormonal; stimuli inducing enzyme secretion are hormonal secretions that act via the hemolymph.

Movements of the alimentary canal that complement the actions of the digestive enzymes have been reported to be under neural control in some insects. In others, no neural connections can be found, and gut movements in these cases are assumed to be myogenic. Hormonal stimuli may also have a great deal to do with the rate of gut movements.

Salivary Glands

Although there may be glands associated with the mandibles and maxillae, the *labial glands* are the ones that most commonly carry on a salivary function. The labial glands (Figs. 3–1 and 3–5) are two in number, lying ventral to the foregut in the head and thorax and occasionally extending posteriorly into the abdomen. They vary greatly in size and shape but are typically described as being *acinar* (i.e., resembling a cluster of grapes). Each "grape," or *acinus*, bears

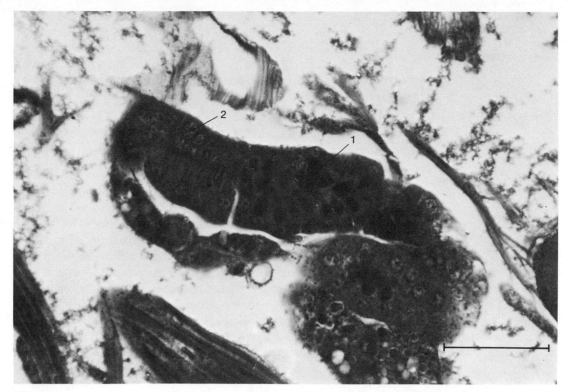

Fig. 3–5

Photomicrograph of a section of a salivary gland in a mosquito. 1, Secretion product; 2, nucleus of salivary gland cell. (Scale line = 50 microns.)

a tiny duct that communicates with other similar ducts, eventually forming a *lateral salivary duct*. The lateral salivary ducts run anteriorly and ultimately join in a *common salivary duct*, which empties, in the case of the labial glands, between the base of the hypopharynx and the base of the labium. This region is called the *salivarium* and in some of the sucking insects forms a salivary syringe that "injects" saliva into whatever is being pierced. The lateral salivary ducts commonly communicate with comparatively large *salivary reservoirs*, as in the cockroaches. Some salivary glands are made up of a number of lobes. In other insects there are no salivary glands.

The secretory products of the salivary glands are generally clear fluids that serve a variety of functions in different insects: (1) they moisten the mouthparts; (2) they act as a food solvent; (3) they serve as a medium for digestive enzymes and various anticoagulins and agglutinins; (4) they form silk in larval Lepidoptera (caterpillars) and Hymenoptera (bees, wasps, and relatives); and (5) they are used to "glue" puparial cases to the substrate in certain flies. The most common enzymes found in insect saliva are amylase and invertase, although lipase and protease also have been found. Aphids secrete a pectinase that aids their mouthparts in the penetration of plant tissues. The spreading factor, hyaluronidase, which attacks a constituent of the intercellular matrix of many animals, has been found in the assassin bug. Anticoagulins and agglutinins have been found in the saliva of bloodsucking insects such as the mosquito and the tsetse fly. Anticoagulins would, of course, facilitate the pumping of blood by blocking the clotting mechanism of the host.

Since salivary enzymes are commonly injected into the tissues of host plants or animals or are applied to the surface of food materials, a certain amount of digestion in many insects is accomplished external to the alimentary canal (i.e., extra-intestinally). Some insects (e.g., larval dytiscid beetles) actually regurgitate digestive juices from the gut and inject them into the prey through their mandibles. In this situation a large portion of digestion is extra-intestinal, the tissues of the prey becoming a soupy material that is drawn into the predatory larva through the mandibles.

Microbiota and Digestion

Under normal circumstances, all insects possess intestinal symbionts, which may or may not contribute to their host's well-being. Among the intestinal microflora and microfauna are found bacteria, protozoa, and fungi of various species. In some cases, as will be discussed later, these microbes may provide substances of nutritive value (e.g., vitamins) for their hosts. In other cases these symbionts may synthesize an enzyme which enables the insect to digest substances that in the absence of such microbes it would be unable to accomplish. Leafhoppers (order Homoptera; family Cicadellidae) harbor yeasts capable of digesting starch and sucrose. Bacteria that ferment cellulose are probably always present in the alimentary tracts of wood-feeding insects. A considerable amount of cellulose fermentation probably occurs in the wood itself, but undoubtedly much also occurs in the guts of certain insect species. Some insects, such as the lamellicorn beetles (stag, bess, and scarab beetles), possess morphological modifications of the gut designed to house these cellulose-fermenting bacteria. In the lamellicorn beetles a portion of the hindgut is enlarged, forming a "fermentation chamber." Wood-feeding termites similarly possess an enlarged sac in the hindgut, but in this case the sac harbors mainly cellulase-secreting flagellate protozoans required for the survival of the termites on a cellulose diet. Certain microbes may not provide an important enzyme but may aid significantly in the processing of the food, for example, the larvae of the blow fly, *Lucilia,* in which bacteria in the gut produce an alkaline state, which aids in the liquefaction of ingested animal tissues. In addition, fungi may aid certain wood-feeding insects in the breakdown of cellulose. It should be mentioned that these intestinal microflora and microfauna also serve as "food" for the host insect.

Insect Nutrition

The inclusion of nutrition with the system responsible for the initial processing of ingested materials seems most logical. Given a source of chemical energy and the basic raw materials, insects, like other heterotrophic animals, are able to synthesize many, if not most, of the more complex molecules necessary for the maintenance of life. However, there are certain molecules which they cannot synthesize but which are necessary in one way or another for survival and reproduction. These molecules, along with those required as raw materials and sources of chemical energy, make up the nutritional requirements of a given insect. Not only are specific molecules required, but the quantities of each are of great importance. Too little of one may result in the impairment of some vital function of the insect. Too much of a given molecule in the diet would result in the excretion of the unneeded portion of it, and since excretion is an energy-utilizing process, this would be metabolically inefficient. Energy-yielding molecules such as sugars and fats of various types are needed in comparatively large amounts in terms of grams per kilogram of body weight per day. In adult insects, substances such as amino acids, purines, and certain lipids are required in somewhat lesser, but still fairly large, amounts, milligrams per kilogram of body weight per day. Immature insects probably require larger amounts of amino acids since they are more actively synthesizing structural protein. Other molecules, in particular vitamins and minerals, are required in comparatively minute amounts, micrograms per kilogram of body weight per day. Specific organic and inorganic molecules required by insects will be discussed below.

Amino Acids.   Amino acids are the building blocks of protein. Different insects, of course, have different requirements, depending upon which amino acids they are capable of synthesizing. As already pointed out, those which they cannot synthesize will fall into the required category. These required amino acids may be present in the food material in a free state or much more commonly in protein that is digested to its component amino acids in the gut. The qualitative and quantitative amino acid composition of a given protein determines its nutritive value.

Carbohydrates.   Carbohydrates are not considered to be essential nutritive substances for most insects, but they are probably the most common source of chemical energy utilized by insects. However, many insects (e.g., many Lepidoptera) do, in fact, need them if growth and development are to occur normally.

Lipids.   Lipids, like carbohydrates, are good sources of chemical energy, in addition to being important in the formation of membranes and synthesis of hormones. It is considered that most insects can probably synthesize most of the necessary fatty acids that make up the larger lipid molecules from carbohydrate and protein. However, some insect species do require certain fatty acids and other lipids in their diets. For example, certain Lepidoptera require linoleic acid for normal larval development. Apparently all insects require a dietary sterol for growth and development. Examples of such dietary sterols are cholesterol and ergosterol.

Vitamins.   Vitamins include a diverse group of compounds which are required in very small amounts for the normal functioning of any animal. They are not used for energy nor do they form part of the structural framework of the insect tissues. Vitamin A is required for the normal functioning of the compound eye of the mosquito *Aedes aegypti* (Brammer and White, 1969). Vitamin A is a fat-soluble vitamin. However, vitamins required by insects are principally of the water-soluble type (e.g., biotin, choline, and nicotinic acid). These compounds are coenzymes, working in conjunction with enzymes in specific metabolic reactions.

Minerals. Like vitamins, various minerals are required by insects in very small amounts. Minerals required by insects include potassium, phosphorus, magnesium, sodium, calcium, manganese, copper, and zinc. Various minerals have been shown to be necessary for normal growth and development in certain insects. Some insects (e.g., the aquatic larvae of mosquitoes, which possess very thin-walled anal papillae; see Fig. 3–18) are able to absorb mineral ions from the water through the cuticle of these structures.

Purines and Pyrimidines. The nucleic acids, DNA and RNA, are composed of purines and pyrimidines. RNA exerts a positive effect on the growth of certain fly larvae.

Water. All insects require a source of water, whether it be from food, by drinking, absorption through the cuticle, or from a metabolic source. Insects vary greatly with respect to amounts of water needed. Those like the mealworm, *Tenebrio molitor,* can survive and reproduce on essentially dry food. Others, for example honey bees and muscid flies, require large amounts of water for survival. The excrement of the mealworm is hard and dry, with almost all the water having been reclaimed by the insect, while the excrement of bees and muscid flies contains large amounts of water.

Microbiota and Nutrition

Experiments involving the surficial treatment of eggs to destroy symbiotic or potentially symbiotic microbes, for example those in food, have shown that some insects require the presence of certain of these microbes for normal growth and development and in some instances survival. In many cases the microbes present are passed from generation to generation. In some insects in which this occurs, specialized cells, *mycetocytes,* and the tissues composed of these cells, *mycetomes,* can be found associated with the gut, fat body, or, appropriately, the gonads. The latter would ensure the infection of any eggs produced, thus furnishing one of the possible means for bridging the gap between generations. Whether hereditary or not, microbes are commonly found in the alimentary canals of insects, often in the various diverticula of the midgut. The kinds of microbes reported to be likely contributors to their host's nutritional requirements include yeasts, other fungi, bacteria, and protozoans. These microbes probably benefit the insect in various ways. Suggestions include the possible fixation of atmospheric nitrogen; synthesis of protein from nitrogenous waste materials; and the provision of vitamins (particularly those of the B group), sterols, and amino acids.

The Circulatory System

The circulatory system in insects is rather different from that of vertebrates and many other invertebrates in that the major portion of the blood or hemolymph is not found within the confines of a closed system of conducting vessels, but instead bathes the internal organs directly in the body cavity, or *hemocoel*. This "blood cavity" is not a true *coelom,* that is, one lined entirely with mesodermal tissue. Although coelomic sacs do occur in the embryo, almost the entire hemocoel is formed from the epineural sinus and lined with ecto- and endodermal tissue. The only conducting tube is the *dorsal vessel*, which is a pulsatile structure and generally extends the length of the insect from the posterior part of the abdomen to just beneath the brain in the head. The blood of insects is commonly called *hemolymph,* a term which implies that it carries out the functions of both blood and lymph, which are distinctly different fluids in vertebrates.

General Characteristics of the Hemolymph

Hemolymph is a clear fluid that is usually colorless or may have a slight greenish or yellowish color imparted by certain pigments. Outstanding exceptions to this are found among some midge (true flies, order Diptera, family Chironomidae) larvae, the hemolymph of which contains the pigment hemoglobin, which imparts a distinct reddish color. The hemolymph makes up from approximately 5 to 40% of the total body weight of an insect, depending on the species. However, the hemocoel is not usually filled with blood and its volume varies with the physiological state of the insect. The pH of the blood is usually slightly acid, in the range between pH 6 and 7. However, in a few insects the pH may be slightly alkaline, pH 7 to 7.5. Buck (1953) states that the pH of the hemolymph may vary intraspecifically up to 0.7 pH unit.

Insect hemolymph is slightly more dense than water, having a specific gravity somewhere between 1.015 and 1.060 and is subject to increase during periods of molting (Patton, 1963). The total molecular concentration in the hemolymph is fairly high, a fact that accounts for the osmotic pressure being somewhat higher than that of mammalian blood. As pointed out by Florkin and Jeuniaux (1964), free amino acids, organic acids, and other organic molecules play a significant role as osmolar effectors in insects, in contrast to vertebrate blood, in which inorganic anions and cations are largely responsible for the osmotic pressure. Like the blood of vertebrates, insect blood can be conveniently separated into two fractions: a fluid portion, or *plasma,* and a cellular portion, the *hemocytes.* In addition to the two basic fractions, several nonhemocytic elements may be found (Jones, 1964). These elements include muscle fragments, free fat body cells, oenocytes, free crystals, spermatozoa, various parasitic organisms (e.g., bacteria, protozoans, and nematodes), tumor cells, and similar entities.

Chemical Composition of the Hemolymph

The chemical composition of the hemolymph shows considerable variation, both qualitatively and quantitatively, among the various insectan groups and is subject to variation in the same species depending upon the physiological state, age, sex, food, and so on, of the organism. A brief listing of some of the major constituents that have been identified follows. The reviews by Florkin and Jeuniaux (1964), Wigglesworth (1965), and Buck (1953) contain extensive information on this subject.

Water. As with all other organisms, in terms of amount, water is the major component of the internal body fluid of insects. A listing given by Buck (1953) contains figures for the water content of hemolymph, which range from 84 to 92% of the total body weight, a range not dissimilar to that found in human blood plasma. Figures from other sources indicate that the percentage of total body weight that is water may be much lower, perhaps less than 50% in certain insects.

Inorganic Constituents. Sodium, potassium, calcium, sulfur, magnesium, chloride, phosphorus, and carbonate comprise the major inorganic materials that have been identified in insect hemolymph. The chloride content is comparatively low relative to that in mammals, and phosphate, calcium, and magnesium are the reverse, occurring in somewhat higher concentration in insects than in mammals. In addition, copper, iron, aluminium, zinc, manganese, and other metallic elements have been found in very small amounts.

Nitrogenous Waste Materials. As the metabolism of proteins and amino acids takes place, nitrogenous waste materials are formed. In insects, these wastes are usually in the form of uric acid, which occurs in a very high concentration in the hemolymph. It is produced mainly in the fat body and is usually excreted by the Malpighian tubules. Other nitrogenous waste materials produced by insects include urea, allantoin, allantoic acid, and ammonia, the latter being formed mainly in aquatic insects.

Organic Acids. Succinate, malate, fumarate, citrate, lactate, pyruvate, α-ketoglutarate, and several organic phosphate compounds are the major organic acids found in insect blood. The first seven are formed during the tricarboxylic acid cycle. It has been suggested that these acids are important in balancing the cations in the blood.

Carbohydrates. Much of the carbohydrate found in insect blood is in combination with protein, forming glycoprotein. A rather surprising discovery has been that α-trehalose, a disaccharide composed of two glucose molecules, is the major blood sugar. Prior to this discovery, trehalose was known only from certain plants and from the cocoons of a species of beetle. However, trehalose is not the major blood sugar in every insect; glucose, fructose, ribose, and others have also been found. Only very small amounts of glycogen have been found in the hemolymph. In addition, glycerol has been found in quite high concentrations in certain insects exposed to cold seasons. In some cases it is thought to lower the freezing point of the hemolymph, thus serving as a kind of insect antifreeze. Salt (1961) discusses this topic at length.

Lipids. According to Wigglesworth (1965), lipids occur in insect blood either as minute fat particles or as lipoprotein (i.e., lipids in combination with protein).

Amino Acids.   The presence of free amino acids in the hemolymph in concentrations higher than that reported in any other animal group is one of the outstanding characteristics of the class Insecta. These free amino acids may be either dietary or synthesized by the insect. Explanations of the function of this exceedingly high amino-acidemia (amino acid in the blood) include the idea that the amino acids are excesses derived from the diet and are being stored in the hemolymph until they can be excreted or that they serve as a reservoir of the raw materials for the construction of new cells during the periods of growth and metamorphosis.

Proteins.   The proteins in insect blood occur in a concentration similar to that in the blood of vertebrates and in somewhat higher concentration than that in the internal fluids of other invertebrates (Florkin and Jeuniaux, 1964). Several different protein fractions have been demonstrated, many of which are in combination with other constituents (e.g., lipoprotein and glycoprotein), but the best defined are those which act as enzymes. A number of enzymes, among them lipase, protease, sucrase, and amylase, have been found in the hemolymph and are most active during metamorphosis. The protein pool in the hemolymph may also serve as a reservoir of raw materials and one of the sources of the free amino acids.

Pigments.   Several pigments have been identified in insect blood. Among them are hemoglobin (a conjugated protein), already mentioned; kathemoglobin, derived from the blood meals taken by the bloodsucking bug, *Rhodnius*; carotene and xanthophyll in plant-eating (phytophagous) insects; and others—riboflavin, fluoroscyanine, and so on.

Gases.   Both oxygen and carbon dioxide may occur in the hemolymph, usually in very low concentrations since the major route of gaseous exchange is via the tracheal system. Exceptions to this are found, of course, in those species which contain the oxygen-carrying pigment, hemoglobin, in their blood.

In summary, the composition of the insect hemolymph is in some respects similar to that of the internal body fluids in other animals. However, there are many blood characteristics that are rather distinctive in insects. Among the more significant of these are the high aminoacidemia, the functioning of amino acids and other organic acids as important osmolar effectors, the unique blood sugar trehalose, the fact that oxygen is usually not carried over great distances by the blood, and the relatively high concentration of uric acid in the blood.

Hemocytes

A number of morphologically distinct cells have been identified in the hemolymph of insects (Fig. 3–6). Some of these hemocytes are found in all groups of insects; others are less common, and some are quite rare, found only in a few or even a single species. Jones (1964) describes nine basic hemocyte types, some of which are considered to

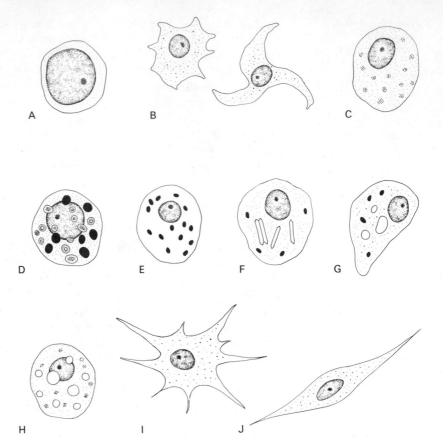

Fig. 3–6

Examples of hemocyte
types. A. Prohemocyte.
B. Plasmatocytes.
C. Granular hemocyte.
D. Spherule cell.
E. Cystocyte. F and G.
Oenocytoid cells.
H. Adipohemocyte.
I. Podocyte.
J. Vermiform cell.

give rise to other forms. A listing of the hemocytes described by Jones
follows. Gupta (1969) provides an extensive and very useful listing of
the synonomy of hemocyte terminologies used by different investigators.

Prohemocytes. These are the smallest blood cells and contain a
large, round nucleus and a small amount of cytoplasmic material,
which gives a basic (basophilic) stain reaction. They have been
identified in all groups of insects and are commonly observed undergoing mitotic division.

Plasmatocytes. Blood cells of this type are widely diverse in form
(polymorphic); some are round, some spindle-shaped, and some have
other shapes. They usually possess a single large nucleus located in
the center and have a granular cytoplasm, which, like the prohemocytes, gives a basic stain reaction. These are the dominant hemocytes
in many species. Gupta (1969) includes prohemocytes in the plasmatocyte category.

Granular Hemocytes. Like the plasmatocytes, granular hemocytes
contain a large, centrally located nucleus. However, this type contains
a large number of inclusions of uniform size that give an acid stain
reaction (acidophilic). They are thought to be derived principally from
prohemocytes, although some may arise from plasmatocytes.

Spherule Cells. Spherule cells contain spherical or sometimes elliptical cytoplasmic inclusions, which commonly obscure the nucleus. The inclusions, like those in granular hemocytes, are acidophilic.

Cystocytes. These are also sometimes called *coagulocytes*. Blood cells of this type have a single, small nucleus with a weakly basophilic or hyaline (colorless; neither basophilic nor acidophilic) cytoplasm with distinct acidophilic inclusions. The rapid breakdown of these cells *in vitro* is associated with plasma precipitation.

Oenocytoids. These are large cells, which may have one or two excentric (not centrally located) nuclei and a weakly acidophilic cytoplasm, which may be filled with a variety of inclusions, depending on the species. They occur in various shapes.

Adipohemocytes. Adipohemocytes are round or ovoid cells which generally have distinct droplets of fat or fatlike substance in the cytoplasm. The cytoplasm also has granular inclusions.

Podocytes. A podocyte is a rare type of hemocyte which is quite large and flattened and bears a varying number of cytoplasmic extensions.

Vermiform Cells. As is well described by the term *vermiform* (in the form of a worm), vermiform cells are long and threadlike, 100 to 300 microns in diameter. They are uncommon.

Origin of Hemocytes

Prohemocytes, plasmatocytes, granular hemocytes, and sperule cells have all been observed undergoing mitotic division. Several investigators consider the prohemocytes to be the basic stem cells from which all the others are derived (Jones, 1964). However, on the basis of *in vitro* transformations of the hemocytes in the American cockroach, Gupta and Sutherland (1966) suggest that the polymorphic plasmatocyte is the basic type of insect hemocyte.

In addition to the transformation of hemocytes in the blood as described above, there have been identified in several orders of insects distinct *hemocytopoietic organs* (blood-cell-producing organs) in which hemocytes may multiply and/or differentiate. These organs are apparently most active during periods of metamorphosis.

Number of Hemocytes

Counts of the total number of hemocytes per unit volume of insect hemolymph have proved to be quite variable, depending upon such parameters as species, developmental stage, various physiological states, and the technique applied to obtain the count. The number of hemocytes per cubic millimeter of hemolymph usually falls somewhere in the range 30,000 to 50,000 (Patton, 1963).

Functions of Hemolymph

Based on what we have said thus far, it is obvious that the hemolymph is a very complex mixture of a variety of materials and different cell types. This complexity is reflected in the many diverse functions carried out by the components of the hemolymph. A brief description of some of the major functions of the hemolymph follows.

Lubricant. The hemolymph, the circulating fluid that bathes the various tissues of the insect's body, serves as a lubricant, allowing easy movement of the internal structures relative to one another.

Hydraulic Medium. Like the hydraulic fluid in an automobile braking system, insect blood is essentially incompressible. This being the case, forces that tend to reduce the blood volume in one portion of the insect's body (e.g., compression of the abdomen) are transferred via the hemolymph to other parts of the body and may effect a change there. A good example of this is found among the members of a group of dipterans (suborder Cyclorrhapha). These flies, when ready to emerge from their puparial cases, which have contained them throughout the period of change (metamorphosis) from immature to adult form, literally pop the preformed lid from the end of the case. They accomplish this by means of the "hydraulic" extrusion of a bladderlike structure, the *ptilinum,* from the anterior portion of their heads. The protrusion of similar structures is also effected by this action. In some, as mentioned in Chapter 2, this hydraulic force may actually oppose muscular contraction in the extension of an appendage. In addition, the expansion of the wings of a newly emerged adult insect may be accomplished at least in part by the forcing of hemolymph into them.

Transport and Storage. Like the body fluids of other animals, the insect hemolymph is involved with the transport of various substances from one tissue or organ to another. These "substances" include nutrients (amino acids, sugars, fats, and so on) absorbed through the gut wall or released from cells that store such materials; metabolic wastes (nitrogenous materials) and foreign materials to be excreted or taken in and stored by certain cells; hormones; and, over usually very short distances, oxygen and carbon dioxide. Any of these may be carried in the plasma, and, in some instances, at least the first three mentioned are carried by certain hemocytes. In addition to transport, the hemolymph serves as an important storage pool for the raw materials for construction of new cells.

Protection. We include under the general heading "Protection" those processes which have been observed to provide some advantage to insects when foreign entities (animate or inanimate) or cell fragments gain entrance to the hemolymph. Following are some of the ways by which the hemolymph affords a degree of protection to insects.

1. *Phagocytosis.* Certain hemocytes, in particular plasmatocytes and granular hemocytes, actively ingest foreign particles of various sorts, bacteria, and cellular debris. This phagocytic role is probably

important in some instances in the protection of the insect from severe infection by microorganisms, but it becomes especially important during the periods of molting and metamorphosis, when many tissues are in a state of disintegration and fragments of cells freely invade the hemolymph.

2. *Encapsulation.* In certain situations, large numbers of hemocytes may actually become layered around an invading entity of a size larger than that which may be phagocytized. This response of the hemocytes may be of value in inhibiting in various ways the activities of an internal parasite.

3. *Detoxification.* Some hemocytes are capable of rendering toxic metabolites and certain insecticidal materials nontoxic.

4. *Coagulation.* Most insects possess "hyaline hemocytes" (coagulocytes included), which are extremely fragile. These cells have been effectively studied only by means of phase contrast microscopy. During coagulation, these cells either produce a fine precipitate or threadlike cytoplasmic extensions that enmesh the cells around them. A recent review by Grégoire (1964) examines this topic in rigorous detail.

5. *Wound Healing.* Hemocytes of various types commonly tend to accumulate at sites of injury and actively phagocytize, coagulate, and produce protective membranes of connective tissue, any one or all of which may aid in the healing of a wound.

6. *Noncellular Protective Factors.* There is evidence that implicates factors other than blood corpuscles in the protection of an insect via the hemolymph. Briggs (1964) discusses these "factors" under the heading "Humoral Reactions" and cites a number of appropriate papers pertinent to this subject.

Formation of Other Tissues.   Certain hemocytes may be capable of forming other tissues; however, there is little evidence available at present to state that this is definitely true.

The Dorsal Vessel and Accessory Pumping Structures

The dorsal vessel (Fig. 3–7) is the principal organ responsible for blood circulation. It lies along the dorsal midline of the insect body, extending from the posterior region of the abdomen to just behind or beneath the brain in the head. It is largely composed of circular muscle, but it may possess longitudinal fibers and may have a connective tissue sheath around the outside. Fine elastic fibers, which arise from the dorsal integument, alimentary canal, somatic muscles, and other structures, serve as suspensors for the dorsal vessel. The dorsal vessel may or may not be extensively tracheated. According to Patton (1963) a pair of lateral nerves run along the heart and give off branches into the heart wall, alary muscles, and lateral vessels when present. These nerves are formed by fibers from both the stomodael system and the ganglia of the ventral nerve cord. The dorsal vessels of some insects are apparently not innervated at all.

It is commonly possible to identify two major divisions of the dorsal vessel (Fig. 3–7): a posterior *heart* and an anterior *aorta*. The heart is usually closed at its posterior end and bears a number of

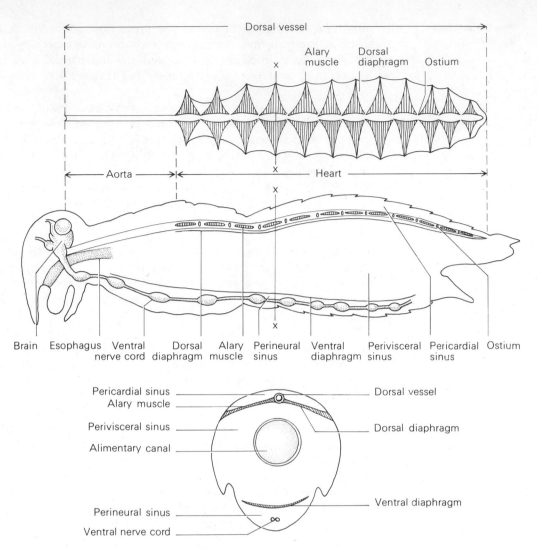

Fig. 3–7

Generalized insect circulatory system. x—x's in top two diagrams indicate cross-sectional plane of bottom diagram.

valves, or *ostia*, the function of which is generally to allow the entrance, but prevent the exit, of hemolymph. Ostia usually occur in laterally oriented pairs but may in some cases be single and located ventrally with respect to the heart. The heart may be constricted between successive ostia, giving it a chambered appearance, but its lumen is in nearly every instance continuous throughout. On either side of the heart in a segmental arrangement are the "wing" or alary muscles, so named because when viewed from either a dorsal or ventral aspect, they resemble wings. These fibroelastic structures are attached laterally to the body wall and vary in number from 1 to 13 pairs, depending on the species of insect. Lateral segmental vessels associated with the heart have been identified in several insects, for example many cockroach species. The heart may extend into the thorax but is generally confined to the abdomen. The aorta extends anteriorly from the heart and opens behind or beneath the brain.

It lacks any valvular openings and in some insects may be variously thrown into one or two vertical loops or arches, lateral kinks, or coils. Also it may have dilated portions along its course.

In addition to the pumping activities of the heart, various accessory pulsatile structures (Fig. 3–8) aid the movement of hemolymph in regions comparatively distant from the dorsal vessel. They have been found at the bases of the antennae, at the bases and within the legs and wings, and within the meso- and metathorax.

Sinuses and Diaphragms

The hemocoel, particularly in the abdomen, is usually separated into two and sometimes three cavities or *sinuses*, which may have a pronounced effect on the circulation of the blood. This compartmentation is produced by the presence of one or two fibromuscular septa. The one that is generally present is the *dorsal diaphragm* or *pericardial septum*. It typically consists of two layers, which enclose the alary muscles associated with the heart. It may be imperforate, but is usually not, containing lateral *fenestrae* or "windows" through which hemolymph can readily pass. This diaphragm divides the dorsal *pericardial* ("around the heart") *sinus* from the *perivisceral sinus* ("around the gut"). In many insects, a second septum, the *ventral diaphragm*, is present. It also contains muscle fibers and is located ventral to the alimentary tract but dorsal to the ventral nerve cord. Like the dorsal septum, it is usually fenestrated around its periphery. The ventral diaphragm separates the perivisceral from the perineural ("around the nerve cord") sinus. Undulation of both the dorsal and the ventral diaphragm may aid appreciably in the circulation of blood.

Mechanism of Circulation

The general mechanism of circulation in insects can be described as follows. The blood enters the heart through the ostia. It is then directed anteriorly by a wave of peristaltic contraction which passes along the dorsal vessel in the direction of the head. The direction of propagation of the wave of peristaltic contraction has been observed to reverse on occasion. Release of the blood from the dorsal vessel occurs in the head. With the aid of the undulatory movements of the diaphragms and the actions of the accessory pulsatile structures, the blood is then circulated throughout the general body cavity and appendages. Ultimately the blood is forced through the fenestrae of the dorsal diaphragm and enters the heart during its relaxation phase. Figure 3–8 shows the generalized pattern of blood flow in an insect. The alternate phases in the heartbeat cycle, contraction *(systole)* and relaxation *(diastole)*, can be measured at a single point along the heart. Figure 3–9 shows a characteristic pattern that results when these measurements are made mechanically. Movement of the curve upward results from contraction and downward, relaxation. A period of rest or *diastasis* occurs between successive beats. Also, toward the end of this rest period a definite expansion may occur. This produces the *presystolic notch* and may possibly be the

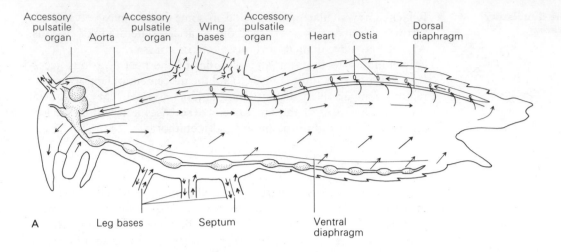

Accessory pulsatile organ | Accessory pulsatile organ | Aorta | Wing bases | Accessory pulsatile organ | Heart | Ostia | Dorsal diaphragm

A

Leg bases | Septum | Ventral diaphragm

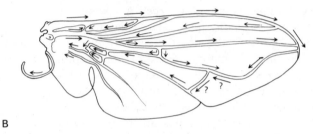

B

Fig. 3–8
Patterns of circulation. A. Insect with a fully developed circulatory system. B. Pattern of circulation in a house fly wing. (B redrawn from West, 1951.)

Fig. 3–9
Mechanical recording of two heartbeats (cardiac cycles) in an isolated cockroach, *Periplaneta americana*, heart.
A, Systolic phase;
B, diastolic phase;
C, relaxation phase with presystolic notch.
(Redrawn with modifications from Yeager, 1938.)

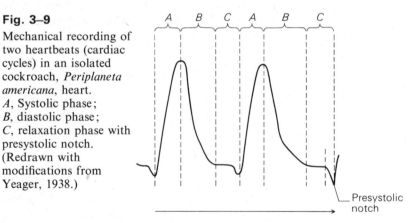

A   B   C   A   B   C

Presystolic notch

result of contraction of the alary muscles or blood being forced from behind into the portion of the heart where recordings are being made.

The rate and amplitude of the heartbeat are affected by a variety of factors. Patton (1963) lists the following: ambient temperature, metabolic rate, developmental stage, and the presence of biologically active chemicals, insecticides being included with the last mentioned. The accessory pulsatile structures are independent of the heart but are influenced by the same factors. Patton summarizes: "In general, any material [or other factor] that increases metabolic rate or acts upon the nervous system will alter either the rate or the amplitude of the heartbeat." The average hemolymph pressure is generally thought to be quite low, although by contractions of regions of the body wall, considerable hydraulic pressure can be brought to bear locally.

Control of Heartbeat

Four types of heartbeat have been described for insects:

1. *Noninnervated myogenic:* The actions of the heart are seemingly independent of nervous modification and control resides within the heart itself.
2. *Innervated myogenic:* The action of the heart is subject to modification by nervous impulse, but rhythmic beating continues to occur when the nervous connections are severed.
3. *Neurogenic:* The actions of the heart are entirely under neural control.
4. *Hormonal:* In certain insects the heart beat is accelerated by a hormone secreted by the corpus cardiacum.

Other Tissues Associated with the Circulatory System

Pericardial Cells. These are comparatively large, usually multi-nucleate cells of mesodermal origin found in association with the heart, alary muscles, and connective tissue (Fig. 3–10A). They are not found circulating in the hemolymph but are stationary in the vicinity of the heart. They possess the ability to absorb particles of colloidal size from the hemolymph. In addition to the pericardial cells, cells in other parts of the insect body have the same ability. According to Wigglesworth (1965), the pericardial cells and those in other locations are probably analogous with the reticuloendothelial system of vertebrates.

Fat Body. The fat body (Fig. 3–10B) is variously distributed in the insect hemocoel depending on the species, but its distribution is the same for all members of a species at a comparable stage of development. It has been described as a loose meshwork of lobes, composed of cells of mesodermal origin and invested in connective tissue strands. Fat body cells function mainly in the storage of various materials, including fat, proteins, and glycogen. Especially large reserves accumulate during the larval stages in insects that later undergo complete metamorphosis. These reserves are probably important in the provision of nutrients and raw materials during the non-feeding pupal stage. Also it is likely to be an important tissue in terms of intermediary metabolism.

Oenocytes. These are highly specialized cells of ectodermal origin found in nearly all insects except certain thysanurans (Fig. 3–10C). They may be arranged segmentally or be dispersed randomly. The function of the oenocytes has not been completely elucidated, although there is evidence that they secrete the lipoprotein, which forms the cuticulin layer of the epicuticle and possibly the free wax on the surface of the integument of cockroaches (Wigglesworth, 1965). Other functions have been postulated but have not been supported by conclusive evidence.

Fig. 3–10

Photomicrographs of tissues associated with the circulatory system in adult mosquito, *Aedes triseriatus*. A. Pericardial cell. B. Fat body. C. Oenocytes. 1, Pericardial cell; 2, heart; 3, midgut epithelium; 4, cuticle; 5, muscle; 6, fat body; 7, oenocytes. (Scale lines = 50 microns.)

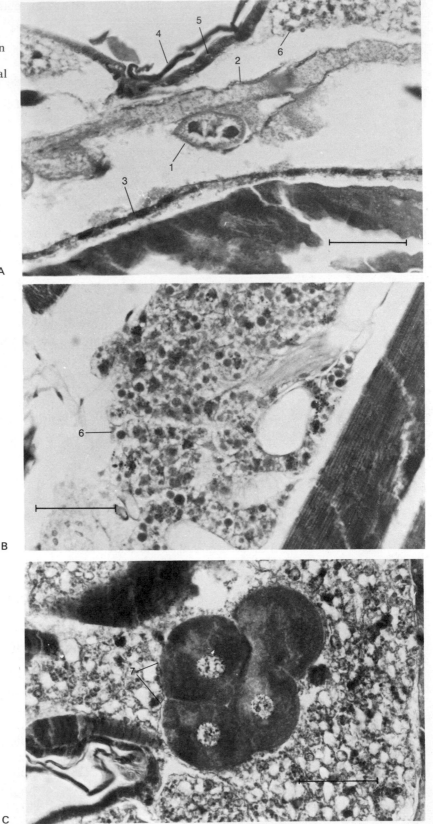

A

B

C

The Ventilatory System

The ventilatory system is involved with the transport of oxygen to the various tissues of the insect body and the transport of carbon dioxide from them. Use of the term *ventilation* to denote this gaseous exchange seems preferable to *respiration,* which in the strict sense is an intracellular process involving the oxidative breakdown of carbohydrate to carbon dioxide and water.

Structure of the Ventilatory System

Tracheae. In a few insects [e.g., many springtails (order Collembola) and certain hymenopterous larvae (ants, bees, wasps, and relatives)] all gaseous exchange occurs via the integument, but in the vast majority it is accomplished by means of a usually quite elaborate system of branching tubules or *tracheae* (Fig. 3–11). In most terrestrial and many aquatic insects these structures communicate with the outside by means of comparatively small openings, the *spiracles.* Internally the tracheae divide and subdivide, their diameters becoming successively smaller and smaller as they probe deeply into the tissues of the insect, eventually giving rise to extremely fine *tracheoles,* the tubes from which the major portion of the oxygen actually enters the cells. Tracheae are of ectodermal origin and are continuous with the integument. In the embryo they originate in each segment independently of one another. However, with the exception of a few apterygote insects, in postembryonic immature and adult stages they do not retain this early developmental arrangement, but become joined by longitudinal trunks so that all parts of the tracheal system are in communication.

From each original developing trachea, branches are given off dorsally to the body wall, heart and aorta, various muscles, and so on; mesially to the alimentary tract and associated structures; and ventrally to the nerve cord, body wall, various muscles, and so on. The tracheae are circular or somewhat elliptical in cross section. Histologically (Fig. 3–12A, B) they are similar to the integument,

Fig. 3–11

Cross section of insect thorax showing the major tracheal branches (diagrammatic). (Redrawn from Essig, 1942.)

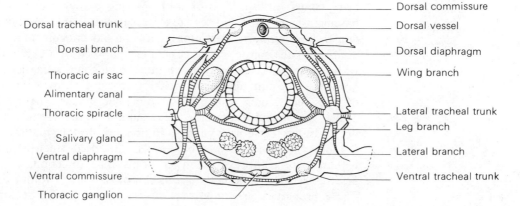

Dorsal tracheal trunk — Dorsal commissure

Dorsal branch — Dorsal vessel

Thoracic air sac — Dorsal diaphragm

Alimentary canal — Wing branch

Thoracic spiracle — Lateral tracheal trunk

Salivary gland — Leg branch

Ventral diaphragm — Lateral branch

Ventral commissure — Ventral tracheal trunk

Thoracic ganglion

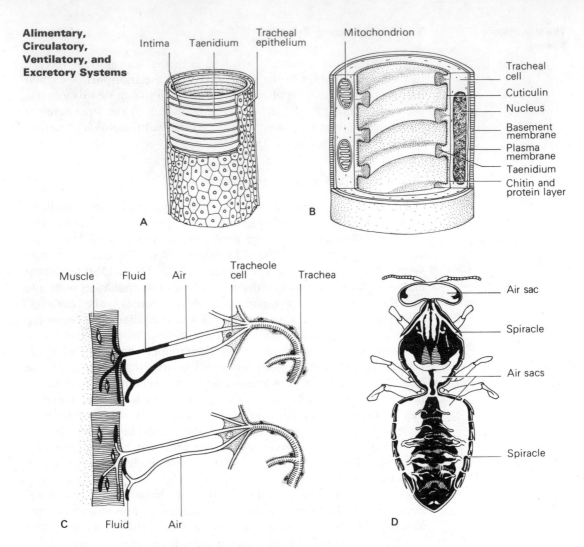

Fig. 3–12

Tracheal structure. A. Portion of trachea. B. Fine structure of trachea.
C. Tracheoles in close contact with muscle. D. Air sacs in the honey bee.
(A and D redrawn from Snodgrass, 1963b; B redrawn from Miller,
1964; C redrawn from Wigglesworth, 1930.)

being composed of a layer of epithelial cells bordered internally by a
basement membrane and secrete a cuticular layer into the lumen, the
intima. The intima is shed along with the old integument with each
molt. Near the spiracles, the cuticle is composed of a cuticulin layer
with a chitinous layer beneath; chitin is lacking in the smaller tracheal
branches. The cuticular lining of the tracheae is thrown into a series
of folds that usually run spirally round the lumen. These folds are
called *taenidia* and lend a measure of support to the tracheae, protecting against collapse with changes in pressure. In the smaller tracheae
the folds of the intima may be annular. Although tracheae are quite
resistant to compression in a transverse direction, they can
be stretched longitudinally a considerable amount without damage.
These properties are important in insects in which the abdomen

becomes greatly distended with food, for example in several blood-sucking forms.

Tracheoles. The tracheoles are the smallest branches of the tracheal system, ranging in size from 0.2 to 1.0 micron in diameter. When all the studies of tracheoles were based on light microscopy, they were thought to lack taenidia, but when examined with the electron microscope, very tiny folds can be identified. The lining of the tracheoles, unlike that of the tracheae, is not shed at molting. The transition between tracheae and tracheoles typically occurs with a *tracheal end cell,* which gives off a number of tracheoles, which are all a part of this cell (Fig. 3–12C). Other tracheae may merge gradually into a tracheole; in some cases the tracheoles arise directly and abruptly from the side of a trachea. The tracheoles may be involved with other tissues in a variety of ways. For example, the processes of the end cells may intertwine and form a sort of tracheal net around certain organs, such as the ovaries and testes, or they may merely lie on the surface of some tissues, such as those of the alimentary tract. Tracheoles may also be interstitial, that is, located between the cells of a tissue. These smallest branches of the tracheal system nearly always end blindly at a diameter of about 0.2 micron.

Although the basic pattern of tracheation is genetically determined, new tracheae and tracheoles can be induced to develop if an insect is reared in an atmosphere with a very low oxygen content. New tracheoles do not develop between successive molts. However, changes in the distribution can occur at the time of molting if there is a demand. For example, if the tracheae and tracheoles in one portion of an insect are destroyed, this section will receive tracheation from an adjacent portion in the next developmental stage.

Air Sacs. Air sacs (Fig. 3–12D) are tracheal dilations of varying size, number, and distribution found mainly in flying insects. They are typically oval in cross section and lack chitin in their cuticular layers. The walls of the cuticular lining lack oriented taenidia, and for this reason air sacs are quite distensible and collapsible. Tracheal air sacs are especially pronounced in those insects which are the most active ventilators. Thus one may actually observe the alternate distension and collapse as air is taken in and released by the insect.

Several functions of air sacs have been discovered; the major one is probably to increase the volume of the tidal air, that is, the volume of air which is inspired and expired. The presence of large air sacs in the body cavity of an insect appreciably lowers the specific gravity and in this way may aid in flight. Air sacs may also provide room for growth of internal organs, as is the case in certain female flies, in which the air sacs provide room for the growth of the ovaries. Other observed functions include aiding in heat conservation in those large insects which must generate high temperatures for flight, assistance in hemolymph circulation, reduction of the mechanical damping of flight muscles by the hemolymph, and formation of the tympanic cavity of the hearing organs of various insects.

Spiracles. Spiracles are formed from the mouths of the ectodermal invaginations, which produce the tracheae. They occur in two basic

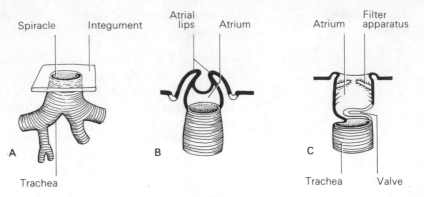

Fig. 3–13
Types of spiracles.
A. Simple, nonatriate
type. B. Atriate spiracle
with lip closure
mechanism. C. Atriate
spiracle with filter
apparatus and valve
closure mechanism.
(Redrawn from
Snodgrass, 1935.)

types, simple and atriate. The simple type (Fig. 3–13A) of spiracle is merely an opening to the tracheal system. The atriate type is formed as a result of the entad migration of the primitive (simple) spiracular opening. Thus in the fully developed atriate spiracle (Fig. 3–13B, C), the opening to the trachea, the tracheal orifice, lies at the bottom of a spiracular chamber or *atrium*. In this type the opening to the outside of an insect is referred to as the *atrial aperture* or *orifice*.

In addition to being quite permeable to oxygen and carbon dioxide, the tracheal system also readily allows the passage of water. This presents a problem for insects since they would naturally tend to lose water very rapidly through the integument or tracheae by evaporation, a phenomenon called *transpiration*. The presence of various types of spiracular closing mechanisms has been one of the evolutionary solutions to this problem. These closing mechanisms are found in most atriate spiracles. Two principal types of closing mechanisms may be found (Snodgrass, 1935): the lip type (Fig. 3–13B), in which folds of the integument form opposing lips, which can be pulled together and effect closure of the atrial aperture; and the valvular type (Fig. 3–13C), which lies at the inner end of the atrium and regulates the size of the tracheal orifice.

Whatever the type of mechanism, closure is ultimately effected by contraction of the associated muscles. Atriate spiracles with the second type of closing mechanism may be lined with tiny hairs, forming a *felt chamber*, or may bear other structures, such as *sieve plates*, which are porous covers over the atrial aperture that probably serve to retard water loss and to prevent airborne particles from entering the tracheae. In addition to these closing mechanisms and accessory structures, glandular tissue is commonly associated with spiracles. In many, especially aquatic, insects these glands secrete a substance that repels water, preventing the wetting of the tracheae. In terrestrial insects the secretions of these glands probably serve as lubricants for the movable parts of the spiracular closing mechanism. In some insects spiracular glands probably also secrete repugnant defensive substances, keep rain water from entering the tracheae, and may serve to improve the seal of the closing mechanism.

Types of Ventilatory Systems

Spiracles, tracheae, air sacs, and tracheoles compose the ventilatory system. Tracheae and tracheoles are always present, but air sacs may be absent and spiracles nonfunctional or absent. Based on the presence, absence, and functional or nonfunctional nature of spiracles, there are principally two types of ventilatory systems, open and closed, with a variety of modifications within each type.

The open ventilatory system (Fig. 3–14A, B) is found among most

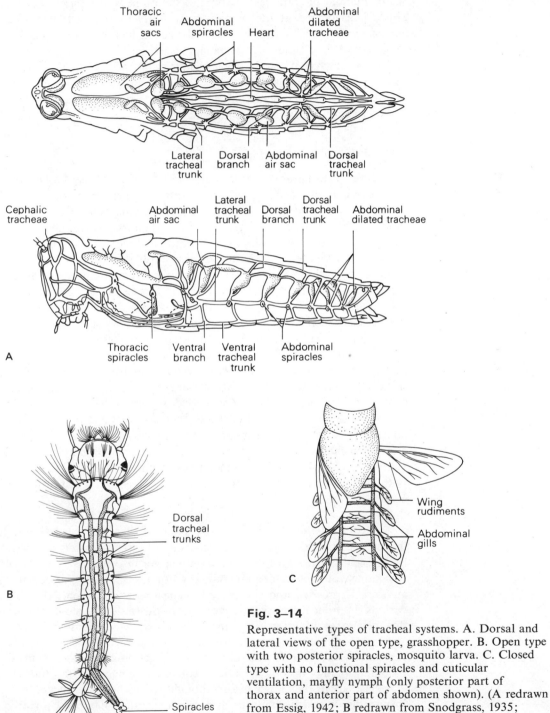

Fig. 3–14

Representative types of tracheal systems. A. Dorsal and lateral views of the open type, grasshopper. B. Open type with two posterior spiracles, mosquito larva. C. Closed type with no functional spiracles and cuticular ventilation, mayfly nymph (only posterior part of thorax and anterior part of abdomen shown). (A redrawn from Essig, 1942; B redrawn from Snodgrass, 1935; C redrawn from Packard, 1898.)

terrestrial insects and many aquatic forms and is characterized by the presence of one or more pairs of functional spiracles. Typically, there are two lateral rows of 10 spiracles each; the first 2 spiracles on each side are located on the meso- and metathorax and the following 8 on the first eight abdominal segments. During the evolution of the various groups of insects there has been a tendency toward reduction in the number of spiracles, in some to the point where there is but a single pair located at the posterior extremity of the abdomen, the *metapneustic* type. This type is well exemplified by the mosquito (Fig. 3–14B) and other dipterous larvae.

The closed type of ventilatory system (Fig. 3–14C) is found in many aquatic and endoparasitic insects. There are no functional spiracles in this type of system; gaseous exchange within the tracheal system occurs directly across the integument.

The Ventilatory Process

Passive Ventilation. Rather elaborate calculations based on measurements of tracheae and on the physical properties of oxygen have shown that simple gaseous diffusion from the outside of smaller insects and from well-ventilated air sacs in the larger ones can supply sufficient oxygen to the body tissues via the tracheoles to maintain life activities. In this text we shall consider simple diffusion as a passive form of ventilation, that is, one in which there are no pumping or other movements that might aid the movement of gases in the tracheae and tracheoles. Use of the phrase *passive ventilation* does not imply that the insects which ventilate exclusively in this manner do not exert any control over the diffusion. Diffusion control is effected by the opening and closing of the spiracles. Thus, in addition to their function in the prevention of excessive water loss from the tissues, the spiracles are able to regulate to some degree the entrance and exit of gases. The spiracles respond to decreased oxygen in the air by remaining open for longer periods of time. Increase in the carbon dioxide content of air produces a similar effect. The muscles of the spiracular closing mechanisms receive nerves from the segmental ganglia of the ventral cord, and there is evidence that opening and closing is under both nervous and chemical control. Also, the possibility of hormonal control exists.

Active Ventilation. It is generally assumed that the ventilation of the smaller tracheae and the tracheoles in all insects is accomplished by simple diffusion. However, in several larger insects passive ventilation alone would be inadequate in many situations. In these insects the air sacs, if present, and the larger tracheae are often ventilated by various rhythmical pumping movements of the body. We shall refer to this type of ventilation as *active ventilation*. Movements that are known to cause the inspiration and expiration of gases include peristaltic waves over the abdomen, telescoping or dorsoventral flattening of the abdomen, and, in some, movements in the thorax. The elasticity of the cuticle is also thought to play a part, especially in inspiration. In addition, muscular movements other than those already mentioned, such as heartbeat, movements of the gut, and so on, may

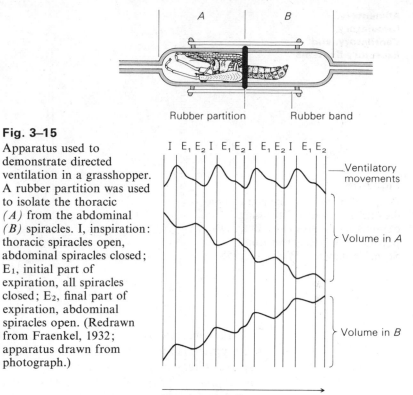

Fig. 3–15
Apparatus used to
demonstrate directed
ventilation in a grasshopper.
A rubber partition was used
to isolate the thoracic
(A) from the abdominal
(B) spiracles. I, inspiration:
thoracic spiracles open,
abdominal spiracles closed;
E_1, initial part of
expiration, all spiracles
closed; E_2, final part of
expiration, abdominal
spiracles open. (Redrawn
from Fraenkel, 1932;
apparatus drawn from
photograph.)

assist in ventilation by pressing against adjacent tracheae. Whatever
type they may be, these pumping movements renew the air in the
tracheae and air sacs by contraction and expansion forces mediated
via the hemolymph. Tracheae that are circular in cross section do not
respond to compressive forces but, as mentioned, are quite extensible.
The tracheae that are oval in cross section and the air sacs are collapsible and hence can serve to increase the volume of tidal air.

In certain insects, the ventilatory movements and opening and
closing of the spiracular valves are coordinated and capable of producing an essentially unidirectional flow of gases through the body,
for example into the thoracic spiracles and out the abdominal ones
(Fig. 3–15). This coordination mechanism has been shown to be
under neural control. As in simple spiracular opening and closing,
oxygen deficiency and carbon dioxide excess can both serve as regulatory stimuli of pumping movements.

Elimination of Carbon Dioxide. The rate of diffusion of carbon
dioxide through air is not too much different from that of oxygen,
but in tissues it diffuses much more rapidly (about 35 times as rapidly).
This being the case, carbon dioxide is much more likely to be eliminated through the tracheal linings and integument than is oxygen to
be absorbed via the same routes. Thus, although most of the carbon
dioxide produced by the respiratory activities of the body cells is
eliminated by means of the tracheoles, at least a small portion of it may
escape through the general body surface of soft-bodied insects and
the intersegmental membranes of the more sclerotized ones.

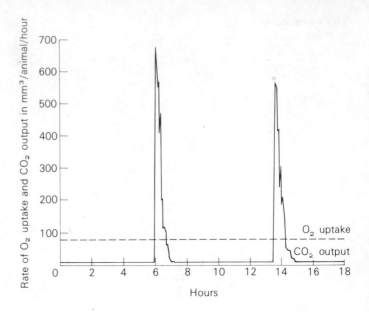

Fig. 3–16

Graph of oxygen uptake and carbon dioxide output in a diapausing cecropia moth, *Hyalophora cecropia,* pupa at 25°C. (Redrawn from Schneiderman and Williams, 1955.)

Many insects have been found to eliminate carbon dioxide through the spiracles in regular "bursts" (Fig. 3–16). Between these bursts, the spiracles remain partially or entirely closed. This phenomenon has been observed particularly in insects that are inactive due to a low ambient temperature or developmental stage. This phenomenon has also been induced experimentally in insects in which it does not normally occur by lowering the temperature and by injury to certain parts of the brain. The ability to release carbon dioxide periodically in this manner allows an insect to keep its spiracles partially or entirely closed most of the time and hence is thought to be an adaptation that favors the conservation of water by diminishing the rate of transpiration.

Oxygen Transport in the Hemolymph. Except over the very short distances between most tracheolar endings and the cells they oxygenate, the hemolymph does not generally function as an oxygen carrier. However, as mentioned earlier certain chironomid larvae have the oxygen-carrying pigment hemoglobin dissolved in their hemolymph. According to Wigglesworth (1965), under ordinary conditions of oxygen tension this hemoglobin is saturated with oxygen and thus does not serve as a carrier, but under conditions of low oxygen tension, it becomes reduced and thus can serve as a carrier. Wigglesworth also mentions the presence of hemoglobin in modified fat body cells of endoparasitic larvae of the bot flies in the genus *Gasterophilus,* which ventilate periodically when they come into contact with an air bubble in the intestine of their host. He points out that the hemoglobin in this situation permits the insects to take in a larger supply of oxygen at any given time than they could otherwise.

In addition, Wigglesworth describes anatomical situations in certain insects in which tracheoles do not seem to be associated with any tissue other than the hemolymph. These, he suggests, may function as "tracheal lungs," serving to aerate the hemolymph.

Ventilation in Aquatic Insects. A large number of insects in several orders spend all or a part of their lives in an aquatic environment. Existence in this environment presents special problems in that an insect must either be able to utilize oxygen in solution or have some means of tapping a source of undissolved oxygen whether it be at an air–water interface (at the water's surface or in the form of submerged bubbles) or from aquatic vegetation. A wide variety of structural and physiological adaptations which solve these problems more or less efficiently has evolved in the insects that now inhabit an aquatic environment.

Fig. 3–17
Ventilatory structures in aquatic insects.
A. Lateral abdominal tracheal gills in a mayfly nymph. B. Terminal abdominal tracheal gills in a damselfly nymph. C. Cuticular gills on the thorax of a black fly pupa. D. Terminal blood gills in a black fly larva. Enlarged gills are diagrammatic. (A and B redrawn from CCM General Biological, Inc., key card; C and D redrawn from Packard, 1898.)

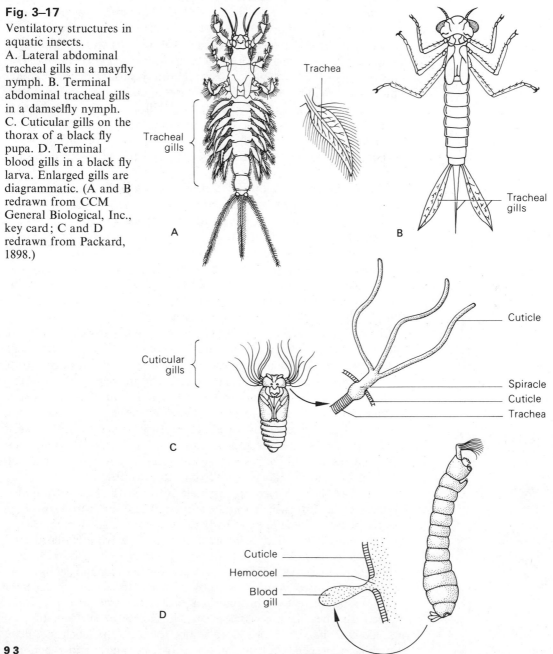

Aquatic insects that have closed ventilatory systems do not make use of an air–water interface but depend entirely on dissolved oxygen. They obtain this oxygen in a variety of ways. Many possess *tracheal gills* (Fig. 3–17A), which are typically evaginations of the integument covered by a very thin cuticle and are well supplied with tracheae and tracheoles. These evaginations may be found anywhere on the body but are commonly found in the abdomen. In some insects they are located at the posterior end of the abdomen (Fig. 3–17B). Thin areas of the integument which are not evaginated may be well supplied with tracheae and tracheoles and function in a similar fashion. Other aquatic insects with closed ventilatory systems possess *spiracular* or *cuticular gills* (Fig. 3–17C). These are filamentous outgrowths consisting mostly of very thin cuticle (about 1 micron thick) which open directly into the tracheae. Another structure, which is at least suggestive of a ventilatory function, is the *blood gill* (Fig. 3–17D). This is a very thin evagination from the body surface or the rectum, which is nearly devoid of tracheae. Many aquatic insects with closed ventilatory systems lack any specialized gill structures and depend upon diffusion of oxygen across the general body surface, i.e. cutaneous ventilation. In fact, all these forms with gills of various sorts probably depend, to a greater or lesser extent, on cutaneous diffusion of oxygen.

Aquatic insects with open tracheal systems obtain oxygen at an air–water interface. Most aquatic insects fall into this group. It may be divided into two categories as follows: those which must surface periodically and depend at least in part on atmospheric oxygen, and those which may remain submerged for an indefinite period of time and are usually somewhat independent of atmospheric oxygen. These two groups are rather indefinite, with intermediates between them. In addition, cutaneous ventilation may under certain circumstances play a role in some of these insects.

Most of the members of the group of aquatic insects that must surface possess hydrofuge structures. These structures are generally associated with particular spiracles and are highly variable from insect to insect. However, all have essentially the same function, the breaking of the surface film of the water and by this action exposing certain spiracles to the atmosphere. They also serve to keep water out of the tracheae when submerged. Hydrofuge structures are usually made up of hairs and are resistant to "wetting" by water, and thus when an insect approaches the surface, the cohesive properties of water cause it to be drawn away from the hydrofuge areas.

Many insects in this group carry air stores in the form of bubbles or films of air into which spiracles open. These air stores are often held in place by a pile of erect hydrofuge hairs, but in some the shape of the body is such that it forms a storage area without the use of hydrofuge hairs. Films or bubbles of air would obviously be a temporary source of oxygen if the insect were forced to remain submerged. How long an insect could survive by utilizing stored oxygen depends on a number of factors, but unless a means for replenishing it from oxygen dissolved in the water exists, the length of survival time is not likely to be very long. In addition, some insects, such as pupal mosquitoes, depend on an air store for buoyancy, and when the air store becomes depleted to the point where the density of the insect plus air store becomes greater than the density of water, the insect is unable to surface and sinks and drowns. Many aquatic insects that carry stores of air are able to replenish the oxygen without surfacing. This is accomplished by the air store acting as a "physical gill." As the oxygen in reserve is used up, a point is reached where the partial pressure of oxygen is less in the air store than it is in the surrounding water. At this point, because there is a much greater tendency for oxygen to diffuse from water to air than for nitrogen to diffuse from air to water, equilibrium tends to be restored by the diffusion of oxygen into the store instead of the diffusion of nitrogen out of it.

Insects that are able to remain submerged indefinitely include those which possess a structure known as a *plastron* and those which obtain their oxygen from submerged vegetation. A plastron is a very thin layer of gas held firmly in place by tiny hydrofuge hairs. Unlike typical air stores, it cannot be displaced by water. The spiracles open into the plastron, and functionally it acts in a manner similar to a physical gill except that it usually does not require repletion by a visit to the surface of the water. Insects that obtain oxygen from submerged vegetation may do so in a variety of ways. Many are able to take bubbles on the surface of plants by means of hydrofuge structures. Others penetrate the tissues of submerged plants by biting into them or by inserting a specialized ventilatory structure into the intercellular air spaces (Fig. 3–18).

Many aquatic forms, whether they ventilate by means of the various modifications of the closed or open system, carry on ventilatory movements of one sort or another. These include such activities as rhythmic undulations of the tracheal gills, moving water in and out of the rectum containing tracheal gills, and directing currents of water over ventilatory structures.

Fig. 3–18
Ventilatory apparatus adapted
for penetration of aquatic plants
in the mosquito *Mansonia
perturbans*. The anal papillae are
involved in osmoregulation.
(Redrawn from Matheson, 1944.)

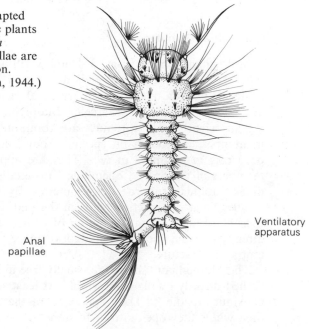

Ventilatory
apparatus

Anal
papillae

Ventilation in Endoparasitic Insects. Endoparasitic insects are those which invade the tissues of their host as opposed to those which feed superficially. The great majority of the insects in this group are endoparasitic only in the immature stages. The environment of these insects presents problems similar to those of the aquatic environment in that the insects have had to adapt to either obtaining oxygen in solution or from the atmosphere or both. In several the tracheal system is nonfunctional and ventilation is cutaneous, gaseous exchange occurring directly between the tissues of the parasite and body fluids of the host. Some endoparasitic insects have tracheal gills rather similar to those found in insects inhabiting aquatic environments, which function in the same way. Other endoparasites depend, at least partly, on atmospheric air, obtaining it either by tubes or other structures which communicate with the tracheal system and which extend out of the host to the atmosphere. Some insects in the group that are endoparasites of other insects actually tap the tracheae of their host and obtain oxygen in this manner. As with most of the aquatic forms, cutaneous ventilation plays a greater or lesser role in most endoparasites.

The Excretory System

The function of the excretory system is the maintenance of a constant internal environment. Since the hemolymph bathes the tissues and organs of the insect body, it largely determines the nature of this internal environment. Thus the excretory system is essentially responsible for the maintenance of the uniformity of the hemolymph. It accomplishes this task by two basic means: the elimination of metabolic wastes and excesses, particularly nitrogenous, and the regulation of salt and water. The Malpighian tubules are the major organs of excretion, although other tissues may play a role in some instances.

Malpighian Tubules

Malpighian tubules (Figs. 3–1 and 3–19) are usually long, slender tubes closed at their distal ends and found in association with the posterior portion of the alimentary canal of insects. They are named after their discoverer, Marcello Malpighi, a seventeenth-century Italian scientist. These tubules are commonly convoluted and vary in number, depending on species, from 2 to 250 or more, usually occurring in multiples of two. They are apparently lacking only in members of the order Collembola (springtails) and the family Aphididae (aphids; suborder Homoptera). By definition, Malpighian tubules lie at the junction between the mid- and hindguts. However, their embryonic origin is a matter of controversy and they may open directly into the midgut or hindgut or more commonly into a dilated ampullar structure.

The Malpighian tubules are usually free in the body cavity and are bathed directly by the hemolymph. At least some are always in close proximity to the fat body and parts of the alimentary canal, other than where they open into it. In some groups of insects, for example

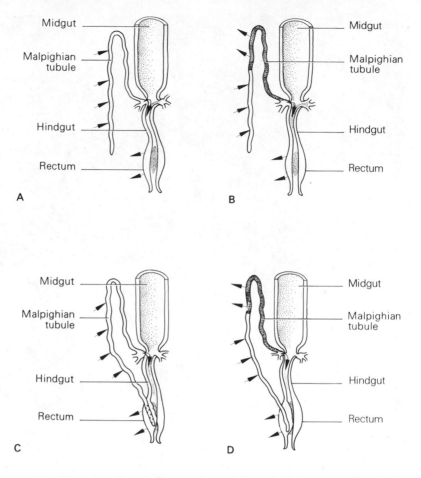

Fig. 3–19
Major types of
Malpighian tubule–
hindgut systems.
A. Orthopteran type.
B. Hemipteran type.
C. Coleopteran type.
Arrows indicate
directions of movement
of substances in and out
of the tubule lumen.
(Redrawn from Patton,
1963.)

most Lepidoptera (butterflies and moths) and Coleoptera (beetles), the distal ends of the tubules are embedded in the tissues surrounding the rectum (Fig. 3–19C, D). This is the *cryptonephridial* ("hidden kidney") arrangement. In some insects, the distal ends of two tubules anastomose, forming a closed loop. In addition, there may be anatomical differences between the tubules within the same insect.

In many insects the Malpighian tubules are capable of a variety of movements. Microscopical investigation reveals that the cause of these movements is the presence of variously oriented muscles intimately associated with the tubes. Apparently only members of the orders Thysanura (silverfish and relatives), Dermaptera (earwigs), and Thysanoptera (thrips) do not have muscles associated with the Malpighian tubules. The function of the tubule movement is currently a matter of conjecture. Suggestions have included such functions as propulsion of the contents of the lumen toward the opening into the alimentary canal and mixing of the luminal contents. The tubules are usually well tracheolated. A basement membrane lies beneath the muscles and surrounds the tubule cells (Fig. 3–20). The tubules are one cell thick, but the circumference may be composed of from one to several cells. The cytoplasm of these cells varies in appearance. It is usually colorless but may have a faint green or yellow appearance. It is generally filled with various refractile or pigmented inclusions and sometimes contains needlelike crystals but may be nearly

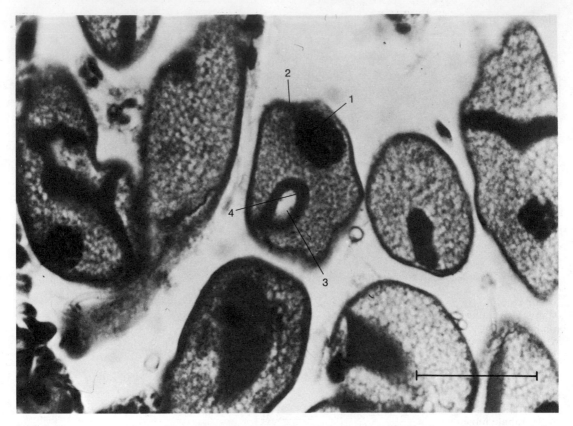

Fig. 3–20

Photomicrograph of sections of Malpighian tubules in the mosquito *Aedes triseriatus*. 1, Nucleus of Malpighian tubule cell; 2, basement membrane; 3, lumen of tubule; 4, striated border. (Scale line = 50 microns.)

clear. The tubule cells communicate with the lumen by means of a finely striated border. The electron microscope has revealed a very large number of mitochondria in the cytoplasm of the tubule cells.

Salt and Water Balance

Different environmental situations pose different salt and water problems for insects. Terrestrial forms are constantly faced with the tendency to lose water through transpiration and are generally dependent on ingested food for the needed water and also salt. Depending on the water content of their diet, the fecal material may be quite watery, as in plant-feeding insects that take in an excess of water, or a dry, powdery pellet from those insects that feed on materials of very low water content. Freshwater insects must excrete the large amounts of water absorbed through the integument and by the gut along with ingested food and at the same time must conserve the inorganic ions. Saltwater forms, like terrestrial insects, must constantly overcome the tendency to lose water, in this case due to an osmotic gradient between the insect and the surrounding medium.

In the majority of insects the regulatory problems outlined in the preceding paragraph are, at least in part, solved through the activities of the Malpighian tubules and the rectum in the hindgut. Various studies have demonstrated that the Malpighian tubules are freely permeable to most small molecules. These molecules may be passed into the lumen of a tubule by simple diffusion set up by a concentration gradient or by an active transport process, as is the case with potassium ions. Secretion of water and ions is not a selective process. However, substances that are required are reabsorbed into the hemolymph either in the proximal portions of the tubules in some insects and/or in the rectum. These reabsorption processes may also be active transport or simple diffusion. It has been suggested that it is easier for organisms to develop ways of reabsorbing needed substances than to develop ways of excreting every type of unneeded substance.

A number of Malpighian tubule–rectal cycling systems have been described. In the simplest situation, found in members of the order Orthoptera (grasshoppers and relatives), the tubules are composed of a single cell type and have only fluid in the lumen (Fig. 3–19A). This fluid passes down the lumen and mixes with the gut contents and as it passes down the hindgut, particularly in the rectum, needed water and ions are reabsorbed into the hemolymph. A more complex cycling system is found in most Hemiptera (Fig. 3–19B). In this case movement of materials into the lumen of a tubule occurs in the distal portion; reabsorption of needed water and ions into the hemolymph occurs in the proximal portion and in the rectum. A third system is that typically found in beetles (Fig. 3–19C), in which the distal portion of a Malpighian tubule is embedded in the tissues surrounding the rectum. This is the cryptonephridial arrangement mentioned earlier. Secretion into the lumen of a tubule apparently occurs in its exposed portion, whereas reabsorption of needed materials likely takes place through the portion embedded in the rectum. It has been suggested that this situation increases the efficiency of the cycling process (Stobbart and Shaw, 1964).

A cryptonephridial arrangement is also typical of members of the order Lepidoptera (Fig. 3–19D) except that in these reabsorption of needed materials into the hemolymph occurs in the proximal portions of the tubules in addition to that in the rectum. Other situations have been described, but those mentioned here should give a general idea of the processes involved. It should be pointed out that factors such as spiracular control, integument permeability, food selection, and habitat selection are also involved with the regulation of salt and water in insects. Also, in certain aquatic insects there exist mechanisms for the uptake of ions other than via the gut. For example, chloride ions are taken into the hemolymph of mosquito larvae by way of four papillae, which surround the anus (Fig. 3–18). This is an active process, occurring against a rather severe concentration gradient. In addition, these papillae are responsible for sodium, potassium, and water uptake (Stobbart and Shaw, 1964).

Nitrogenous Excretion

Nitrogenous products of various types tend to accumulate in the hemolymph as a result of protein and amino acid metabolism. These materials are generally of no use to an insect and may be toxic. This being the case, they must either be eliminated or stored in an inert situation until they can be eliminated. Uric acid is the major nitrogenous waste product formed and excreted, making up 80% or more of the nitrogenous end products found in the urine of most terrestrial and many aquatic insects. It does not require a large amount of water for its elimination and is thus quite appropriate in insects in which water conservation is a problem. On the other hand, ammonia is the major nitrogenous waste formed by many aquatic insects and by blow fly larvae. However, it is highly toxic and requires large amounts of water for its elimination. Other nitrogenous products that have been identified in insect urine are allantoin, allantoic acid, and urea. These arise from the breakdown of uric acid and are usually present in very small amounts. Amino acids are sometimes found but usually in very small quantities.

The Malpighian tubules are the major organs involved in the excretion of nitrogenous materials, but other tissues may be involved to a greater or lesser extent depending upon the species concerned. As mentioned earlier, the nitrogenous wastes may be stored by certain tissues until they can be eliminated. This most commonly occurs in the fat body in the *urate cells* (Wigglesworth, 1965) but also in certain other tissues.

Insect Urine

The nature and composition of insect urine is, not surprisingly, highly variable. Its physical appearance varies from the dry powdery material egested by terrestrial insects that inhabit dry environments to a clear fluid in those in which water conservation is not a problem (e.g., plant feeders and aquatic forms). Its chemical composition is dependent upon the dietary substances that are in excess of an insect's needs, those which are absorbed into the hemolymph by the gut but are not usable by the insect, the nitrogenous wastes formed as a result of protein and amino acid metabolism, and the excesses of salts and water that may occur.

Selected References

GENERAL
    Bursell (1971); DuPorte (1961); Patton (1963); Smith (1968); Snodgrass (1935); Wigglesworth (1965).

ALIMENTARY SYSTEM AND NUTRITION
    Brammer and White (1969); Brues (1946); Day and Waterhouse (1953a, 1953b, 1953c); Gelperin (1971); House (1961, 1965a, 1965b); Trager (1953); Waterhouse and Day (1953).

CIRCULATORY SYSTEM
    Arnold (1964); Briggs (1964); Buck (1953); Florkin and Jeuniaux (1964); Goodwin (1966); Grégoire (1964); Gupta (1969); Jones (1962, 1964); McCann (1970); Salt, G. (1970); Salt, R. W. (1961); Wigglesworth (1959b); Wyatt (1961).

VENTILATORY SYSTEM
    Buck (1962); Edwards (1953); Miller (1964).

EXCRETORY SYSTEM
    Craig (1960); Patton (1953); Stobbart and Shaw (1964).

4

The Nervous, Endocrine, and Muscular Systems

The Nervous System

The many, diverse activities of the various systems of an insect are coordinated in large part by the nervous system. This system is composed of elongated cells, or *neurons*, which carry information in the form of electrical impulses, from external and internal *sensilla* (sensory cells) to appropriate *effectors*. The nature and location of the stimulated sensilla determine the nature of the response. The basic effectors are muscles and glands. However, there are other effectors (e.g., the light-producing organ of the firefly). The nervous system and effector organs enable an insect to adjust continually to changes (stimuli) in both the internal and external environment and behave in such a manner as to ultimately subserve the maintenance of the individual and the survival of the species.

Structure and Function of the Nervous System

The Neuron.   The basic functional unit of the nervous system is the nerve cell or *neuron* (Fig. 4–1A). A neuron may be described as a thin-walled tube which varies from much less than 1.0 mm to more than 1 meter (in larger animals) in length and has a diameter between 1.0 and 500 microns. Insect neurons are usually comparatively small in diameter, somewhere on the order of 45 to 50 microns for the largest and much less for some of the smaller. Typically a neuron consists of a *cell body* or *soma* and one or more long, very thin *fibers* or *axons*. A branch of an axon is called a *collateral*. Associated with the cell body or near it are tiny branching processes, the *dendrites*. Similar branching processes, the *terminal arborizations,* are found at the end of an axon. Axons are invested by the *Schwann* or *glial cells,* which may have a variety of functions, among them being the active transport and storage of nutrients and the laying down of a connective tissue sheath around ganglia, the *neural lamella.*

Many insect nerve cells are invested in a *myelinated sheath.* Neurons may be *unipolar, bipolar,* or *multipolar* (Fig. 4–1A, B, C). In unipolar neurons, a single stalk from the cell body connects with the axon and a collateral. In bipolar neurons the cell body bears an axon and a single, branched or unbranched dendrite, while multipolar neurons have an axon and several branched dendrites. The individual neurons are not continuous with one another, but the finely branching terminal arborizations of an axon come into extremely close associa-

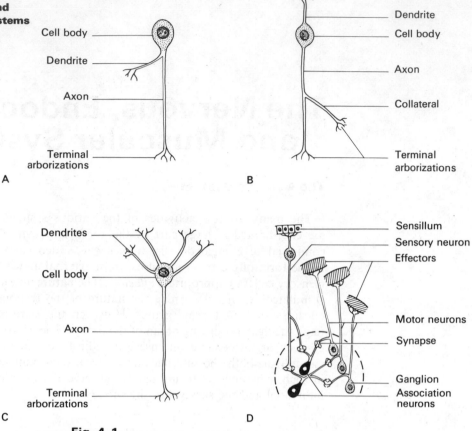

Fig. 4–1

Types of neurons. A. Unipolar. B. Bipolar. C. Multipolar. D. Relationship among sensory, motor, and association neurons. (Redrawn from DuPorte, 1961.)

tion with the dendrites of another. A very small but measurable distance, however, always lies between them. The region of close association between terminal arborizations and dendrites is called a *synapse* and the space between the arborizations and dendrites is called the *synaptic cleft* (Fig. 4–1A).

Neurons may be a part of a ganglion (Fig. 4–2), which is an aggregation of neural material. The term *ganglion* literally means a "swelling." There are four basic histological divisions of a ganglion: (1) an outer connective tissue layer, the *sheath* or *neural lamella*; (2) cell bodies of neurons; (3) tracts of fibers; and (4) the *neurophile,* which is composed of intermingled fibrous processes and neuroglial elements. Bundles of axons constitute *nerves.*

Types of Neurons. The insect nervous system is composed of basically three types of neurons (Fig. 4–1D): (1) *sensory*, or *afferent*; (2) *motor*, or *efferent*; and (3) *association*, or *internuncial*.

Sensory or afferent neurons are usually bipolar and are located peripherally in the insect. The distal process, or dendrite, is associated with a sensory structure of some type; the proximal process may either enter a ganglion and connect with motor or association neurons

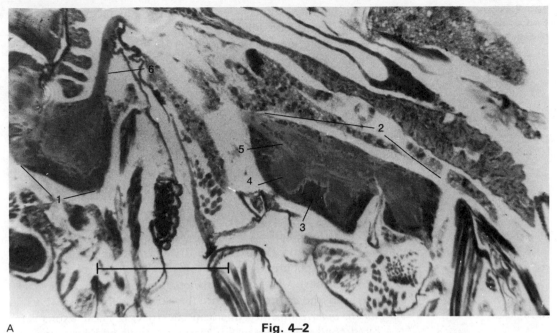

A

Fig. 4–2
Photomicrographs of sagittal sections of neural tissue
in a mosquito. A. Section of a mature pupa. B. Section
of the prothoracic ganglion with special silver nerve
stain. 1, Brain; 2, thoracic ganglia; 3, cell bodies of
neurons; 4, neuropile; 5, fiber tract; 6, nerve; 7, neural
lamella. (Scale lines: A, 250 microns; B, 50 microns.)
(Tissue prepared by S. M. Meola.)

B

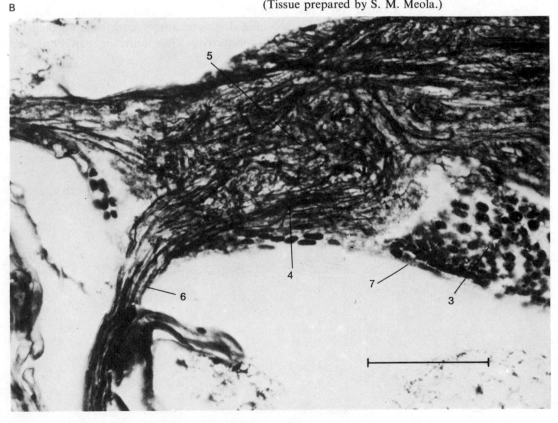

or be connected with other neurons, which are distributed over large areas of the central nervous system. Some sensory neurons may be bipolar or multipolar with their distal processes, or dendrites, branching, sometimes quite elaborately, over the inner surface of the integumental wall, through perforations in the integumental wall, or over the alimentary canal while their axons enter the ganglia of the central nervous system.

Motor or efferent neurons are unipolar. The cell body lacks dendrites and is located in the periphery of a ganglion. The bundles of axons from the cell bodies form the motor nerves that activate muscles; the collateral of each neuron enters the neuropile, and its terminal arborizations connect with those of association or sensory neurons.

Association or internuncial neurons also have their cell bodies located in the periphery of a ganglion and may synapse with one or more other association neurons, sensory neurons, or motor neurons. Some association neurons are quite large and are connected to "giant axons" which have very large diameters (approximately 45 microns) and may run the entire length of the ventral nerve cord (e.g., in *Periplaneta americana*, the American cockroach). These axons serve as a rapid conduction system for alarm reactions and are associated with a variety of sensory-motor combinations in different insects. Another group of association neurons is located in the *mushroom bodies* of the brain, which are thought to be the "higher centers" in the central nervous system, controlling the most complex insect behavior.

Central Nervous System. The insect central nervous system (Fig. 4–3) is composed of a double chain of ganglia connected by lateral and longitudinal connectives. The anterior ganglion, the

Fig. 4–3

Generalized insect central nervous system.

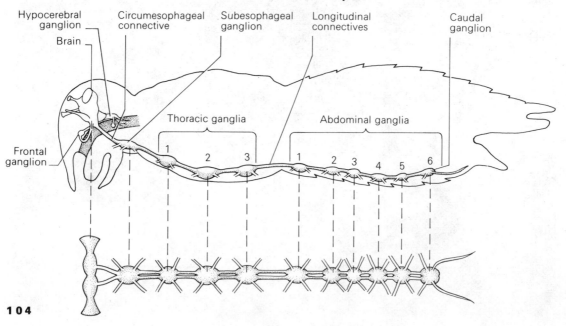

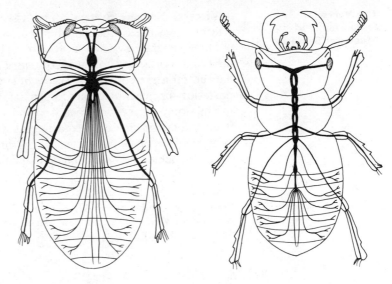

Fig. 4–4

Variation in the concentration of ventral chain ganglia in two species of beetles. (Redrawn from Packard, 1898.)

brain, is very complex and is located dorsal to the foregut in the head. It is usually connected to a ganglion ventral to the foregut by *circumesophageal connectives.* This is the *subesophageal ganglion,* which is also highly complex, being composed of three fused ganglia. It innervates sense organs and muscles associated with the mouthparts, salivary glands, and the neck region. In addition to innervating the mouthparts, the subesophageal ganglion in many insects has an excitatory or inhibitory influence on the motor activity of the whole insect. Posterior to the subesophageal ganglion are typically three segmental *thoracic ganglia,* each containing the sensory and motor center for its respective segment. In some insects [e.g., adult Diptera, Hymenoptera, and some Coleoptera (Fig. 4–4)], the thoracic ganglia may be more or less fused longitudinally, forming what appears as a single neural mass in the thorax.

In the more primitive insects each of the first several abdominal segments contains a ganglion, the first 8 in apterygote insects, 7 in dragonfly and damselfly nymphs, and 5 or 6 in grasshoppers and their relatives. However, there has been a tendency toward the reduction in the number of abdominal ganglia; for example, several adult flies have only one, which is more or less fused with the large, single thoracic ganglion. In those insects which possess abdominal ganglia, the posteriormost or *caudal ganglion* is always compound and furnishes the sensory and motor nerves for the genitalia. This ganglion is therefore intimately involved in the control of copulation and oviposition.

The insect brain (Fig. 4–5) is a very complex structure, apparently formed by the fusion of three separate primitive ganglia and is hence composed of three paired lobes as follows: (1) the *protocerebrum,* which innervates the compound eyes and ocelli and contains two groups of association neurons, the mushroom bodies; (2) the *deutocerebrum,* which innervates the antennae; and (3) the *tritocerebrum,* which connects the brain to the visceral nervous system, innervates the labrum, makes up the circumesophageal connectives, and may receive sensory nerves from the head capsule.

Visceral Nervous System. The visceral nervous system is often referred to as the sympathetic nervous system of insects. It is made up of three separate systems as follows: (1) *stomodael,* associated with the brain, aorta, and foregut; (2) *ventral sympathetic,* associated with the ventral nerve cord; and (3) *caudal sympathetic,* associated with the posterior segments of the abdomen.

The stomodael system (Figs. 4–5 and 4–6) arises during embryogeny from the dorsal or dorsal and lateral walls of the stomodaeum and eventually becomes connected to the brain. Its various components typically include a *frontal ganglion,* which lies on the dorsal midline of

Fig. 4–5

Brain and stomodael nervous system of the grasshopper *Dissosteira carolina;* A. Anterior view. B. Lateral view. (Redrawn with slight modifications from Snodgrass, 1935.)

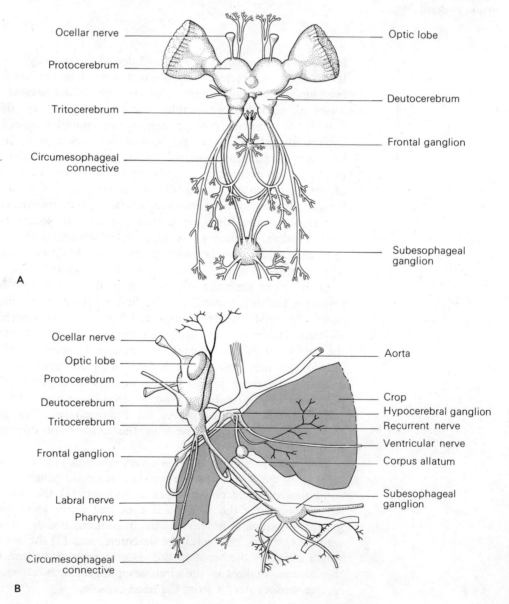

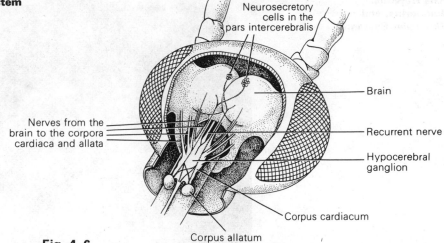

Fig. 4–6

Stomodael nervous system and endocrine tissues in an orthopteran species (semidiagrammatic). (Redrawn from Engelmann, 1970.)

the foregut just anterior to the brain. The frontal ganglion connects with the brain by bilateral nerves. The *recurrent nerve* arises medially from the frontal ganglion and extends beneath and posterior to the brain. Beyond this point there is considerable variation in the stomodael system among different kinds of insects. Therefore, the description that follows will pertain only to one of the more common arrangements. The recurrent nerve ends posteriorly in a *hypocerebral ganglion,* which may give rise to one or two *gastric* or *ventricular nerves,* which continue posteriorly and eventually terminate with a *ventricular ganglion.* Also associated with the hypocerebral ganglion are two pairs of neural–glandular bodies, the *corpora cardiaca* and the *corpora allata.* The stomodael system apparently exerts some control over the movements of the gut and the heart and possibly, in certain instances, labral muscles, mandibular muscles, and salivary ducts. The corpora cardiaca and corpora allata are involved with hormone secretion and are thus considered as parts of the endocrine system.

The ventral sympathetic system is associated with the ganglia of the ventral nerve cord. From each segmental ganglion a single median nerve arises and divides into two lateral nerves, which supply the spiracles of the segment in which they are located. These nerves may be lacking altogether.

The nerves of the caudal sympathetic system arise from the caudal ganglion of the ventral chain and supply the posterior portions of the hindgut and the internal sexual structures.

The Endrocrine System

The endocrine system is involved with the secretion of hormones. These diverse substances complement the coordinating activities of the nervous system. They are generally secreted into the hemolymph from a rather well defined tissue and circulate throughout the insect body. However, they exert their influence quite specifically upon target

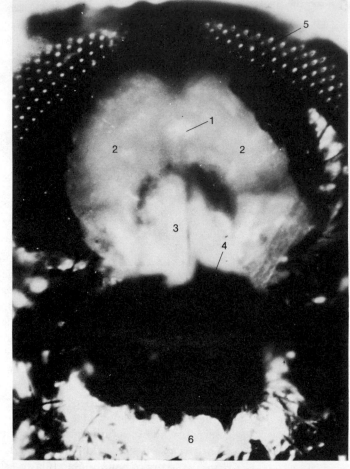

Fig. 4–7
Neurosecretory cells in the brain of the mosquito *Aedes taeniorhynchus*. A. Cells exposed by retraction of posterior part of head capsule. B. Neurosecretory cells removed from brain. Highly magnified. 1, Neurosecretory cells; 2, brain; 3, pharyngeal muscles; 4, retracted portion of head capsule; 5, compound eye; 6, anterior portion of the thorax. (Courtesy of A. O. Lea.)

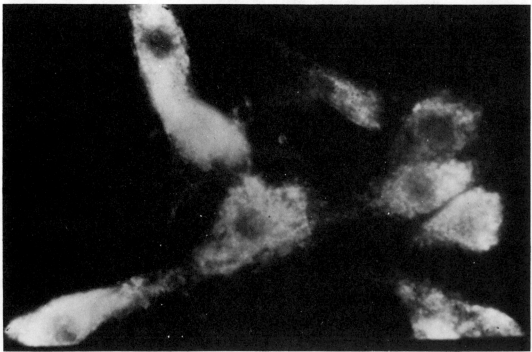

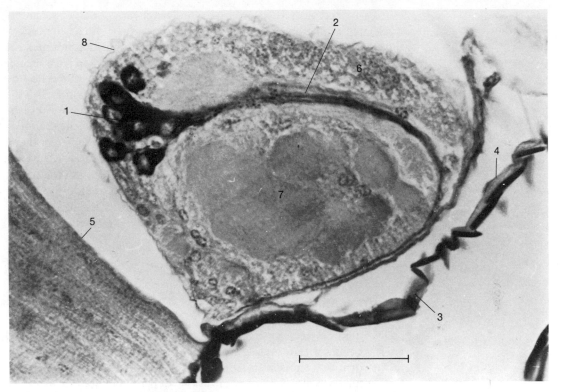

Fig. 4–8

Photomicrograph of neurosecretory cells exposed by staining with paraldehyde fuchsin (PF). Sagittal section of the brain of a mosquito, showing neurosecretory cells and axons running anteriorly and then posteriorly. The axons eventually reach the corpora cardiaca (see Fig. 4–6). 1, Neurosecretory cells; 2, axons of neurosecretory cells; 3, cuticle of pharyngeal pump; 4, recurrent nerve; 5, pharyngeal pump muscle; 6, cell bodies of neurons; 7, neuropile; 8, neural lamella. (Scale line = 50 microns.) (Courtesy of S. M. Meola.)

organs or tissues. Like the nervous system, hormones help the insect to adjust to external and internal environmental changes.

The best identified endocrine structures and those whose function is at least partially understood are the corpora cardiaca, the corpora allata, and the *thoracic glands* (Fig. 4–6; see also Fig. 8–26). In addition, *neurosecretory* cells are located in groups in the proto-cerebral part of the brain (Fig. 4–7). Neurosecretory cells are also intimately associated with other neural tissue and are modified neurons, the difference from ordinary neurons being that they show extensive secretory activity. They are usually identified by the use of specific stains; for example, they stain deeply with paraldehyde fuchsin (Fig. 4–8). They are also characterized by the presence of electron-dense granules when examined with the electron microscope. Neurosecretory cells are found in many locations other than the protocerebrum. For example, they have been identified in other parts of the brain, in the subesophageal ganglion, the thoracic ganglion, and the corpus cardiacum.

The locations of the corpora cardiaca and corpora allata have already been described and specific functions will be discussed under appropriate sections of this text. The thoracic glands are irregular masses of tissue of ectodermal origin which are usually intimately associated with tracheae. They may or may not be innervated. They have been found to secrete *ecdysone*, the hormone which initiates the molting process. As will be discussed in more detail when we consider development (Chapter 8), their secretory activities are stimulated by hormones released by the neurosecretory cells in the brain.

Several endocrine functions have been identified in insects. Examples of some of the functions include stimulation of the secretory activities of glandular tissues other than that which secreted them, influencing color changes, regulation of the excretion of water, control of heartbeat rate and amplitude, control of certain metabolic activities, such as the maintenance of carbohydrate level in the hemolymph and protein synthesis, involvement with sclerotization and melanization of the cuticle, control of growth and metamorphosis, and possible control of circadian rhythms.

The Muscular System

The final conversion of chemically stored energy into mechanical energy occurs within muscle tissue. This mechanical energy is usually in the form of a shortening in length, or *contraction*. Muscle contraction in turn produces a variety of results, depending upon the location, attachments, and the degree of stimulation of the muscles which

Fig. 4–9

Photomicrograph of a sagittal section of a mosquito showing the indirect flight muscles in the thorax and intersegmental muscles in the abdomen. 1, Longitudinal flight muscles; 2, tergosternal flight muscles; 3 intersegmental muscle. (Scale line = $\frac{1}{2}$ millimeter.) (Tissue prepared by S. M. Meola.)

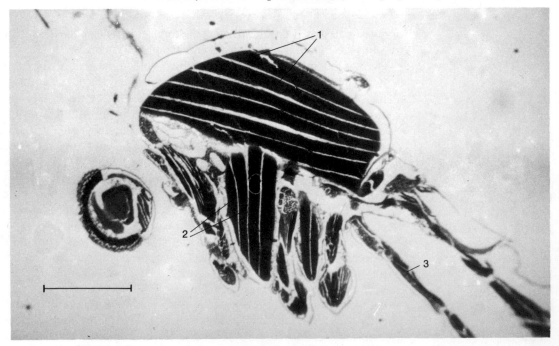

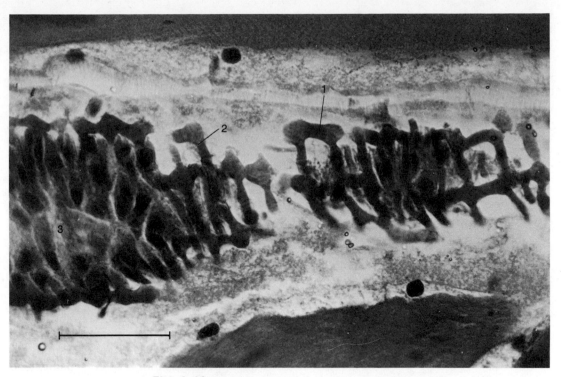

Fig. 4–10

Photomicrograph of a section of the midgut of a pupal mosquito, *Aedes triseriatus*, showing visceral muscles. 1, Longitudinal muscle; 2, circular muscle; 3, gut epithelial cells. (Scale line = 50 microns.)

contract. It is responsible for most forms of locomotion, the maintenance of posture, and movements of the viscera, as for example the peristaltic propulsion of food along the alimentary canal. Quantitatively speaking, muscle is probably the most abundant tissue in the higher animals.

Muscles are commonly classified as striated, those containing definite transverse lines or striations, and smooth, those which lack these striations. Insect muscles are all of the striated variety, although in some types, the striations are extremely difficult to discern. They are typically colorless or grayish in appearance, but may be tinged with yellow, orange, or brown, as is the case for wing muscle in many insects. In view of the small size of insects, one might assume that they have comparatively few muscles. Actually, the reverse is true; larger insects possess perhaps two or three times the number of individual muscles as, for example, does man.

Based on their general location, insect muscles can be divided into two groups, *skeletal* and *visceral*. Skeletal muscles (Fig. 4–9) are attached at both ends to regions of the integument and are those associated with the maintenance of posture and the various movements of the skeleton. Visceral muscles occur in regular meshes of circular and longitudinal strands or in irregular networks surrounding various internal organs, such as the alimentary canal (Fig. 4–10), the ovaries, and the Malpighian tubules. Visceral muscles, which surround the various internal organs, usually attach to other visceral muscles; skeletal muscles attach to the integument.

Skeletal Muscles

Skeletal muscle in most cases has a more-or-less stationary attachment area, the *origin,* and a movable attachment area, the *insertion.* Areas of attachment may be either directly on the body wall (Fig. 4–11A–D) or on the inner surface of an apodeme (Fig. 4–11E). A number of different means exist by which muscle is attached to integument, but in all cases where it is present, the outer membrane of a muscle fiber, the *sarcolemma,* is continuous with the basement membrane of the hypodermal layer. Typically fibrillar structures, the *tonofibrillae,* which are derived from the hypodermal cells, extend through or between the hypodermal cells and are associated intimately with the cuticle (Fig. 4–11B–D). However, in some cases, muscle fibres may attach directly to unmodified hypodermal cells (Fig. 4–11A). Lai-Fook (1967) considers the ultrastructure of developing muscle insertions in insects.

Groups of Skeletal Muscles. The skeletal muscles of the head carry on three basic functions: (1) movements of the entire head, (2) mouthpart movements, and (3) antennal movements. Muscles whose origins are on the anterior part of the prothorax and insertions on the tentorium and various parts of the head capsule are responsible for head movement. The head appendages, mouthparts, and antennae are moved both by *extrinsic muscles,* which originate on the inner surface of the head capsule or tentorium and insert within a given appendage, and by *intrinsic muscles,* whose origins and insertions are entirely within a given appendage (Fig. 4–12A).

Thoracic skeletal muscles are, of course, those involved mainly with the major locomotor appendages, the legs and wings. The legs (Fig. 4–12B), like the mouthparts and antennae, are operated by both

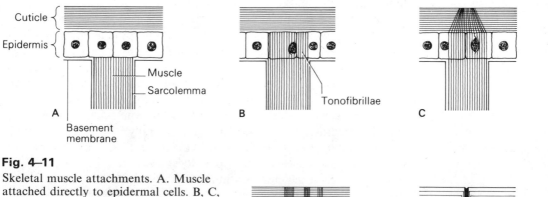

Fig. 4–11
Skeletal muscle attachments. A. Muscle attached directly to epidermal cells. B, C, and D. Muscles attached to cuticle by means of tonofibrillae. E. Muscle attached to apodeme. (Redrawn from Richards, 1951.)

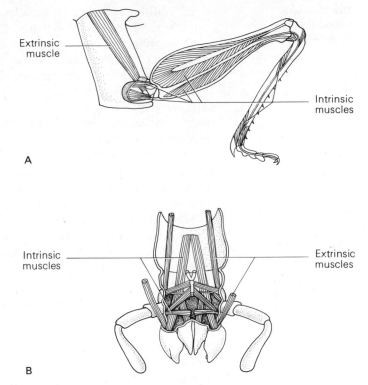

A

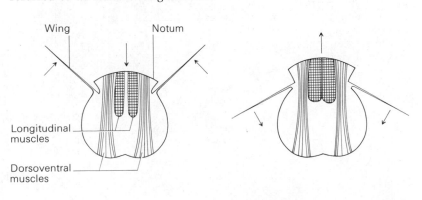

B

Fig. 4–12
Extrinsic and intrinsic muscles. A.
Metathoracic leg of a grasshopper.
B. Labium of a cricket. (Redrawn
from Snodgrass, 1935.)

extrinsic and intrinsic muscles; the extrinsic ones are responsible for
leg movement and the intrinsic ones for movement of the individual
segments. The origins of the various extrinsic leg muscles are on the
terga, pleura, and sterna of each thoracic segment. In the typical
wing-bearing segment are well-developed longitudinal muscles which
run between the phragmata of the pterothorax (Figs. 4–13 and 4–9),
and dorsoventral muscles that attach to the tergum and sternum of
each segment. Contraction of the longitudinal muscles causes arching
of the tergum, which, owing to the construction of the thorax, causes
the depression of the wings. The action of the dorsoventral muscles
is antagonistic to the longitudinals, contraction resulting in the depression of the tergum and a consequent elevation of the wings. Since the
action of these two sets of muscles on the wings is indirect, they are
referred to as *indirect wing muscles*.

Fig. 4–13
Cross section of the
mesothorax showing the
action of the indirect
flight muscles. A. Notum
depressed, wings
elevated. B. Notum
arched and elevated,
wings depressed. Arrows
indicate direction of
movement. (Redrawn
from Snodgrass, 1963b.)

A   B

In addition to indirect muscles, there are muscles that attach directly to the bases of the wings. Direct muscles are responsible for various wing movements during flight and for the flexing of the wings over the abdomen in those insects which possess this ability. In some insects direct muscles provide the main force of propulsion. Their action will be considered in more detail when we discuss flight. The wing muscles are, of course, well developed only in the pterothorax, being reduced or nonexistent in the prothorax. Besides the muscles associated directly or indirectly with leg and wing movements, there are muscles that run between the pleura and terga or sterna and lateral intersegmental muscles. Also, since there are spiracles located in the thorax, muscles associated with their closure mechanisms are present.

The skeletal musculature of the abdomen is somewhat simpler than that of the head and thorax. The most prominent abdominal muscles are the longitudinal intersegmental or segmental ones (Fig. 4–14). These occur both in association with the terga and the sterna and run between the successive antecostae. Also present are lateral abdominal muscles, which may be oblique in orientation, the *oblique sternals,* but typically are dorsoventral in orientation, the *tergosternals.* These may be inter- or intrasegmental. Spiracular muscles are also present. In addition, special muscles are involved in the various movements of the copulatory structures, ovipositor, and cerci.

For more detailed information on the musculature of insects, the reader should consult Snodgrass (1935, 1952), Richards and Davies (1957), and the various individual papers that deal with the morphology of specific insects or groups of insects.

Basic Units of Contraction

The Muscle Fiber. The basic structural unit of a muscle is the *muscle fiber* (Fig. 4–15). Typically, it is composed of an outer membranous layer, the *sarcolemma,* which ensheaths a bundle of tiny fibers, the *myofibrils,* each of which has a cross-sectional diameter of about 1 micron. Also within the sarcolemma lie numerous cell nuclei

Fig. 4–14
Skeletal musculature of the ventral portion of the abdomen (diagrammatic).
A. Dorsal view.
B. Lateral view.

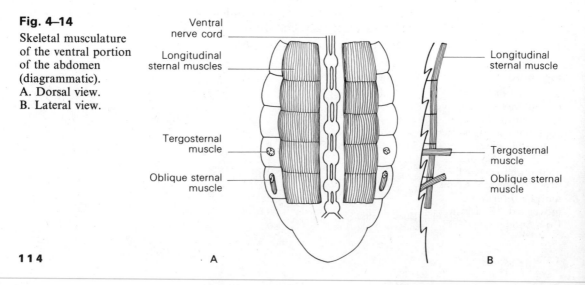

Ventral nerve cord

Longitudinal sternal muscles

Tergosternal muscle

Oblique sternal muscle

Longitudinal sternal muscle

Tergosternal muscle

Oblique sternal muscle

A

B

Table 4–1
Histological Types of Insect Muscle

	Larval	Skeletal	Fibrillar	Tubular
Occurrence	Particularly larval Diptera	Skeletal muscles of most insects	Indirect flight muscles of Hymenoptera, many Diptera, many Coleoptera; tymbal muscles of cicadas	Adults of higher Diptera and Hymenoptera, some Coleoptera
Color	White	White	Yellowish	White
Nuclei	In sarcoplasm around the myofibrils	Scattered immediately beneath sarcolemma or throughout the fiber	Obscure	Centrally located in a core of sarcoplasm which extends the entire length of the fiber
Myofibrils	Surrounded by thick plasm layer; centrally located; striations sometimes obscure	Tightly packed; much more pronounced than in larval muscle	In individual bundles bound together by tracheoles	In bundles surrounding the central core
Sarcolemma	Present	Present	Absent; sarcosomes (giant mitochondria) scattered between myofibrillar bundles	Present

and mitochondria. The myofibril is the basic unit of contraction. The chemical composition by weight of a muscle fiber is approximately 20% protein; the remaining 80% is primarily water, plus small quantities of salts and metabolic substances (Huxley, 1965).

Insect Muscle Fibers. On the basis of color, arrangement of the myofibrils (sarcostyles), presence or absence of the sarcolemma, and other characteristics, four histological types of insect muscle fibers have been described (Table 4–1, Fig. 4–15).

Physiology of Insect Muscle

The individual insect skeletal muscle fibers are organized into discrete morphological entities, the *muscle units*. The fibers composing each unit are enveloped by a tracheolated membrane. Muscle units, in turn, compose the various skeletal muscles. According to Hoyle (1965) a muscle commonly contains 10 to 20 muscle fibers, although some may contain as few as 1.

In addition to the morphological muscle units, muscles can also be divided into *motor units*. A motor unit may be a single muscle unit or all the muscle units in an individual muscle, and it is defined as the total of muscle units innervated by a given motor axon (Hoyle, 1965).

Insect skeletal muscle units commonly receive more than one motor axon. This situation is called *polyneural innervation*. In the majority of cases the number of axons that innervate a given motor unit is two. However, a third has been identified in the jumping hindlegs of *Romalea* sp. and other locusts. In a typical muscle unit (Fig. 4–16) one axon is a fast one, that is, one which elicits a rapid twitch. The other axon is a slow one and induces the tension involved with muscle tonus and slow, smooth contractions. The term *tonus* refers to the continuous tension applied by a muscle as opposed to alternate contraction and relaxation. Muscles in the tonic state support an insect in a given stance. For example, in soft-bodied larvae such as caterpillars they are responsible for the maintenance of the rigidity of the exoskeleton, which is necessary particularly during locomotion. The third motor axon, when present, has been suggested to function as an inhibitor of the effects of the other two fibers, but the evidence for this is sparse.

Muscle Power. Great feats of strength have been ascribed to insects. For example, it has been said of the flea that if he were the size of a man he would have super powers, for example be able to leap over tall buildings. This is a common fallacy. It is true that some

Fig. 4–15

Histological types of insect muscle fibers.
A. Larval, honey bee.
B. Skeletal, beetle.
C. Tubular, honey bee.
D. Fibrillar, honey bee.
(Redrawn with modifications from Snodgrass, 1925.)

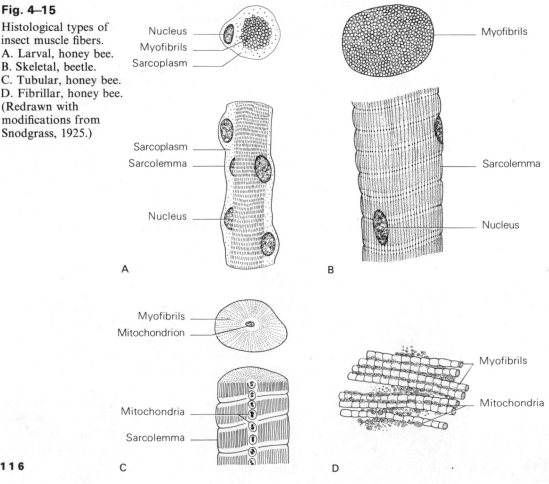

Nucleus
Myofibrils
Sarcoplasm

Myofibrils

Sarcoplasm
Sarcolemma

Sarcolemma

Nucleus

Nucleus

A

B

Myofibrils
Mitochondrion

Myofibrils

Mitochondria

Mitochondria
Sarcolemma

C

D

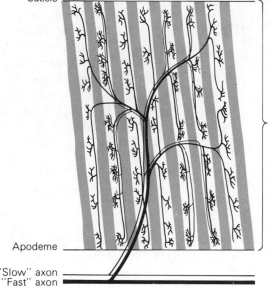

Cuticle

Muscle
unit

Apodeme

"Slow" axon
"Fast" axon

Fig. 4–16

Innervation of a typical muscle unit
showing fast and slow axons. (Redrawn
from Hoyle, 1965.)

insects can leap comparatively great distances and some are capable
of lifting or moving objects many times heavier than themselves.
However, it is not true that if these insects were to increase in size, their
strength would increase correspondingly. The power of muscle
varies directly with its cross-sectional area, while the volume or
mass of a body varies with the cube of its linear dimensions. As an
insect body would increase in size, its volume and mass would
increase at a greater rate than a cross-sectional area of muscle, and
hence the muscles would become relatively weaker. Thus we can say
that insect muscles are relatively more powerful than those of larger
animals only because of their comparatively small size.

Actually when the absolute muscle power (maximum force a muscle
can apply per square centimeter) of vertebrates and insects is compared, it is found that there are no great differences (Wigglesworth,
1965).

Physiological Types of Insect Muscle. Insect muscles can be
divided into two types as follows: *resonating* (*asynchronous*) and
nonresonating (*synchronous*). The resonating type of muscle in insects
is the fibrillar type described earlier. In appropriate mechanical situations it is capable of undergoing several successive contractions when
stimulated with a single nervous impulse. This type of muscle is almost
exclusively associated with flight, but it is associated also with the
halteres of the true flies and with the sound-production mechanism
of the cicada. Nonresonating muscles include the other types of muscle
fibers described earlier and are associated with movement other than
that produced by fibrillar muscle. The differences between the resonating and nonresonating types of muscle will become more clear when
we discuss the flight of insects.

Energy for Muscle Contraction. The elaborate biochemical processes that produce the high-energy adenosine triphosphate (ATP)

occur in the sarcoplasm of muscle. The phosphate for replenishing ATP utilized in muscle contraction is furnished by the phosphogen, arginine phosphate, which has been identified in a number of insects. Creatine phosphate is the phosphogen found in vertebrate muscle. Carbohydrates are commonly the primary fuel for muscle contraction, although many insects utilize fat and occasionally protein or amino acids. For example, the amino acid proline serves as the primary flight fuel in the tsetse fly. For further information on this topic the student is referred to the reviews by Gilmour (1965) and Sacktor (1965).

Selected References

NERVOUS SYSTEM

Bullock and Horridge (1965); Huber (1965); Stevens (1966); Treherne and Beament (1965).

ENDOCRINE SYSTEM

Highnam and Hill (1970); Wigglesworth (1970).

MUSCULAR SYSTEM

Hoyle (1965); Huxley (1965); Maruyama (1965); Sacktor (1965).

5

Sensory Mechanisms;
Light and Sound Production

Sensory Mechanisms

In Chapter 4 we discussed the system responsible for the collection, integration, and interpretation of information from the external and internal environment (the nervous system) and the systems that enable an insect to respond and hence adjust to environmental changes (the muscular and endocrine systems). We shall now consider those structures which carry on the actual collection of environmental information, that is, the various sense organs or *sensilla* (singular sensillum). Basically, the function of any sense organ is to receive some form of energy (stimulus) from the environment and subsequently set off a chain of events that ultimately result in a nerve impulse (Dethier, 1963). A large portion of the energy "sensed" by insects is in the form of various mechanical changes, either gross changes such as the bending of a hair or the stretching of a portion of the body, or molecular movements in the form of sound waves propagated through a solid, liquid, or gaseous medium. The sensation of these changes falls under the general heading of *mechanoreception*.

Another form of energy perceived by insects is ". . . potential energy existing in the mutual attraction and repulsion of the particles making up atoms" (Dethier, 1963). The perception of this form of energy is referred to as *chemoreception*. If the molecules happen to be water, *hygroreception* is the appropriate term. In other instances, the energy stimulating a given sense organ may be in the form of electromagnetic waves (or photons), as, for example, light and heat. The sensation of these forms of energy will be considered under the headings *photoreception* and *thermoreception*.

Morphology of Sense Organs

The majority of sense organs are composed of two types of cells: a *receptor cell* or cells, and *accessory cells*. The receptor cells are usually bipolar neurons which carry out the actual detection of stimuli and generation of the nervous impulse. They are modified epithelial cells, which during the development of an insect, send out a process (the axon) which eventually communicates with the central nervous system. The accessory cells surround the receptor cells and are usually involved with or actually secrete the specialized cuticular structures which make up the most obvious parts of a sense organ. In addition to the type of sense organ just described, there are multipolar receptor neurons which have no contact whatsoever with the cuticle. These multipolar neurons are associated with the various muscles, the alimentary canal, and entad surface of the integument.

Sense organs can be classified morphologically on the basis of the differences in associated cuticular structures. It is thought that all sensilla were originally derived from setae and hence are homologous structures. However, photoreceptors do not readily fall into this grouping and will therefore be discussed separately when we consider photoreception.

Since the various sensilla are considered to be derived from setae, the hair sensilla (*sensilla trichoidea*) should be the first ones to be discussed. You will recall from Chapter 2 that each hair is formed by two cells, the "hair-forming" *trichogen cell,* which is surrounded by the "socket-forming" *tormogen cell.* The addition of a single or several bipolar receptor cells to this picture produces the basic *trichoid sensillum* (Figs. 5–1A and 5–2). Other closely related sensilla include those with bristlelike processes (*sensilla chaetica*), scalelike processes (*sensilla squamiformia*), and peglike or conelike processes (*sensilla basiconica*; Fig. 5–2A). A further modification consists of processes such as those just described, which are sunken in shallow pits (*sensilla coeloconica*; Figs. 5–1B and 5–2) or comparatively deep pits (*sensilla ampullacea*; Figs. 5–1C and 5–2).

Two other types, *campaniform sensilla* (*sensilla campaniformia*; Figs. 5–1D and 5–2B) and *placoid sensilla* (*sensilla placodea*; Figs. 5–1E and 5–2), lack any hairs, pegs, cones, or bristles. The campaniform variety appear as shallow, round or oval, pits and in section can be seen to consist of a bell-shaped cuticular cap innervated by a single receptor cell. The placoid variety appear as platelike structures which are made up of a round or oval cuticular plate surrounded by a narrow membranous ring. In contrast to the campaniform type, a placoid sensillum is innervated by a number of receptor cells. A type of sensillum that is rather dramatically different from those already described

Fig. 5–1

Types of sensilla related to the trichoid type (diagrammatic). A. *Sensillum trichodea.* B. *Sensillum coeloconicum.* C. *Sensillum ampullaceum.* D. *Sensillum campaniformia.* E. *Sensillum placodeum.*

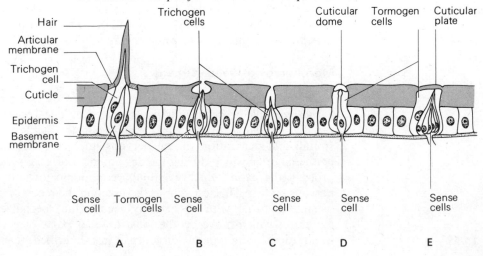

Fig. 5–2

Scanning electron micrographs of sensilla on the antennal flagellum of a worker honey bee. 1, *Sensillum trichodea;* 2, *sensillum basiconica;* 3, *sensillum campaniform;* 4, *sensillum placodea;* 5, *sensillum coeloconicum* or *sensillum ampullaceum.* (Scale lines = 10 microns.) (Courtesy of Alfred Dietz and Walter J. Humphreys.)

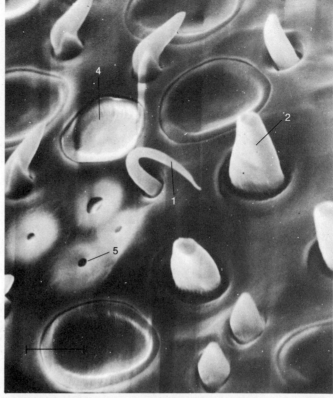

A

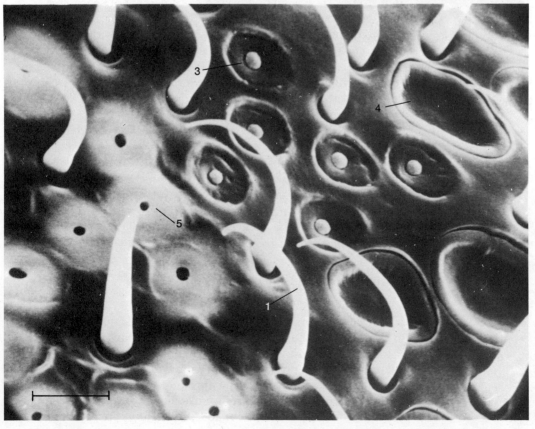

B

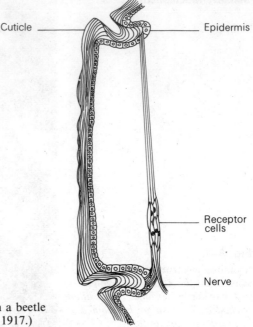

Cuticle — Epidermis

Receptor cells

Nerve

Fig. 5–3
Simple chordotonal organ in a beetle
larva. (Redrawn from Hess, 1917.)

is the *scolopophorous* or *chordotonal organ* (Fig. 5–3). This kind of
sensillum usually occurs in bundles and consists of a bipolar neuron,
which is usually stretched between two internal integumental surfaces.
Johnston's organ is a good example of this type of sensillum.

It should be pointed out that there is considerable variation within
any given morphological type of sensillum both with regard to the
appearance of the associated cuticular structures and to the numbers
of receptor neurons. In addition, in many instances, a morphological
type does not always imply a particular function, since a given sensillum may have different functions in the same insect or may contain
two or more receptors which receive different forms of stimuli. For
example, a single hair on the labellum of the blow fly, *Phormia*, may
have two chemoreceptor cells and a mechanoreceptor cell associated
with it (Dethier, 1963). However, there are certain sensilla which
always seem to be associated with the same general function. For
example, the campaniform type has always been found to be a
mechanoreceptor which is stimulated by deformation of the cuticle.
The same is true for the chordotonal sensilla, which are sound,
vibration, and stretch receptors.

Methods Used to Study Insect Sense Organs

Before we consider the various senses, it would be appropriate to
first discuss in general terms some of the methods that have been used
to study the structure and function of sense organs.

Earlier work with insect sense organs generally consisted of a combination of two approaches: behavioral and morphological. Based
upon their gross appearance and appearance in stained sections under
the light microscope, structures were identified as sensory or not. This
approach did not, of course, enable investigators to be certain of the

specific function of a given sense organ, although based on other work they could speculate quite accurately in some instances. The use of behavioral criteria allowed the process to be carried somewhat further since innate or learned responses to various stimuli could be taken advantage of in different ways. For instance, an insect may be naturally attracted to, repelled by, or respond in some other way to a given stimulus. Once this fact is established, a structure or structures suspected of receptor activity can be removed or blocked in a variety of ways to determine if they are, in fact, active with regard to the stimulus being tested.

When a specific receptor or specific receptors are identified, further experiments can be made to determine their sensitivity by varying the concentration or intensity of the stimulus. An excellent example of this kind of work has been carried out on the tarsal receptors of the blow fly (Dethier, 1955). The tarsal sense organs are "taste" or contact chemoreceptors and are sensitive to various sugars. When a solution of a "tastable" sugar is applied to them, the fly extends its labellum in response. Using this labellar response, the sensitivities to various concentrations of several sugars have been worked out.

These earlier approaches are still used today, but have been complemented by the much more detailed observation possible with the electron microscope, both transmission and scanning, and by the use of highly sophisticated electrophysiological techniques. The latter have enabled investigators to make exceedingly critical studies of the functions of the sensory structures at the cellular level.

Mechanoreception

Organs that are sensitive to the actions of stretching, bending, compression, torque, and so on, applied to the integument or some internal organ are the *mechanoreceptors*. These sense organs are responsible for the maintenance of posture, stability during locomotion, and body position with respect to gravity. In addition, many are designed such that they enable the insect to detect sound waves or vibrations in a solid substrate. In certain instances they provide information as to the state of certain internal organs (e.g., the alimentary canal).

We shall approach the topic of mechanoreception by a consideration of the various mechanical senses insects are known to possess. These senses are as follows: *tactile* (touch), *proprioceptive,* and *sound* or *vibration.*

The Tactile Sense

The hair sensilla are the external receptors involved in the sense of touch. Movements of these hairs caused by contact with another surface are carried via their basal portions to the associated receptor cells. At this point mechanical energy is *transduced* (changed) into electrical. In other words, the movement of the base of a hair initiates a nervous impulse or a train of impulses.

Many hair sensilla are strictly mechanoreceptive in function and have only a single receptor cell. However, others may possess a number of other kinds of receptor cells (e.g., chemical) in addition to the mechanoreceptive ones. An example of the latter was given earlier—certain hairs on the labellum of the blow fly, *Phormia* sp.

According to Dethier (1963) the mechanosensitive hairs studied to date appear to fall into two groups: *velocity sensitive* and *pressure sensitive*. In the first group neural impulses are generated only in the presence of constantly changing stimuli, as would be the case, for example, in an insect during flight. Not surprisingly, these hairs have been found on the anterior edges of wings of various insects. Hair sensilla that fall into the second category initiate a steady train of nervous impulses when statically deformed. This situation would occur, for example, when an insect was at rest or on a solid substrate.

Hair sensilla are located in a number of different regions of the insect body, where they are able to come into contact with and hence be deformed by a variety of surfaces. For example, they are commonly found on the legs, mouthparts, and antennae, all of which frequently come into direct contact with the substrate or other surfaces. They are also found on the cerci and here may function in initiating an escape response resulting from air movement or something suddenly touching these appendages and consequently bending the associated hairs. In the cockroach the nerves from these hairs are ultimately connected with the "giant axons," which allow very rapid transmission of the nervous impulse that initiates a sudden scurrying off, the alarm reaction, in these insects. Hair sensilla in these locations are usually of the velocity-sensitive type. In addition to being found in locations in which they come into contact with other than the insect body itself, many hair sensilla are located between joints, body segments, and other areas in which there is direct contact between two body surfaces. Their function in these situations is proprioceptive, and they are generally of the pressure-sensitive type.

The Proprioceptive Sense

Proprioceptive organs are those which are stimulated by changes in various parts of the insect body. The changes may be in length, tension, and so on. These receptors provide the insect with continuous information as to the position of the various body parts and the tensions of the various muscles. It follows, then, that these types of sensilla are of critical importance in the maintenance of "proper" orientation of the body parts with respect to one another, or of the entire body with respect to gravity both in the stationary and moving insect. You will find examples of this later in the section on locomotion. The role of proprioceptors in orientation is complemented by the actions of other receptors (e.g., tactile and photoreceptors).

A number of different types of sensilla function as proprioceptors. Among these are clusters of tactile hairs or hair plates, campaniform sensilla, stretch receptors, chordotonal organs, and Johnston's organ.

Hair plates are very common in insects and appear as clusters of tiny trichoid sensilla. They may be looked upon as tactile receptors, which respond to the insect "touching" itself since they are usually

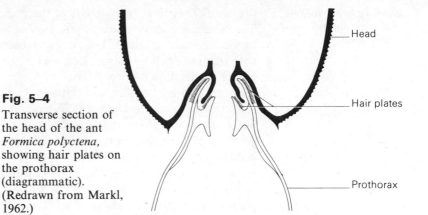

Fig. 5–4
Transverse section of
the head of the ant
Formica polyctena,
showing hair plates on
the prothorax
(diagrammatic).
(Redrawn from Markl,
1962.)

Head

Hair plates

Prothorax

located in appressed or overlapping areas of the body. For example, in the ant they are located between certain joints of the antennae, in the neck region between the head and thorax (Fig. 5–4), at the bases of the coxae and the trochanters, in the petiolar region between the first and second abdominal segments, and in the ventral gaster region between the second and third abdominal segments.

Since a given position of one body part relative to another would "stimulate" particular hairs of specific plates to specific degrees, a definite complex sensory pattern would impinge on the central nervous system for every possible body position and movement. These hair plates in the ant are then important in the maintenance of "proper" posture, whether the insect is stationary or moving. In the praying mantis, hair plates in the neck region function in the process of prey capture. As the mantis turns its head, visually following potential prey, the changing pattern of impulses from the hair plates is critical in determining the accuracy of the strike with its grasping forelegs. If the operation of the hair plates is experimentally interfered with by denervation or immobilization of the head, the accuracy of the strike is seriously impaired or destroyed altogether (Roeder, 1963).

Campaniform sensilla serve as compression and stretch receptors and are located only in areas on the integument that are exposed to strains of various sorts. They are concentrated particularly in areas where compression and stretching occur as a result of muscular activity, for example in the legs, wings, halteres of flies, ovipositor, and the bases of the mandibles. The role they play in association with the legs, wings, and halteres will become clear when we discuss locomotion.

The multipolar neurons associated with muscles, the alimentary canal, and entad surface of the integument, which were mentioned earlier in this chapter, have been found to act as stretch receptors. They respond with a nervous impulse when the tissue in which they are embedded is subjected to a change in length. They have been found in dragonfly nymphs and members of the orders Orthoptera (grasshoppers and relatives), Hymenoptera (ants, bees, wasps, and relatives), and Lepidoptera (moths and butterflies) (Dethier, 1963).

Chordotonal sensilla, as explained earlier, usually occur in bundles and are usually stretched between two internal integumental surfaces. They have been found in virtually every insect examined for their presence. They have been found in the pedicel of the antennae in all insects studied, in the mouthparts, wing bases, halteres, legs, and in the abdominal segments. They have also been found closely associated with tracheae, pulsatile structures, and in the hemocoel. Early investigators ascribed an auditory function to them, and, in fact, many of them are auditory in function. However, a number are known to be proprioceptive and, as Dethier (1963) points out as very likely, "... all not associated with tympanic membranes or grouped to form subgenual and Johnston's organs . . . will eventually be proved to be proprioceptors." Several different proprioceptive functions have been suggested for them, for example sensation of body orientation, passive body movements, and muscular movements. Their close associations with tracheae, pulsatile organs, and the various hemocoelic cavities suggests that they may respond to changes in intertracheal air pressure and in blood pressure.

In virtually every insect studied, a specialized group of sense organs, which are quite similar in structure to chordotonal sensilla, are found in the pedicel (second segment from the base) of each antenna (Fig. 5–5). These structures are attached to the pedicellar wall and to the membrane between the pedicel and the third antennal segment and are in a radial arrangement. They make up the Johnston's organ. This structure varies in complexity depending on the insectan species and reaches its greatest development in two families of flies (Culicidae, mosquitoes; Chironomidae, midges). In these two

Fig. 5–5
Photomicrograph of a sagittal section of Johnston's organ in the mosquito *Aedes triseriatus*. 1, Nerve; 2, scolopophores; 3, cell bodies of neurones; 4, pedicel. (Scale line = 50 microns.)

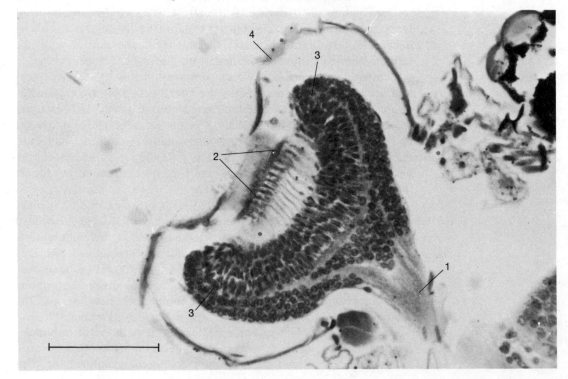

Dipteran families it completely fills the pedicel. Johnston's organ is known to function as a proprioceptor in some insects, although it has different functions in others (e.g., sound reception in mosquitoes and midges). An example of an insect in which it has a proprioceptive function is the honey bee. During the flight of this insect, Johnston's organ responds to movements of the antennal flagellum and in this way provides the bee with a measure of the stream of air passing over it. The amplitude of the wingbeat is regulated in response to this measurement (Wigglesworth, 1965).

Another example is the use of the antennal movements and hence Johnston's organ by certain aquatic Hemiptera (true bugs). Some of these (*Corixa* and *Naucoris*) swim with the dorsal side up, while others swim with their dorsal sides down (*Notonecta* and *Plea*). In either case, the proper body orientation during swimming is maintained because they are able to sense when their dorsum is up or down. This is accomplished by the buoyant action of a small air bubble trapped between the ventral part of the head and each antenna. Any change in the position of the insect results in a change in the direction of the buoyant force of the bubble relative to the insect and hence results in a movement of the antennae which in turn causes a change in the sensory patterns generated by each Johnston's organ.

Sound Perception

For the purposes of this discussion, we shall define sound as longitudinal waves of kinetic energy propagated through a continuous medium (gas, liquid, or solid). As a matter of convenience the perception of sound will be divided into two senses, the sense of vibration and the sense of hearing. The sense of vibration will then entail the perception of sound via the substratum, and hearing, the perception of sound via air or water.

The perception of sound by insects is important to their survival in a number of ways. Many of the stimuli that impinge on an insect from the external environment are in the form of sound. Some of these sounds are produced by other insects of the same or different species (see the section on sound production in this chapter) and other sounds come from a variety of environmental sources. Sound detection may be of value in sensing a potential danger situation, a potential mate, prey animals, other members of the same species (as might be the case where individual territories are maintained), and so on.

Sensilla Involved in Sound Perception. Only two basic types of sensilla have definitely been shown to be involved in sound reception, trichoid sensilla and specialized organs composed of chordotonal receptors. However, others, such as campaniform sensilla or stretch receptors, may well be suspect. We shall consider first the organs composed of chordotonal sensilla and then the trichoid sensilla.

Those sound-sensitive structures composed of chordotonal sensilla are the *tympanic organs, subgenual organs,* and *Johnston's organ,* which was described earlier. Most tympanic organs are composed of the same fundamental parts. However, the degree of development of these parts varies from group to group. The basic structures (Fig.

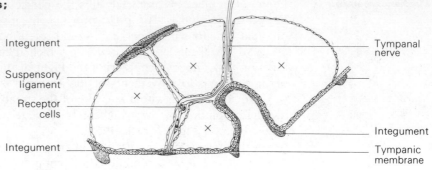

Fig. 5–6
Section of a metathoracic tympanal organ of a noctuid moth. × indicates a
tracheal air sac. (Redrawn from Roeder, 1959.)

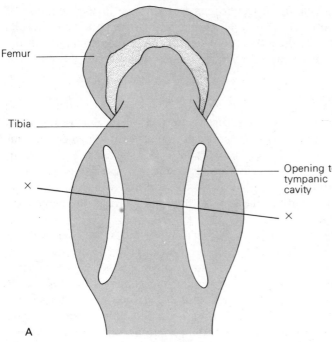

Fig. 5–7
Tympanic organ on the fore tibia of a
long-horned grasshopper. A. Anterior
view of the tibia. B. Transverse section
at the level indicated approximately by
the line ×—× in A, epidermal cells
omitted. (B redrawn with modifications
from Schwabe, 1906.)

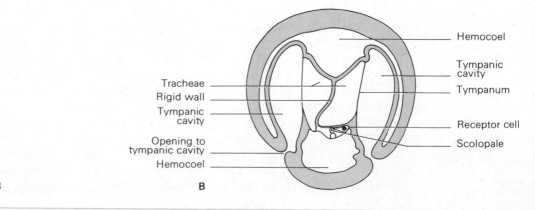

5–6) involved are a thin integumental area (the *tympanum*) and a group of chordotonal sensilla attached directly or indirectly to the entad surfaces of the tympanum. Usually a tracheal air sac is closely associated with the tympanum and sensilla. In some insects (e.g., male cicadas) the tracheal air sac may serve to amplify certain frequencies. The number of chordotonal receptors varies from 2 in moths in the lepidopteran family Noctuidae to 1,500 or more in cicadas (Dethier, 1963).

Tympanic organs have been identified in a number of different locations in a variety of insects. In the order Orthoptera (grasshoppers and relatives), they are found on the tibiae of the forelegs in the families Tettigoniidae (long-horned grasshoppers; Fig. 5–7A, B) and Gryllidae (crickets) and on either side of the first abdominal segment in members of the family Acrididae (short-horned grasshoppers). Tympanic organs occur in the metathorax of Noctuid moths and in the abdomen in geometrid and pyralid moths. In cicadas (suborder Homoptera) these structures are located in the abdomen. In addition, tympanic structures have been identified in a few members of the order Hemiptera (e.g., in waterboatmen, family Corixidae). Examples of specific functions of hearing via the tympanal organs will be discussed in Chapter 7.

Subgenual organs (Fig. 5–8) are groups of chordotonal sensilla which are located in the basal portion of the tibial leg segment. They are not associated with any joints. They vary considerably in degree of development from group to group, being somewhat weakly developed in the true bugs, more developed in the members of the orders Lepidoptera and Hymenoptera, and most highly developed in the beetles and the true flies.

Vibration Perception. One needs only to sharply tap a surface on which an insect is resting to be convinced that insects are able to perceive vibrations through the substrate. According to Swartzkopff (1964), among the insects in which vibration sensitivity has been measured, those showing the greatest sensitivity have been found to possess subgenual organs. These organs are generally thought to be

Fig. 5–8

Subgenual organ of an ant exposed by section of the tibia. (Redrawn with modifications from Schön, 1911.)

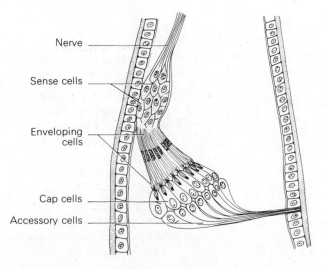

Nerve

Sense cells

Enveloping cells

Cap cells

Accessory cells

especially adapted to the perception of vibrations. Subgenual organs have not been found in the less sensitive species, the trichoid and small chordotonal sensilla in the legs being the vibration receptors. In *Locusta* (order Orthoptera, family Acrididae) trichoid sensilla on the sternites are sensitive to substrate vibration (Dethier, 1963).

Hearing. Tympanic organs, trichoid sensilla, and Johnston's organ have all been shown to subserve sound perception in a number of insects. The tympanal organs are, as far as known, always used for hearing. However, as you have seen earlier in this chapter, trichoid sensilla and Johnston's organ have been demonstrated to have other functions.

The sounds received by tympanic organs are used by nocturnal noctuid and related moths to detect the ultrasonic emissions used by echolocating, predatory (insectivorous) bats. Detection of these sounds stimulates avoidance behavior of various sorts (see Chapter 7). In the orders Orthoptera and Hemiptera tympana are the receivers of the sounds produced by members of the same species and are particularly involved with sexual behavior. In the waterboatmen, *Corixa* (order Hemiptera), males produce waterborne sounds that are perceived and responded to by the females. Evidently males also hear each other produce these sounds, since they chirp as long as the tympanic organs are intact. These are the only insects shown to be sensitive to waterborne sounds (Dethier, 1963). Since tympanal organs occur bilaterally on some part of the insect body, they facilitate localization of a sound source through differential stimulation.

Certain trichoid sensilla on exposed regions of the body have been shown to be sensitive to airborne sounds in a number of insects. Sensilla on the cerci of Orthoptera (e.g., cockroaches) are especially sensitive, and their stimulation by sound may elicit the characteristic "alarm" reaction mentioned earlier.

Johnston's organ in the antennae of some species of mosquitoes and midges has been shown to be a sound receiver. The males are able to detect the sounds produced by the rapidly beating wings of the females. Their response to these sounds can be elicited in a male by using a tuning fork that sounds at an appropriate frequency.

Chemoreception

As defined earlier, chemoreception is the process by which the "potential energy existing in the mutual attraction and repulsion of the particles making up atoms" is detected (Dethier, 1963). Thus chemoreceptive organs are responsive to direct contact with chemicals. Chemical cues from the environment are useful to insects in several ways, for example, food (host plant or animal, prey, decaying organic material, and so on), mediation of caste functions in social forms, mate location, identification of noxious stimuli that are a potential threat to survival, selection of oviposition site, habitat selection, and others.

In general terms the chemoreceptive activities of insects may be divided into three "chemical senses": distance chemoreception, or *olfaction*; contact chemoreception, or *gustation*; and "general" or "common" chemical sensitivity. Distance chemoreception is mediated by chemoreceptors that are responsive to molecules or ions of a chemical in the gaseous state at comparatively low concentrations. These receptors are considered to be highly sensitive and may show a high degree of specificity with regard to the kind of chemical that initiates a response. Contact chemoreceptors are excited by direct contact with molecules or ions of a chemical in solution at a concentration usually somewhat higher than olfactory chemostimuli. Generally these receptors are less sensitive than the distance chemoreceptors and are commonly associated with feeding activities. The "general" chemical sense involves receptors that are comparatively insensitive except to relatively high concentrations of a stimulating chemical. These receptors are much less discriminating than either contact or distance chemoreceptors and are usually associated with an avoidance or escape response. Actually no specific general chemical receptors have been positively located and nonspecific effects on neurons may be involved (Hodgson, 1964).

Although the classification of the "chemical senses" presented above is quite useful, it is not without difficulties. For example, the differentiation between distance and contact chemoreception must be qualified by pointing out that actually in both types the molecules or ions of the stimulating chemicals always come into direct contact with the receptor cell membranes. Compounding the problem, in aquatic and subterranean insects the same receptors may respond to chemicals either in the gaseous state or in aqueous solution.

A wide variety of the morphological types of sensory structures may be involved with chemoreception. Among those for which there exists at least some evidence of olfactory activity in certain insects are the sensilla trichodea, sensilla basiconica, sensilla placodea, and sensilla coeloconica. Sensilla trichodea and sensilla basiconica have been identified as contact chemoreceptors in a number of insects.

Proved or suspected contact chemoreceptors have been found in several parts of the insect body. They probably exist in the mouthparts of all insects, for example the hypopharynx and epipharynx in caterpillars, the tips of the maxillary and labial palps and in the buccal cavity of the cockroach *Periplaneta* sp., and possibly the buccal cavity (cibarium) of mosquitoes (Day, 1954). Chemoreceptors probably also exist in the foregut; for example, sensilla in the pharynx of the house fly apparently have to do with the control of the passage of ingested food.

In honey bees and certain wasp species, contact chemoreceptors are located on the distal segments of the antennae, enabling the insects to differentiate between sweetened and unsweetened water (Wigglesworth, 1965). The distal portion of the tibia and tarsi of the forelegs of many insects bear contact chemoreceptors. In many butterflies, true flies, and bees the mouthparts are extended in response to stimulation of the fore tarsi with sugar water. Contact chemoreceptors may be located in the ovipositors of parasitic Hymenoptera, which deposit their eggs directly into host insects. Similarly, distance (olfactory) receptors have been identified in the antennae and mouthparts of a variety of insects and in the ovipositor of at least one.

Chemoreceptive sensilla may occur in tremendous numbers in certain insects. For example, in the antennae of male polyphemus moths (*Antheraea polyphemus*) there are approximately 70,000 sensilla with about 150,000 sense cells. Male honey bees (drones) have approximately 30,000 placoid sensilla on each antenna, while members of the worker caste have about 6,000 (Wigglesworth, 1965).

Considerable work has been done with the contact chemoreceptive abilities of the honey bee. This insect is capable of differentiating the qualities sweet, bitter, acid, and salt (von Frisch, 1950). The honey bee and man have been compared with regard to their sugar-tasting abilities. According to Wigglesworth (1965), ". . . out of 34 sugars and related substances tested, 30 appear sweet to man, only 9 to the honey bee; all these nine being present in the natural food of the bee and capable of being metabolized by it." Honey bees are a little more sensitive to bitter substances such as quinine. Another substance, acetylsaccharose, is exceedingly bitter to man but not to the honey bee. This substance has been suggested for addition to cane sugar so that the sugar might be sold more cheaply to beekeepers and not used as human food (Wigglesworth, 1965).

Von Frisch (1950) established two thresholds for honey bees relative to the sugar sucrose (cane sugar). The first he called the *threshold of acceptance*, which he defined as the minimum concentration of sugar which the bees could be induced to ingest. This threshold was established at about 40% when many of the bees' foraging plants were in bloom and approximately 5% in the fall when flowers were scarce. The second threshold von Frisch measured was the *threshold of perception,* that is, the minimum concentration of sucrose that the bees could perceive. This threshold, if accurately determined, would be expected not to vary with the scarcity or abundance of flowers. By starving the bees for several hours and then determining the minimum concentration of sucrose which they would accept, this threshold was established to be somewhere between 1 and 2%. Contact chemoreception has also been extensively studied in a number of true flies (*Phormia* sp., *Calliphora* sp., *Musca* sp., *Drosophila* sp., and others). In those instances, both behavioral and electrophysiological techniques have been utilized to establish thresholds and to gain insight into the basic mechanisms involved.

As with contact chemoreception, the distance chemoreceptive abilities of the honey bee have also been widely investigated. One would, of course, expect olfactory abilities to be great in these insects since their life depends to a large extent on flowering plants, and it would be of obvious advantage for a bee to be able to visit repeatedly a particular kind of flowering plant that was currently in bloom. Evidently, the olfactory abilities of honey bees and man are similar in terms of threshold concentration of various scents; however, honey bees seem to have a greater ability to discriminate among many different scents (Wigglesworth, 1965).

Olfactory cues no doubt play an important role in the lives of many insects, and the distance chemoreceptive abilities of some are fantastically acute. For example, the male silkworm moth *Bombyx mori* reacts to the sex attractant produced by the female at a concentration as low as 100 molecules of attractant per cubic centimeter of air (Wilson, 1970).

Thermoreception

Based on definite behavioral responses, it is well established that many insects are sensitive to changes in temperature. For example, honey bees trained to visit warm places are able to detect temperature differences as small as 2°C (Wigglesworth, 1965). In some insects heat sensitivity seems to be somewhat generalized over the entire body. In others, specific locations have been identified. For example, in the true bug, *Rhodnius* sp., the antennae are extremely sensitive to small differences in air temperature. The sensilla presumed to be involved are the thick-walled sensilla present in very large numbers on the antennal segments. *Rhodnius* sp. is a bloodsucking insect and in this and other bloodsucking species (e.g., mosquitoes, lice, bedbugs) perception of warmth is extremely important in host finding. All insects will move away from high temperatures, but this is probably a generalized sensitivity with no particular sensilla being involved. Some insects are evidently able to perceive the radiant heat of the sun or other light source; for example, stink bugs will turn their dorsal sides toward a light source when the ambient temperature is low. By thus exposing the largest surface to the light, they are able to receive the maximum possible radiant heat.

Hygroreception

As with thermoreception, that insects are able to perceive moisture in the air is well known due to specific behavioral responses. Spring-tails, like other small, soil-dwelling insects, are very sensitive to moisture, both in the air and the substratum. They are attracted to regions of high humidity. Other insects, such as earwigs and meal-worm beetles avoid very moist areas. Some insects, the honey bee and others, can perceive water from a distance. The specific body regions and sensilla that are sensitive to moisture have been identified in only a very few insects. In the human louse, *Pediculus humanus*, antennal "tuft organs" composed of several small hairs have been shown to be specific hygroreceptors (Wigglesworth, 1941).

Photoreception

Photoreception may be defined as the ability to perceive energy (light) in the visible or near-visible (near ultraviolet) range of the electromagnetic spectrum. In order for an organism to perceive light, a pigment capable of absorbing light of a given wavelength must be present, and there must be a means of producing a nervous impulse as a result of this absorption.

Many different kinds of environmental information are available to an insect in the form of light stimuli. For instance, an insect may perceive to a greater or lesser extent, form, pattern, movement, distance, color, relative brightness, the polarization plane of light, light versus dark, and the length of a light period.

Generally speaking, three types of photoreceptive structures have

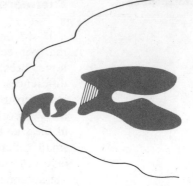

Fig. 5–9

Photoreception in the house fly
larva; cephalopharyngeal skeleton
with pocket that contains
photoreceptive cells, indicated by
hatching.

been found in insects. These are the *compound eyes, lateral ocelli,*
and *dorsal ocelli.* Each of these kinds of photoreceptive structures
will be considered in turn. Refer to Chapter 3 for information regarding the location and numbers of compound eyes and ocelli. In the
larvae of certain higher dipterans (true flies) are found specialized
photoreceptive organs which do not readily fit into any of these three
groups of photoreceptors. These consist of photosensitive cells (Fig.
5–9), which are located in small cavities in the anterior end of a larva
(the larvae or "maggots" of higher Diptera do not have well-defined
heads). The negative phototactic response (i.e., the tendency to move
away from a source of light) is presumably accomplished by the larva
orienting its direction of movement such that the light-sensitive cells
receive minimal stimulation. This orientation would obviously be with
the anterior end of the larva away from the light source, with most
of the body interposed between the light source and the sensitive cells.

In addition to definite organs associated with light perception,
many insects apparently possess a light sensitivity over the general
body surface. This is evidenced by the fact that certain insects will give
definite responses to light even when the operation of the photoreceptive structures listed above is disrupted in some way. For
example, cockroaches continue to demonstrate a preference for dark
situations even after being totally blinded. Similarly, following
decapitation mealworm larvae (*Tenebrio* sp.) continue to avoid light.

Compound Eyes

The compound eyes (Fig. 5–10A) are the major photoreceptive
organs of adult insects. They are two in number (with certain bizarre
exceptions) and are located on either side of the head. Each is composed of a number of individual sensory units or *ommatidia* (Fig.
5–10B). Externally, these ommatidia are marked by hexagonal cuticular facets. The facets, and hence the ommatidia, may vary in number from a very few to several thousand depending upon the kind of
insect.

Structure of the Compound Eye. Before we can appreciate how,
at the present level of knowledge, compound eyes work, we must first
have a more detailed understanding of their structure than has
already been presented. To this end, it is useful to consider an indivi-

dual ommatidium (Fig. 5–10C, D) as being divisible into two parts: the *dioptric apparatus,* which acts as the "lens," and the *receptor apparatus,* in which the events leading to the initiation of a nervous impulse occur.

The dioptric apparatus is composed of the *cornea,* the *crystalline cone,* and *corneal pigment cells.* The cornea is a cuticular structure and is continuous with the cuticle of the integument. It has the shape of a planoconvex lens, the convex portion forming the outer surface. In many insects a hexagonal array of very small conical projections (approximately 0.2 millimicron from tip to base and center to center) may be found on the outer surface of the cornea. These "corneal nipples" are thought to act as an antireflection coating, which reduces the reflection from the air–cornea interface (Miller, Bernard, and Allen, 1968). It is suggested that these "nipples" may also serve in insects active during light periods as camouflage by cutting out mirror-like reflections from the cornea which might attract predators. In insects active at night, they may help in some way to increase the sensitivity of the eyes.

The crystalline cone lies immediately beneath the cornea and is composed of a translucent material. The darkly pigmented corneal pigment cells are usually located on the periphery of the crystalline

Fig. 5–10

Compound eye structure (diagrammatic). A. Head with compound eye. B. Four ommatidia removed and magnified. C. Apposition ommatidium. D. Superposition ommatidium.

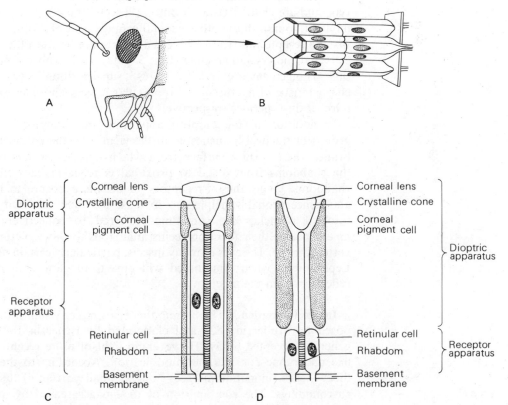

cone, except in very primitive insects [e.g., thysanurans (silverfish and relatives)], in which they lie beneath the cornea. They are considered to be the cells that originally secreted the cornea.

The receptor apparatus (Fig. 5–10C, D) is composed of six or seven *retinular* (nerve) *cells* arranged in one (usually) or two layers. If arranged in two layers, one is proximal to the other. Like the crystalline cone, the group of retinular cells is usually surrounded by rather darkly pigmented cells. Usually each retinular cell gives rise to an axon that passes through a basement membrane and enters the brain. In addition to giving rise to an axon, each retinular cell contributes to the formation of a centrally located retinal rod, or *rhabdom*. The contribution of each retinular cell to the rhabdom is called a *rhabdomere*. The rhabdomeres have essentially replaced the dendrites and are considered to be the receptive surfaces of the retinular cells. Use of the electron microscope has revealed that the rhabdomeres are made up of tiny, closely packed fingerlike projections (microvilli) from the retinal cells. These microvilli project at right angles to the long axes of the retinal cells. It is generally thought that the microvilli contain the light-absorbing pigment(s) directly involved with photoreception. The pigment *retinene,* which has been identified in the eyes of many animals, has been found in insects.

There are a variety of different ways the dioptric apparatus and the receptor apparatus are associated with one another. Generally, however, based on the association of these ommatidial components eyes fall into one of two rather broad categories. The retinal cells may lie immediately beneath the crystalline cone (Fig. 5–10C). Eyes with the ommatidial components so arranged are of the "apposition" type and are characteristic of diurnal insects (those active during the daylight hours). In the other general arrangement of dioptric apparatus and receptor apparatus there is a clear space between the retinular cells and the crystalline cone (Fig. 5–10D). Eyes with ommatidia of this arrangement are referred to as "superposition" eyes and are characteristic of nocturnal or crepuscular insects (those active during dark or dusk periods, respectively).

In addition to the dioptric apparatus and receptor apparatus, groups of tracheal branches lie in the vicinity of the basement membrane. These form a surface from which light which has traversed the rhabdoms from distal to proximal is reflected back along the rhabdoms, giving these "receptors" a double exposure to the light and hence probably helping to increase the light sensitivity. These tracheal branches are sometimes referred to collectively as the *tapetum* since their function seems analogous to this structure in the vertebrate eye. The eyes of many insects, particularly certain nocturnal Lepidoptera, when illuminated will appear to glow as a result of reflection from the tapetum.

Image Formation.   The *mosaic* theory of insect vision intially proposed by Muller in 1826 and elaborated by Exner in 1891 is still generally accepted today. However, in light of more recent work, it has undergone considerable modification. According to the mosaic theory, each ommatidium "sees" only a small portion of the insect's surroundings. The combination of the images sensed by individual ommatidia supposedly together forms a composite or mosaic view of the external environment. This situation is somewhat analogous to looking at the surroundings through a handful of soda straws. Only a small part of the total view is seen through any one straw, but the combination of these "small parts" gives a mosaic image of the surroundings (Wigglesworth, 1964a).

According to the mosaic theory there are two basic types of compound eyes, *apposition* and *superposition*. Apposition eyes (Fig. 5–11A) are composed of ommatidia of the apposition type and are thus, as mentioned earlier, characteristic of many diurnal insects. Similarly, superposition eyes (Fig. 5–11B, C) are made up of superposition ommatidia and are characteristic of many nocturnal or crepuscular insects. In the apposition type of eye there is little or no movement of the pigment in the pigment cells surrounding the crystalline cone in response to changes from light to dark or vice versa. Thus the pigment remains rather uniformly distributed. However, in superposition eyes, there is considerable pigment movement in response to a light–dark change. In a lighted situation (Fig. 5–11B), the pigment in the pigment cells tends to migrate proximally producing the light-adapted condition. On the other hand, in a dark situation (Fig. 5–11C), the pigment migrates distally (dark-adapted condition).

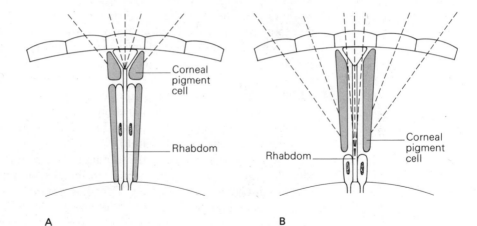

A

B

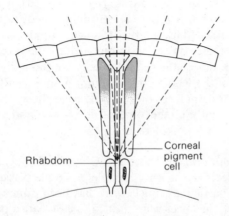

C

Fig. 5–11

Image formation in compound eyes (diagrammatic). A. Apposition type. B. Superposition type, light adapted. C. Superposition type, dark adapted. Dashed lines represent rays of light. See text for explanation.

In the apposition eye and the light-adapted superposition eye, only light rays entering parallel or close to parallel to the long axis of an individual ommatidium are carried to the rhabdom. Oblique rays are absorbed by the pigment, which thus has the effect of optically isolating adjacent ommatidia from one another. The image formed in this situation is an *apposition image* since the light reaching a single rhabdom has entered only via the dioptric apparatus of the same ommatidium. In the dark-adapted superposition eye, the distal movement of the pigment has the effect of removing the optical isolation between adjacent ommatidia. In this instance, the light rays reaching the rhabdom of a given ommatidium enter via several ommatidia. The image formed under these circumstances is referred to as a *superposition image*. One would not expect a superposition image to be as sharply defined as an apposition one because of the increased amount of light reaching each rhabdom. Also the formation of a sharply defined superposition image would require that all the light rays reaching a single rhabdom impinge at exactly the same point. It follows, then, that eyes in which pigment optically isolates individual ommatidia would probably have a greater resolving ability than those in which pigment does not form an optical boundary. Exner theorized that the dark-adapted condition in a superposition eye increases the light sensitivity since it enables more light to reach each rhabdom. The adaptation of the eye to light or dark situations is probably also partly due to a change in the retinal cells or in the central nervous system.

In recent decades Exner's theory has come under attack from several points of view. For example, it seems that superposition images are not formed in many, if not most, insects with superposition eyes. In addition, there is evidence that the resolving ability of eyes is, in fact, not decreased by the distal migration of pigment. However, Exner's idea that distal pigment migration (dark adaptation) in superposition eyes increases sensitivity still holds.

Recent additions to the understanding of insect vision include the idea that the crystalline tracts found in the eyes of many insects may act as light guides in that once light has entered them it does not escape but is carried directly to the rhabdoms. Besides acting as a light guide, a crystalline tract surrounded by pigment cells is thought to act as a longitudinal pupil (Miller, Bernard, and Allen, 1968). When pigment envelopes the tract (light-adapted), it absorbs some of the light; however, when the pigment migrates distally and no longer envelopes the tract (dark-adapted), much more light is transmitted to the rhabdom. Other recent additions include the possibility that in at least certain instances the retinular cells are the individual functional units instead of the whole ommatidium. Also, recently, images which are formed proximal to the retinal image have been discovered. They are essentially of the superposition type and their significance is unclear.

Perception of Form and Pattern.   Considerable behavioral evidence exists that insects are at least capable of simple perception of pattern and form. In particular, much work has been done with honey bees. For example, honey bees were trained to associate a particular pattern of colored paper with the presence of sugar water in a cardboard box.

Fig. 5–12

Figures used to study
form perception in honey
bees. (Redrawn from
Hertz, 1929.)

Subsequently, they were presented with several boxes marked with
pieces of colored paper cut in the pattern to which they had been
trained and others with a rather similar pattern. None of these boxes
contained sugar water. It was judged that the bees could discriminate
between these two patterns since only a small number visited the
boxes marked with the second pattern (von Frisch, 1950). Other
work, again with training, to a particular pattern has demonstrated
that bees can discriminate between the shapes in the upper row of
Fig. 5–12 and those in the lower row, but cannot distinguish between
the shapes within each row. This suggested that the bees perceive
form on the basis of the "brokenness" of pattern (von Frisch, 1950).
It is supposed that broken or interrupted patterns produce a flickering
visual impression as the bee passes during flight. It is not surprising,
therefore, that bees tend to visit flowers that are being shaken by the
wind more readily than those which are not.

It is well known that bees, ants, and wasps are able to locate their
nests on the basis of various landmarks. In addition, flying bees have
been shown to be able to distinguish right and left, before and
behind, and above and below (Wigglesworth, 1965). However, the
form and pattern perception abilities of insects are considered to be
less than those of man.

Intensity and Contrast Perception. Optomotor responses have
been used to study intensity perception. An optomotor response is
behavior stimulated by a moving visual pattern. When an insect flies,
the apparent movement of the surroundings tells it the direction and
rate of movement. Certain orientation movements, maintenance of
flight, and changes in velocity of flight and landing may be in response
to a change in the rate or direction of the moving visual pattern. In the
laboratory, a moving visual pattern can be produced by surrounding a
stationary insect with a cylinder on which are painted vertical stripes.
Rotation of the cylinder gives the insect the sensation that it is moving.

In the study of intensity perception, optomotor responses to patterns of rotating stripes have been used to determine the ability of
insects to discriminate between different levels of intensity of adjacent
stripes. Measurements indicate considerable variation among different
insects. Discrimination of different intensities apparently depends on
whether those being compared are relatively high or low. For example,
in the house fly, at low intensities the more intense light must be 100
times brighter than the less intense, while at high intensities the
magnitude difference may decrease to 2.5 times (Wigglesworth, 1965).
The brightness of the background may influence intensity percep-

tion. For example, the hummingbird hawk moth will favor a dark disc on a white background over an equally dark disc on a gray background. In other words, the moth "prefers" the situation in which there is greater contrast. It is interesting that under natural circumstances these moths enter dark crevices to spend the night.

Movement Perception. As was the case for form and pattern perception, there is considerable behavioral evidence for the perception of movement. In fact, certain responses are elicited only by movement. For example, dragonfly nymphs will not attack prey with their labial jaws unless it is moving. The optomotor response elicited by the moving pattern of stripes is based on the perception of movement by an insect.

Distance Perception. It is not difficult to think of several instances in which it would be essential for an insect to possess the ability to judge the distance of an object from itself. An excellent example would be in prey capture. To catch prey, particularly on the wing, distance perception must be especially acute. One needs only to watch a dragonfly capture its prey in flight to be convinced. Binocular vision must be involved since the ability to judge distance accurately is lost when one compound eye is blocked in some fashion. However, unlike man, the eyes of insects are fixed and depth perception depends, not on convergence of the eyes on a fixation plane, but upon the equal, simultaneous stimulation of corresponding retinal points. The distance of an object will then be determined by the location of the object relative to the points of intersection of projections of axes of the corresponding ommatidia (Dethier, 1953). A schematic representation of how depth perception is accomplished by a dragonfly is shown in Fig. 5–13. For this means of depth perception, the insect must, of course, face the object perceived.

Color Vision. The range of the electromagnetic spectrum perceived by insects is from about 253 millimicrons (near ultraviolet) to approximately 700 millimicrons (infrared; Dethier, 1963). Although there is considerable variation in the sensitivities of different insects to different wavelengths, in general terms insects are particularly

Fig. 5–13
Depth perception in a dragonfly nymph. The lines represent the visual axes of selected ommatidia. The distance and position of objects are determined by the points of intersection of the visual axes. The extended labium is shown in grey. Potential prey at point *A* is within reach of the extended labium, but not at points *B, C, D,* and *E*. (Redrawn from Baldus, 1926.)

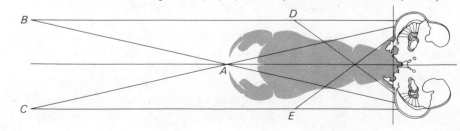

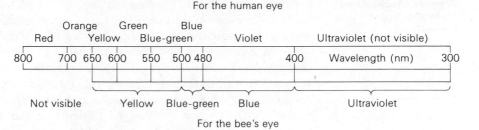

Fig. 5–14
Comparison of the
spectra perceived by
humans and honey bees.
(Redrawn from von
Frisch, 1950.)

sensitive in the ultraviolet and blue-green regions of the spectrum. Some insects, such as the nocturnal stick insect, *Dixippus* sp., are apparently colorblind. In those insects which do possess color vision, part of an eye may be color sensitive and another part colorblind [e.g., in the waterboatman, *Notonecta* (order Hemiptera, true bugs)]. Evidence exists that there is more than one type of color receptor in the cockroach eye (Dethier, 1963). The dorsal area of the eye is more sensitive to ultraviolet than blue-green light; the reverse is true in the ventral area. The spectral sensitivity of an insect may vary with its physiological state. For example, cabbage butterflies (*Pieris brassicae*) seem to prefer blue or yellow flowers; gravid females ready to oviposit seem to prefer green and blue-green.

Both electrophysiological and behavioral techniques have been used to establish the color-perceiving abilities of insects. Different wavelengths have been tested to see if they elicit a particular electro-retinogram (ERG) pattern or a specific behavioral response. Both the various optomotor responses and training experiments have been utilized. For example, von Frisch (1950) used training techniques to demonstrate the color-perceiving capacities of the honey bee. He first placed a dish of honey on a blue card where bees could get to it. After several hours the bees were "trained" to associate the color blue with the presence of honey. The bees were then presented with a fresh (unscented) blue and red card where only the blue one had been previously. The bees visited the blue card and ignored the red one. Although this experiment shows that bees can distinguish between blue and red, it does not prove that they actually perceive color, since red and blue differ in relative brightness. To determine whether the bees were distinguishing brightness or color, several cards of different gray shades between white and black and a blue card were placed where only the blue card had previously been. In addition, to discount the possible role played by scent the cards were covered with a glass plate. In this situation the bees still visited only the blue card; therefore, they are able to perceive color.

Similar results were obtained when the bees were trained to orange, yellow, green-violet, and purple. However, the color red was confused with the black and dark gray cards; therefore, the bees are red-blind. Further experiments in which bees were presented with several different-colored cards showed that they were unable to distinguish certain colors from others. As a result of these and other experiments, it has been demonstrated that bees can accurately distinguish only 4 colors as opposed to the 60 or so that can be distinguished by the human. For instance, the portion of the spectrum in which we can recognize several distinct colors between orange and green appears yellow to bees, and so on (Fig. 5–14). These characteristics of color vision in the honey bee have very interesting ramifications when we consider the relationship between bees and flowers.

We have mentioned that the spectral range in which insect eyes are generally most sensitive is in the ultraviolet and blue-green regions. We also established that bees are red-blind. However, there are insects that are highly sensitive to wavelengths in the red region. For example, certain butterflies are capable of recognizing red flowers or red models of flowers. Also, the firefly, *Photinus* sp., is able to perceive flashes of light up to 690 millimicrons, which is well into the red region of the spectrum.

Polarized Light Perception.   That insects could recognize the direction of polarization of light was first discovered in honey bees and ants. These insects were found to be able to detect the polarization pattern of the sky, which varies with the position of the sun and enables them to determine direction. This directional ability is, of course, important in finding the hive or nest after a foraging or hunting trip. According to von Frisch, it is also of importance in the orientation of the bees communication dances used to inform other members of the hive as to the location of food. Since the discovery of the ability to detect the polarization plane of light in ants and bees, it has subsequently been found in all insects examined for it.

Stemmata

Structurally the stemmata or lateral ocelli are quite variable. Some types are seemingly very similar in structure to an individual ommatidium of a compound eye. This is particularly evident in the ocelli of larval butterflies and moths (Fig. 5–15A). In these insects each ocellus consists of a cornea, a crystalline body, and a number of retinal cells forming a rhabdom. In other insects the ocelli are very different from ommatidia. In these insects a single corneal lens overlies several groups of retinal cells which form several rhabdoms (Fig. 5–15B).

The stemmata function in the manner of eyes. In various insects they have been shown to be involved with color, form, and distance perception.

Although a detailed pattern of the external surroundings is not likely to be perceived at any one time by an insect possessing only a few stemmata, the movement of the head back and forth, "scanning," may allow much greater detail perception than would otherwise be possible. For this type of activity to be effective, the insect must be able to convert spatial patterns into temporal patterns. In other words, the external scene is viewed as a sequence of events in time.

Dorsal Ocelli

Dorsal ocelli vary somewhat in their structure but are generally similar to the second type of lateral ocellus described in the preceding section. They have been shown to be light sensitive but are apparently not important in image perception. They are generally thought to be "stimulatory organs," which increase the sensitivity of the compound eyes to light since when they are blocked in some manner, reactions of the insect to light are diminished somewhat.

Light Production

The production of light by living organisms, *bioluminescence*, has fascinated scientific investigators and curious lay observers for centuries. However, it has been only during the last hundred years that a fundamental understanding has begun to develop. The phenomenon of bioluminescence has been described in several groups of plants, microbes, and animals. For example, it has been found in the clam *Pholas dactylus,* the crustacean *Cypridina hilgendorfii,* several marine annelids, a number of insects, marine dinoflagellates, and several fungi and bacteria. Luminous bacteria commonly share a symbiotic

Fig. 5–15
Types of stemmata (diagrammatic). A. Rhabdom centrally located.
B. Several rhabdoms distributed beneath the corneal lens.

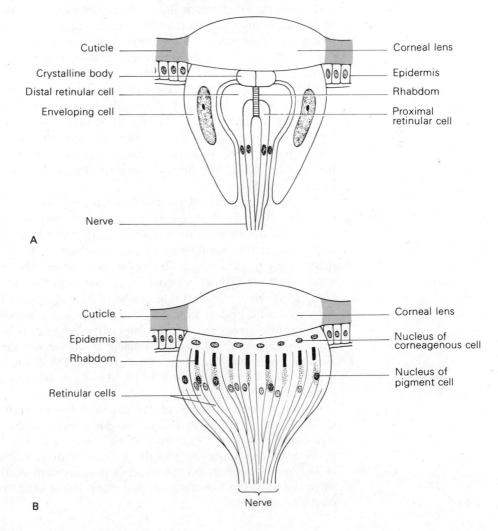

relationship with animals. For example, they are responsible for the luminescent body regions of several marine fish.

Among the insects, those with specific light-producing mechanisms are found in the orders Collembola (springtails), Homoptera (cicadas, leafhoppers, and relatives), Coleoptera (beetles), and Diptera (true flies). Bioluminescence in other insects has been found thus far to be due to the presence of bioluminescent bacteria.

When certain species of springtails (e.g., *Achorutes muscorum*) are stimulated, luminescence occurs over the whole body (McElroy, 1964).

A number of instances of bioluminescence have been reported for the Homoptera, but only one is well established. This is in the species *Fulgora lanternaria*. Luminescence of the head in this species has been observed only when males and females are together and evidently has something to do with mating behavior.

Bioluminescence has been described in more beetles than any other group of insects. Several families are known to be involved, among them Lampyridae (fireflies), Elateridae (click beetles), and Drilidae. Members of the family Lampyridae are also referred to as "lightning bugs" and have been the most extensively studied. The light-producing organ is in the abdomen of these insects and may occur in both sexes or in females only, and in the larval stage. Luminescent larvae in this family are often called "glowworms." Not all species of lampyrids are luminescent, but in those which are, the light flashes are involved in mating activities, serving to attract members of the opposite sex. The flashes are quite species specific. A well-known American species of fireflies is *Photinus pyralis,* the mating behavior of which has been extensively studied. In the family Elateridae, those in the genus *Pyrophorus* are especially well known. These beetles have luminous green spots on the prothorax and orange ones on the abdomen beneath the elytra. The orange spots are thus visible only when the insects are in flight.

Pyrophorus are particularly significant from a historical standpoint since they were the first insects in which the biochemical nature of the light-producing reaction was studied. An investigator by the name of DuBois in the late nineteenth century observed that if the light-producing organs were ground up in cold water, the homogenate would glow for a short period of time and then the glow gradually faded away. He found that if he added an extract obtained by boiling light-producing organs in water to the cold-water homogenate, the glow was briefly restored. DuBois later carried out similar experiments with a luminescent clam and named this reaction the *luciferin–luciferase reaction*. The active principle in the hot-water extract in fireflies much later turned out to be the heat-stable substance adenosine triphosphate (ATP).

A rather bizarre example of the luminous beetles is in the family Drilidae among larvae in the genus *Phrixothrix*. These larvae are commonly known as "railroad worms" because of the 11 pairs of green luminous spots on the thorax and abdomen and the pair of red luminous spots on the head. The movement of these larvae when stimulated to luminesce gives the impression of a string of railroad cars at night.

The only examples of truly luminescent dipterans are found among larvae of the families Platyuridae and Bolitophilidae. The members of both these families are called *fungus gnats*. The larvae of the New Zealand glowworm, *Bolitophila luminosus,* occur in well-shaded humid areas, for example the environmental situation found in certain caves. The light-producing organ has been found to be modified portions of the Malpighian tubules, which are located near the end of the abdomen. According to McElroy (1964), when these fly larvae occur in caves, they lay their eggs in a mucous glue on the ceiling by silken sheaths up to 2 feet long. These sticky threads serve to trap small flying insects upon which the larvae prey. From the description of the appearance of large numbers of these larvae suspended from the ceiling and luminescing, one gains the impression that it is a beautiful sight. Luminescent larvae of the species *Platyura fultoni* have been found in moist environments in the Appalachian Mountains.

The morphology of the light-producing, photogenic organs is extremely complex and variable and will not be considered here. Similarly, the biochemical aspects are also very elaborate and far from being completely understood. For the reader who wishes to delve further into this subject the reviews by McElroy (1964) and the article by McElroy and Seliger (1965) will serve quite well for an introduction to the very extensive literature on this topic.

Sound Production

A wide variety of insects are able to produce distinct sounds. These sounds are commonly correlated with well-developed organs of hearing and often play an important role in various types of behavior. The sounds are produced by several different mechanisms, which will be considered in a general way in this section.

A useful classification of sound-producing mechanisms has been presented by Haskell (1961). Haskell's classification is as follows: (1) sounds produced as a by-product of some usual activity of the insect, (2) sounds produced by impact of part of the body against the substrate, and (3) sounds produced by special mechanisms.

The sounds of many insects which are produced as a result of some usual activity are those which are essentially accidental in that no specifically adapted structures are involved. Sounds of this type include those which are by-products of flight. These sounds are produced by wingbeats, vibration of the thoracic sclerites, the wings striking one another, and similar mechanisms. Other sounds in this category include those produced as a result of movement during copulatory behavior, cleaning, feeding, and so on (Haskell, 1961). In many instances the sounds are adventitious in that they serve no specific function. On the other hand, some sounds that fall under this heading do have behavioral significance. A good example of this behavioral significance is found among the mosquitoes, where the sound produced by the wingbeat of the flying female elicits a mating response in males. Another possible example is the "piping" of queen honey bees, which is thought by many to result from the vibration of the thoracic sclerites. This sound is produced when a colony possesses a number of virgin queens and has been suggested to be a sound of challenge (Butler, 1963).

A number of insects are known to produce sound by tapping the substrate with some part of the body. A commonly cited example is that of the "death watch" beetles (*Anobium* and *Xestobium*, family Anobiidae), the larvae of which tap the walls of their galleries in wood (unfortunately sometimes furniture) with their heads, producing a characteristic tapping sound. Other insects, which have been shown to produce sounds in similar manners, include the subterranean termite (*Reticulitermes flavipes*), book lice (order Psocoptera), and several members of the order Orthoptera.

Under the category of sounds produced by special mechanisms, Haskell (1961) includes frictional mechanisms, vibrating membrane mechanisms, and mechanisms directly involving air movement.

Frictional mechanisms are highly variable morphologically and are found among several groups of insects. Although these mechanisms are structurally diverse, they consist of similar parts. Frictional mechanisms (Fig. 5–16A, B), are located in areas where two surfaces may be rubbed together. One of these surfaces, the "file," bears a row of regularly spaced ridges and the other bears the "scraper," a ridge or knoblike projection. When the file and scraper are rapidly rubbed together a sound is produced, the quality of which is based on the rate of rubbing, the spacing of the ridges on the file, and the resonance characteristics of the surrounding cuticle. Frictional mechanisms have been described in several orders, in different life stages, and involving many different combinations of body parts. The production of sound by means of a frictional mechanism is sometimes called *stridulation*.

The vibrating membrane or tymbal mechanisms have been found only in members of the orders Hemiptera and Lepidoptera. Although tymbal mechanisms have been described in several homopterans, the best-known examples are found among the cicadas (Fig. 5–16C). In these insects the tymbal organs are paired structures on the dorsolateral surface of the probable first abdominal segment. The tymbal organs of Lepidoptera have been found in certain arctiid moths (tiger and footman moths) and others. In the moths these organs are paired and are located on either side of the metathorax. In addition, tymbal mechanisms have been identified in the hemipteran family Pentatomidae.

The third category of special mechanisms, those directly involving the movement of air, is the smallest and least understood. Haskell (1964) points out that many cases where it has been described are dubious. However, he does cite one well-established example of such a mechanism in the death's head hawk moth, *Acherontia atropos*. In this moth a sound is produced by the forcible inhalation and exhalation of air through the proboscis by means of the pharyngeal muscles. Evidently the passage of air over the epipharynx produces the sound. Air released forcibly through the spiracles may result in sound production in certain Diptera, Hymenoptera, and Orthoptera. The cockroach *Gromphadorhina* is capable of producing an audible hissing sound via the spiracles when disturbed.

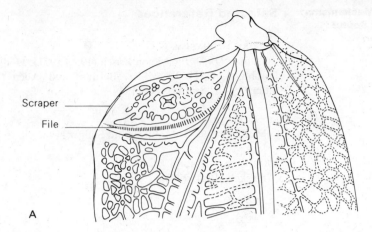

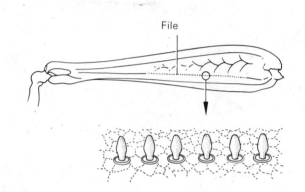

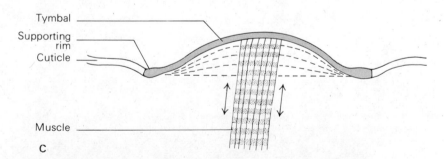

Fig. 5–16

Examples of sound-producing mechanisms. A. File and scraper on underside
and posterior edge, respectively, male katydid, *Neduba carinata*. The file of
one wing and scraper of the other are rapidly rubbed together, producing
a chirping sound; B. Hind femur of a short-horned grasshopper,
Stenobothrus sp., with a portion of the file enlarged. In sound production
the file is rapidly rubbed against a thickening (the scraper) of the basal
portion of the forewing. C. Section of a tymbal of a cicada. The dashed
lines and arrows indicate the pattern of movement of the tymbal during
sound production. (A redrawn from Essig, 1942; B redrawn from
Comstock, 1940; C redrawn with modifications from Haskell, 1961.)

Selected References

SENSORY MECHANISMS

Dethier (1953, 1963); von Frisch (1950, 1971); Hodgson (1964); Mazokhin-Porshnyakov (1969); Miller, Bernard and Allen (1968); Roeder (1963); Slifer (1970); Swartzkopff (1964).

LIGHT PRODUCTION

McElroy (1964); McElroy and Seliger (1962).

SOUND PRODUCTION

Haskell (1961, 1964).

6

Locomotion

The ability to change position within the environment is of essential importance to the survival of all nonsessile organisms. Escape from predators, food gathering, dispersal, mate finding, adjustment to temperature and humidity changes, all depend to a greater or lesser degree upon the ability of an organism to move about. Insects were originally terrestrial organisms that subsequently invaded both the aerial and aquatic environments. They are the only group of invertebrates that contains members capable of flight.

In this chapter we want to consider in general terms the ways in which insects move within their environment.

Terrestrial Locomotion

Walking and Running

Walking and running are accomplished by the six thoracic legs. Actually these limbs, unless modified for some function other than walking, serve two basic purposes: they suspend and support the insect body off the ground, and they exert the necessary forces to propel the insect. The body literally "hangs" close to the ground, resulting in a low center of gravity and thus a high degree of stability. Also, in some insects the tip of the abdomen is in contact with the ground and provides additional stability. The use of walking legs as a mode of terrestrial locomotion is found in the adults of nearly all flying and nonflying insects and many nymphal and larval forms as well. In many insects walking legs are the only source of locomotion or the only source used to any great extent. For example, apterygote insects such as silverfish and proturans are solely walkers and runners; cockroaches, although they possess functional wings, seldom take to the air, preferring to take advantage of their excellent walking and running abilities. In several beetles the elytra are fused and hence they are grounded for life. Larval and nymphal forms that possess walking legs are entirely dependent upon them for locomotion since they do not bear wings.

Although several insects are rather sluggish in their walking and, in fact, may use their legs more as organs for clinging to a surface than actual locomotion, many are quite rapid and agile runners. One who has ever suddenly turned on the light in a cockroach-infested room will readily attest to their running ability—the cockroaches', that is—unless the observer happens to be squeamish about insects. The fastest speed measured for *Periplaneta americana* is 2.9 miles per

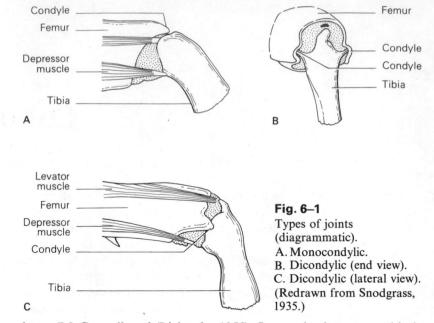

Fig. 6–1
Types of joints
(diagrammatic).
A. Monocondylic.
B. Dicondylic (end view).
C. Dicondylic (lateral view).
(Redrawn from Snodgrass,
1935.)

hour (McConnell and Richards, 1955). In an absolute sense this is a considerably slower speed than many vertebrates are able to attain. However, in relation to body size, it is remarkably fast. Insects also have the ability to accelerate rapidly over short distances and to change direction rather suddenly. The speed of walking and running has been shown to be proportional, within certain limits, to temperature.

Functional Morphology of the Insect Leg

In the discussion of the insect skeleton the leg has already been described as being composed of a series of segments which articulate with one another. The nature of these intersegmental articulations is of paramount importance in the capabilities for movement of the entire leg. Two basic types of articulations are found: *monocondylic* and *dicondylic*. A condyle is a prominence of exoskeleton upon which an adjoining segment articulates. Hence a monocondylic joint (Fig. 6–1A) is one that contains a single condyle. This type of joint allows considerable movement and has been said to be the nearest single-articulation equivalent to the ball-and-socket joint found in vertebrate animals. A dicondylic joint (Fig. 6–1B, C) consists of two condyles and restricts movements to a single plane.

Hughes (1965) describes the insect leg as being composed of four main regions: (1) the coxa, which articulates with the thorax by a dicondylic joint or sometimes a single pleural articulation; (2) the trochanter and femur are fused and the trochanter articulates with the coxa by means of a dicondylic joint; (3) the tibia forms a dicondylic joint with the femur and monocondylic joint with the proximal tarsomeres; and (4) the tarsus, which is composed of tarsomeres joined by monocondylic joints. Hughes points out the dicondylic joints between the coxa and thorax and between the femur and tibia are oriented such that they enable movements in different planes and therefore allow the entire leg to move in all directions about the articulation with the body. This system of two dicondylic joints provides the equivalent of the ball-and-socket joint.

It was mentioned in Chapter 4 that leg movements are accomplished both by intrinsic muscles, which originate within the leg itself, and extrinsic muscles, which are attached between the leg and the thoracic wall. The movement of the leg segments relative to one another is possible because the muscles cross one or more joints between their origins and insertions.

Patterns of Leg Movement During Walking and Running

The sequence of movements or gait of the legs during walking and running has been a topic of considerable interest. Attempts to detect details of gait by direct observation are extremely difficult, if not impossible, because of the general rapidity of the leg movements. Even if one were able to discern the gait of a slow-moving insect, the results would not necessarily apply to the same insect moving more rapidly or to other species. The use of cinematography has afforded an excellent solution to this problem. Insects may be filmed at whatever speed or under whatever conditions the experimenter desires, and then the film may be projected at reduced speed and photographs of different frames may be produced.

The order of sequence of leg movement during walking seems to be fairly constant at a given speed for a given insect. However, variations do occur at different speeds and between different species. The classical description of forward walking of insects is that of alternating tripods of support (Fig. 6–2A). The first and third legs on one side and the middle leg on the other advance while the other three legs remain stationary and provide a tripod of support. The cycle then repeats itself with the previously stationary three legs becoming the moving ones. An insect progressing in this fashion follows a zigzag course. Many deviations from this classical de-

Fig. 6–2
Walking patterns
A. Classical alternating
triangles of support.
B. Typical pattern.

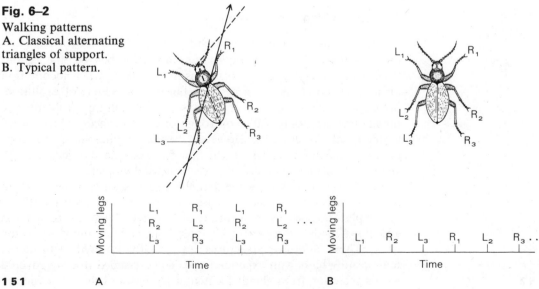

151 A B

scription are known; thus it is useful only in a very general way. Probably the most common pattern is a modification of the alternating triangles of support where the three legs of each triangle do not move simultaneously but in sequence (Fig. 6–2B). In addition, accidental or experimental amputation of one or more legs immediately results in a new coordinated pattern of walking suitable for the remaining legs. Also, it should be pointed out that all insects do not necessarily always use all six in walking, as, for example, the praying mantis, which walks using only the meso- and metathoracic legs. The reviews of Hughes (1965) and Wilson (1966) consider the topic of gait in detail.

Coordination of Legs During Walking and Running

The walking and running of insects may be resolved into two basic components, the movements involved with each leg and the coordination of movements of all the legs. It is pretty well established that each thoracic segment contains afferent and efferent pathways necessary for the individual movement of the two legs associated with it. However, the coordination of all the thoracic legs is intersegmental, as evidenced by the fact that section of the nerve cords disrupts walking. This coordination lies within the thoracic ganglia since decapitation (removal of brain) does not disrupt the ability of an insect to walk. Presumably coordinated walking depends on nervous signals from individual legs acting via the thoracic ganglia in such a way that they influence the timing of the stepping movements of adjacent legs. Changes in walking patterns with the experimental amputations of different legs can then be explained by the difference in the patterns of nervous signals from the remaining legs. Although, as previously mentioned, decapitation does not affect the coordination of walking, it may result in increased locomotor activity. As a result, the brain is thought to exert a degree of inhibitory influence on locomotor activity.

Aids to Walking and Running

The propulsive force applied by the legs would be of little use to an insect were it not for a certain amount of friction between the leg and the substrate. This problem becomes more critical when locomotion on an inclined surface is necessary. During the course of evolution, insects have devised a number of ways to cope with this problem. The tarsal claws generally suffice on rough or dirty surfaces, which afford ample points for grasping. However, there are a number of situations in which the tarsal claws would fail, for example the very smooth surface of an inclined piece of glass or a window pane.

Yet it is common knowledge that many insects, such as the common house fly, are able to walk with ease on such surfaces. They are able to accomplish this because of a variety of adhesive structures: the pulvilli and tarsal pads described in Chapter 2 or pads at the base of the tibia. These adhesive structures are usually covered with dense mats of tiny hairs with expanded tips. The expanded tips are covered with a secretion from glands located at the bases of the hairs and are the parts of the adhesive pads that come into direct contact with the substrate. Apparently, molecular forces among the expanded tips, the glandular fluid, and the substrate account for the adhesion of the pad. Since it is the tiny hairs that are ultimately responsible for the "clinging" ability, they are commonly referred to as *tenent hairs*. A number of explanations have been advanced to explain the action of these hairs, but the one advanced by Gillett and Wigglesworth (1932) seems to be the most generally applicable one. He points out that the tenent hairs which comprise the climbing organ (Fig. 6–3) on the distal end of thc tibia of *Rhodnius* (order Hemiptera, family Reduviidae) are wedge shaped at their tips (Fig. 6–3C). only the hindmost part coming into contact with the substrate. When the insect moves in an anterior direction (toward X, Fig. 6–3C) or an external force (e.g., gravity) is pulling the insect in an anterior direction, the secretion between the hairs and the substrate acts as a lubricant, reducing the friction. However, if the insect is moving or being forced to move in the opposite direction (posterior or toward Y, Fig. 6–3C), the surfaces of the hair tips come into very close contact with the substrate and "seizure" occurs, the surfaces being held together by the adhesive forces of the molecules of the glandular

Fig. 6–3

Climbing organ of *Rhodnius prolixus* (Hemiptera; Reduviidae). A and B. Proximal portion of tibia and tarsus showing position of climbing organ. C. Tenent hairs which comprise the climbing organ. (See text for explanation.) (Redrawn from Gillett and Wigglesworth, 1932.)

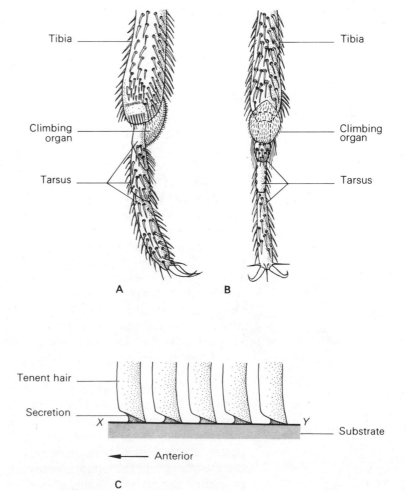

secretion. This is borne out by the fact that *R. proxilus* can walk up a pane of carefully cleaned glass at an incline of 80° but slips at an angle of 22° when walking down the same pane. The holding power of adhesive pads is well illustrated by the fact that one bug in the same family as *R. prolixus* can support a tension of greater than 50 grams (Wigglesworth, 1965).

Jumping

Under this heading we include all the means, other than wings, by which insects are able to propel themselves through the air. Many insects are capable of jumping to a greater or lesser extent, but only in certain groups has this become a pronounced specialization. The suggestion has been made that perhaps wings first appeared on a jumping insect that used the jumping ability for a propulsive force and early wings merely as gliding surfaces. Most jumping insects depend on modified hindlegs for their jumping, although some make use of other specialized mechanisms.

As just mentioned, most jumping insects jump with their legs. They all have enlarged, muscular femora on the hindlegs, the extension of which produces the force necessary to propel the insect. Examples of insects that jump in this way are the members of several orthopteran families (e.g., grasshoppers and katydids), flea beetles, and fleas. Relative to their size the magnitude of leaps these insects are capable of is rather astounding. For example, Wigglesworth (1964a) points out that the trajectory of a jumping flea, which may reach a height of 8 inches and cover a distance of 13 inches, if proportionally carried out by a human being would result in a jump of 800 feet! However, as explained in Chapter 4, since strength of contraction is a function of cross-sectional area of muscle (linear dimension squared) and increase in mass a function of volume (linear dimension cubed), a flea the size of a man would have relatively much more mass per unit cross-sectional area of muscle than a normal-sized flea and would thus probably still be able to jump only 8 inches high. Grasshoppers have been reported to leap up to 30 cm high over a distance of 70 cm (Hughes, 1965).

Members of the order Collembola (springtails), click beetles (family Elateridae), and cheese maggots (family Piophilidae, order Diptera) have developed rather unique methods of jumping. Springtails (see Fig. 2–42B) are minute wingless insects that possess a fork-shaped structure, the *furcula*, on the posterior part of the abdomen and a structure called the *tenaculum* on the anteroventral portion of the abdomen. The furcula is bent anteroventrally and its tip caught and held firmly by the tenaculum. When jumping, tension is built up in the muscles of the furcula and it is then suddenly released by the tenaculum, allowing the furcula to "snap" posteriorly against the substrate. The reaction to this posteriorly directed force propels the insect into the air.

Click beetles utilize a similar principle, an elongated spine on the posterior part of the prosternum being inserted into a receptacle on the mesosternum. The jump produced by the sudden release of this prosternal spine is elicited only when the beetle is on its back and hence serves as a righting mechanism. Cheese maggots use their entire body in jumping. The last abdominal segment is held by the mouth hooks, muscular tension is built up within the body, and the last abdominal segment is suddenly released against the substrate, propelling the insect.

Crawling

Crawling, sometimes referred to as creeping, is generally associated with the locomotion of larval insects, particularly those which propel themselves by means other than the six thoracic legs alone. However, many larval and most nymphal insects do, in fact, move in the same or a similar manner to adults. Some of these hexapodous immature forms, for example several beetle larvae, are aided in their locomotion by eversible *pygopodia,* which arise from the terminal abdominal segment. A pygopodium is everted by means of hemolymph pressure and aids the thoracic legs in the progression of the insect.

The immature forms of butterflies and moths (caterpillars) and certain wasps (family Tenthredinidae) bear, in addition to the three pairs of thoracic legs, accessory legs, or *prolegs,* on certain of the abdominal segments. The movement of the prolegs depends on the integrated activity of the body musculature of the abdomen. Since these insects are typically soft-bodied, hemolymph pressure is important in maintaining a "hydrostatic skeleton." The pressure of the hemolymph is maintained by the turgor muscles of the insect body. Puncturing one of these insects results in an immediate shriveling, owing to the contraction of these turgor muscles. This indicates that they respond to a reduction in hemolymph pressure by contraction and gives some insight as to how they function in an intact larva. In addition to the turgor muscles, there are locomotor muscles that fail to contract when a larva is punctured and are involved with the actual movement

Fig. 6–4

Caterpillar walking. Arrow indicates direction of peristaltic waves and direction of progression. (See text for explanation.) (Redrawn from Barth, 1937.)

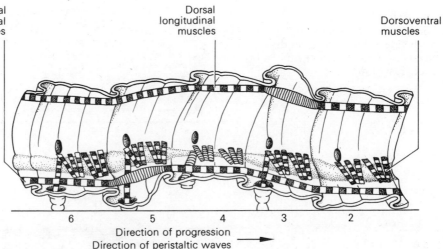

Ventral longitudinal muscles

Dorsal longitudinal muscles

Dorsoventral muscles

6   5   4   3   2

Direction of progression
Direction of peristaltic waves →

of the prolegs. These muscles are arranged in longitudinal, dorso-ventral, and transverse patterns.

Walking patterns in larvae with prolegs have been best described in caterpillars. Usually single peristaltic waves of contraction pass anteriorly from the posterior end of the insect. As each wave passes, the two legs of each segment always move simultaneously. Each wave involves three main phases. Hughes (1965) describes these three phases as follows (Fig. 6–4):

> In the first, the dorsal longitudinal and dorsal intrasegmental muscles contract, together with the large transverse muscles. This results in the segment becoming shortened dorsally and con-sequently its posterior end becomes inclined forward so that the segment behind is lifted from the ground. In the next phase the segment contracts dorsoventrally and its feet are released simul-taneously from the substratum. After the legs have been moved forward, the ventral longitudinal muscles contract so as to bring the segment down toward the ground and the feet become fixed. The wave of perstalsis passing over the body is therefore not limited to a single segment but involves a simultaneous con-traction of muscles in at least three segments. Contraction starts at the last segment with the release of the terminal appendages. They are lifted and placed on the ground at varying distances forward.

This description is, of course, a rather simplified one, since other sets of muscles (e.g., the transverse muscles) are no doubt involved. In addition, it should be pointed out that the contractions of the turgor muscles must be in coordination with the contractions of the loco-motor muscles. An interesting variation of this type of crawling is shown by the larvae of geometrid moths. These larvae are often referred to as *inch worms* or loopers. In these larvae a considerable portion of the trunk is out of contact with the substrate as they move along (Fig. 6–5).

Terrestrial larvae that have no legs whatsoever, for example fly maggots, depend entirely upon peristaltic waves of contractions for their progression. Many possess spines or various other structures which aid them by increasing the friction between their bodies and the substrate. In most apodous larvae, the waves of peristaltic contrac-

Fig. 6–5
Locomotion of "inchworm," geometrid moth caterpillar. *d* indicates distance advanced in one cycle. (Redrawn from Snodgrass, 1961.)

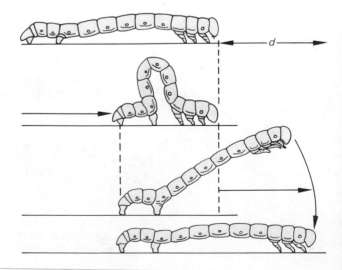

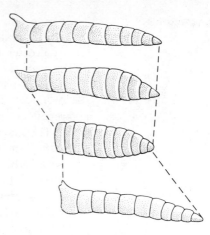

Fig. 6–6
Locomotion in a crane fly larva.
(Redrawn with modifications from
Kevan, 1963, after Gilyarov.)

→ Direction of progression

← Direction of peristaltic waves

tion, like those in larvae with prolegs, pass from posterior to anterior, that is, in the direction of progression. However, in certain ones adapted to burrowing through the soil [e.g., burrowing crane fly larvae and bibionid larvae (order Diptera)], the waves of contraction proceed from anterior to posterior, or in the direction opposite to progression. With this method of progression, the insect is able to use its posterior as a stable, unmoving point and push the anterior portion of its body through the soil (Fig. 6–6).

Aquatic Locomotion

For convenience we shall examine the locomotion of aquatic insects from the standpoint of two general aquatic situations. The first will be those adaptations which enable insects to propel themselves on the surface of the water. The second will be those adaptations by which insects "swim" beneath the water's surface.

Surface Locomotion

To fully appreciate the various adaptations for surface locomotion in insects, we must first consider some of the characteristics of the water's surface and its interaction with other surfaces. At temperatures and pressures characteristic of the environments in which insects are found, the molecules of water are strongly attracted to one another, a phenomenon known as *cohesion*. In the absence of attractive forces upon the molecules, a droplet of water tends to contract such that it has the smallest possible surface area. Theoretically this would result in a sphere. The molecules of the gases that compose air have little attraction for water molecules, and since they are not as tightly packed as water molecules, they have little attraction for one another. Hence any surface of water in contact with air tends to contract. This force of contraction is called *surface tension*, and its presence makes the water's surface behave somewhat like a thin, elastic membrane. Some surfaces or surface coatings attract water molecules, and when this attractive force is equal to or greater than that between water molecules themselves, the water spreads out on the surface, which is said to be wettable or *hydrophilic* (water loving). Other surfaces have the opposite effect, exerting little or no attractive force on water molecules. Such surfaces are water repellent or *hydrophobic* (water fearing).

The magnitude of the forces associated with surface tension seems rather small when we think of large animals, which have a comparatively small ratio of surface to volume and hence comparatively little surface area per unit body mass. However, surface-tension forces are quite significant in small organisms, such as insects that have relatively large ratios of surface to volume and thus a comparatively large amount of surface area per unit body mass. Since, in general, the insect cuticle is hydrophobic, many small insects are able to be supported against the force of gravity by the forces of surface tension. If, however, the cuticle of an insect that is supported by surface tension is made hydrophilic by the application of a "wetting agent," such as a detergent, it will immediately sink or become trapped by the water.

Surface-dwelling insects take advantage of the properties of surface tension and hydrophilic and hydrophobic surfaces in a variety of ways. Many secrete a waxy material which coats their tarsi, making them hydrophobic and allowing the insect to walk on the water's surface. The surface-dwelling bugs (order Hemiptera, several families) are able to either walk on the water like other insects would walk on the ground (e.g., the water measurer, *Hydrometra,* family Hydrometridae) or use the middle legs as synchronous oars, rowing from place to place (e.g., water striders, family Gerridae, and others). The tarsi of these insects do not break the surface of the water (Nachtigall, 1965). The *collophores* and tarsal claws of certain springtails (e.g., *Podura aquatica*) are hydrophilic while the rest of the integument is hydrophobic. This enables them to anchor themselves to the water and yet easily move from place to place by means of their furculae. Beetles in the genus *Stenus*, family Staphylinidae, secrete a substance from anal glands which lowers the surface tension of the water behind the insect. As a result they are drawn forward by the effects of the contractive forces of the water in contact with their bodies. Nachtigall (1965) states that these beetles are able to attain speeds of 45 to 70 cm/sec.

Subsurface Locomotion

Many insects live beneath the surface of the water. These insects can be conveniently divided into three groups: (1) those which utilize appendages in swimming, (2) those which move by various undulations of the body, and (3) those which are able to produce jets of water through the anus. It should be pointed out that many of these insects utilize air stores of various sorts for buoyancy and maintenance of equilibrium. In addition, in many, particularly the rapid-swimming, hard-bodied forms, the body is flattened and tapered, a shape that provides a minimum of friction during movement through the water.

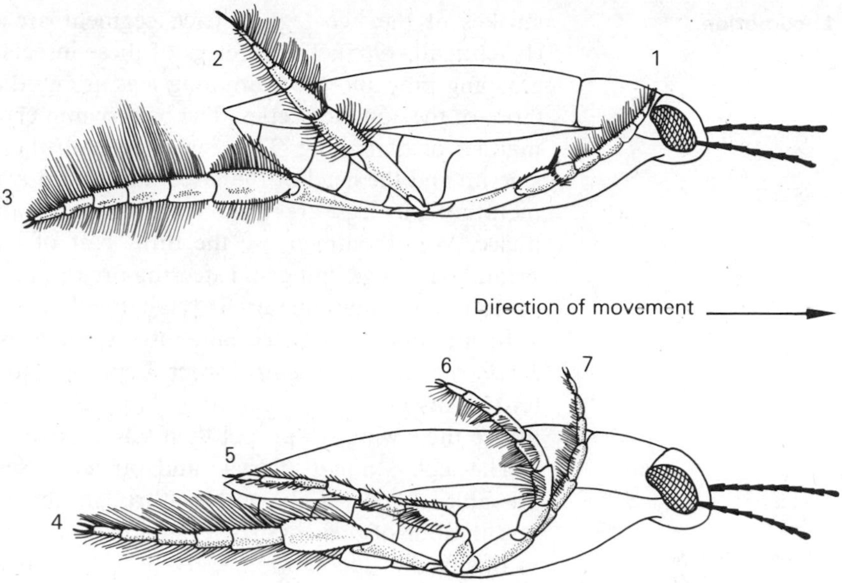

Direction of movement ⟶

Fig. 6–7
Swimming movements of a predaceous diving beetle *Acilius sulcatus*. Numbers indicate the temporal sequence of leg positions during the swimming stroke. Top, thrust stroke; bottom, recovery stroke. Note the positions of the hairs at various points in the stroke cycle. (Redrawn from Nachtigall, 1960.)

Insects that utilize appendages in swimming include representatives from the orders Coleoptera (Dytiscidae, predaceous diving beetles; Gyrinidae, whirligig beetles; Haliplidae, crawling water beetles; Hydrophilidae, water scavenger beetles; and others), Hemiptera (e.g., Notonectidae, backswimmers; Corixidae, waterboatmen; Belostomatidae, giant water bugs; and others), Trichoptera (caddisflies), Neuroptera (dobsonflies, lacewings, and relatives) and even Lepidoptera. Other than those insects which merely crawl about on the bottom of a body of water or on submerged vegetation, the vast majority of the insects in this category have legs which are either flattened like oars, covered with "swimming hairs," or commonly both. For example, in whirligig beetles in the genus *Gyrinus* the rowing legs are so flattened that the broad side has five times the area of a comparable round leg (Nachtigall, 1965).

The swimming legs of adult dytiscid beetles bear large numbers of hairs. These hairs (Fig. 6–7) are movable and when the insect swims they open out, increasing the surface area applied against the water during the thrust stroke. They fold down on the legs during the recovery stroke. The rowing or swimming legs are used in various patterns. The whirligig beetles use the oarlike meso- and metathoracic legs. Their forelegs are long and slender and are used for capturing prey. These beetles are the ones commonly seen darting to and fro, the hydrophobic dorsum of their bodies breaking the water's surface. Beetles in the family Dytiscidae may use all three pairs of legs in locomotion. However, in the larger members of this family the forelegs are used for catching prey, the mid- and hindlegs for propulsion. Hydrophilid and haliplid beetles use all three pairs of legs for propulsion, but in contrast to the beetles just mentioned, the thrust and recovery strokes of legs on the same thoracic segment are in opposite phase.

Among the aquatic Hemiptera members of the families Nepidae (waterscorpions), Naucoridae (creeping water bugs), and Belostomatidae utilize the last two pairs of legs in swimming and the rowing strokes of the two legs of each segment are out of phase by 180° (Nachtigall, 1965). The forelegs of these insects are generally used for grasping prey and the swimming legs are modified in ways similar to those of the aquatic beetles. The backswimmers have a rather different manner of swimming. They swim beneath the water with the ventral side up and the head of the insect directed downward. The meso- and metathoracic legs are used for swimming and move in the same phase. Waterboatmen use the third pair of legs for propulsion, the second pair as grasping and steering organs, and the first pair to scoop up food-containing materials (Nachtigall, 1965).

In addition to legs, certain insects, such as the adult females of *Hydrocampa nymphaeata* (order Lepidoptera) and a few hymenopterous insects (*Polynema* and *Limnodite,* family Proctotrupidae), utilize their wings to propel themselves through the water.

The legless aquatic larvae and pupae of several dipterous insects are able to propel themselves through the water by twitches and undulations of their bodies. Figure 6–8 shows some of the variations in this type of locomotion. Certain of these insects take advantage of the buoying force of air stored in "air bladders" (e.g. *Chaoborus* larvae) or captured gas spaces resulting from the shape of the body (e.g., mosquito pupae).

The nymphs of most dragonflies and the nymphs of the mayfly genus *Chloeon* are able to draw water into the hindgut and if alarmed expel it with considerable force. The expulsion of this water propels the insect several centimeters. Nachtigall (1965) states that speeds of 50 cm/sec can be attained. When expelling the water forcibly in this fashion, the thoracic legs are held close to the body, which has the effect of reducing the friction between the water and the insect.

Fig. 6–8
Swimming movements of legless aquatic larval and pupal diptera.
(Redrawn from Nachtigall, 1963.)

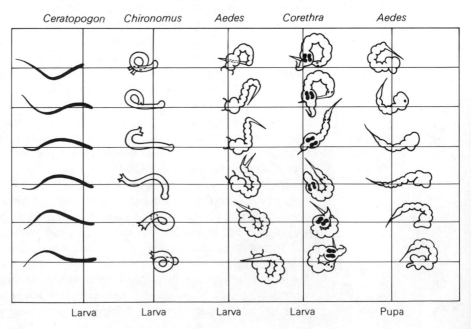

Ceratopogon	*Chironomus*	*Aedes*	*Corethra*	*Aedes*
Larva	Larva	Larva	Larva	Pupa

Fig. 6–9
Hypothetical insects with bilateral extensions of the nota in the "glider" stage of insect evolution. (Redrawn from Snodgrass, 1958.)

Aerial Locomotion

As mentioned earlier, insects, alone among the invertebrates, possess the ability to fly. This ability is perhaps one of the most important reasons for their tremendous success relative to the rest of the animal kingdom. Flight has enabled insects to take advantage of environmental situations that are virtually untouchable by their non-flying rivals.

During the course of insectan evolution, wings and their control and propulsive mechanisms have been superimposed on segments originally designed for walking. This is different from the vertebrates, in which walking appendages and the associated nerves and muscles have become the organs of flight. The wings of insects apparently arose as bilateral expansions of the nota of the thorax during the Devonian or Mississippian periods (see Chapter 10). Fossil insects with winglike expansions of the nota, *paranotal processes*, have been found in rock strata representing these periods. Presumably these expansions first served as surfaces that enabled insects to glide (Fig. 6–9), perhaps from trees or high vegetation to the ground or from a running jump. It is not too difficult to imagine a subsequent gradual development of an articulation and neuromuscular arrangement by which these insects could vary the tilt of the developing wings and thereby exert a degree of control of their glides. The next step would have been the development of an improved basal articulation and neuromusculature which would allow the wings to exert both controlling and propulsive forces.

The skeletal structure, thoracic musculature, and insect wings have already been discussed in rather general terms. The reader may want to review appropriate parts of Chapters 2 and 4 before reading this section. We now want to consider the functional morphology of wings in more detail.

Functional Morphology of Wings

As discussed in Chapter 2, the wings of insects show considerable variation with respect to several different characteristics. One characteristic in particular deserves elaboration at this point. There has been a tendency during insectan evolution toward the reduction of the wings to a single unit. The pressures in this direction have likely been due to the inefficiency of a hind pair of wings operating in the turbulence produced by the movements of the forewings. The wings of certain groups of insects, such as the members of the order Orthoptera and the lower neuropterans, Isoptera (termites), and others, operate in this fashion, and these insects are rather poor fliers as a result.

Other flying insects demonstrate a number of evolutionary solutions to this problem. The wings of dragonflies and damselflies, although they operate separately, are quite efficient because the sequence of wingbeat is reversed, the hindwings moving before the forewings during each cycle. Along the lines of developing a single unit, there have been two basic solutions. One solution has been the development of a variety of mechanisms which enable the fore- and hindwings to be coupled to one another (see Fig. 2–40A–C). Correlated with the presence of wing-coupling mechanisms is a reduction of the size of the hindwings. Coupling mechanisms are found among the Hymenoptera, Trichoptera, Hemiptera, many Lepidoptera, and others. The other evolutionary solution has been the loss or extreme modification of one pair of wings so that they no longer function as propulsive organs of flight. For example, in the true flies, the hindwings have become the halteres. The forewings of beetles have similarly lost their function as organs of flight.

Except for members of the orders Odonata (dragonflies and damselflies) and Ephemeroptera (mayflies), flying insects are able to fold the wings posteriorly over the abdomen at rest and have sclerites at the wing bases which can be interpreted by means of a typical or generalized plan (Fig. 6–10), as was explained in Chapter 2 for insect structure in general. The basal or axillary sclerites make the various wing movements possible and lend considerable strength to this region of excessive wear and tear. In addition, the cuticle in the articular region is largely composed of the protein resilin, which is responsible for the rubberlike flexibility. Members of the primitive

Fig. 6–10
Generalized wing base. (Redrawn from Snodgrass, 1935.)

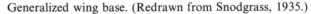
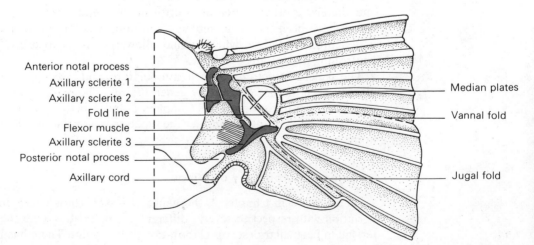

Anterior notal process
Axillary sclerite 1
Axillary sclerite 2
Fold line
Flexor muscle
Axillary sclerite 3
Posterior notal process
Axillary cord

Median plates
Vannal fold
Jugal fold

orders Odonata and Ephemeroptera have no wing-flexing mechanisms. The articulation of the wing allows movement of significant amplitude in the dorsoventral plane. For example, in certain species of wasps the wings are able to move over an arc of 150° (Wigglesworth, 1965).

Flight and Its Control

In the section on muscles in Chapter 4 it was explained that both "direct" and "indirect" muscles were associated with or influenced wing movement. Probably the muscles attached directly to the wings were the original sources of the major propulsive force of wing motion. Direct muscles are still dominant in cockroaches and very important in other insects of the order Orthoptera as well as Odonata and many Coleoptera (Pringle, 1957). In other orders the main propulsive forces are furnished by the indirect muscles of each wing-bearing thoracic segment. In these insects the indirect muscles are quite large relative to the direct ones, although the direct muscles are always important in wing motion. There is a direct correlation between the weight of flight muscle as a percentage of body weight and the strength of flying. For example, according to Wigglesworth (1965): "In the relatively weak flying orthopteran *Oedipoda* the flight muscles comprise only 8% of the total body weight, but in strong fliers they make up far greater proportions: *Musca* 11%, *Apis* 13%, *Macroglossa* 14%, *Aeshna* 24%." The common names of the four genera listed are fly, bee, sphinx moth, and dragonfly, respectively.

The actual patterns of wing movements in flying insects have been analyzed by means of a variety of techniques. Earlier methods for tracing the pattern of wingbeat included attaching light-reflecting devices to the wingtips, which would make the trajectory of the wingtip visible and observation of wing position in dead insects. High-speed cinematography has also proved quite useful. Another important technique has been the use of the strobe light, a device that can produce regularly intermittent light flashes over a wide range of rates. At appropriate rates of flashing the wing is illuminated regularly at various points in its cycle, which makes the wing appear to be moving very slowly. This enables one to directly observe the changes in wing tilt and so on as they occur. When the flash frequency is a whole-number multiple of the wingbeat frequency, the wing is caught in the same position during each cycle and appears to stand still. This is one of the methods for determining wingbeat frequency. Recording and analysis of the pitch of the sound produced during wingbeat is another important method for determining frequency of wingbeat. Highly sophisticated mechanical and electronic recordings of neuromuscular activities during flight have also been used extensively.

In modern insects, the wings are responsible for both lift and propulsion and usually of little importance in gliding. They have been described as producing the same effects on an insect's body as a propeller on an airplane. Naturally for bilateral appendages to produce these effects, the movements must be exceedingly complex. According to Pringle (1957), the movements involved consist of "elevation and depression, promotion and remotion (fore and aft

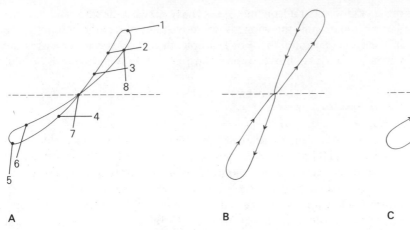

Fig. 6–11

Wingbeat patterns traced by the tip of a wing during flight. Dashed lines indicate longitudinal axis of body. A. Forward flight. Numbered lines indicate position and inclination of the wing at different stages of the wingbeat cycle. B. Hovering. C. Backward flight. (A redrawn with modifications from Mangan, 1934; B and C redrawn with modifications from Stellwaag, 1916.)

movement), pronation and supination (twisting) and changes of shape by folding and buckling."

To add to the complexity, apparently the velocity of wing motion varies in different parts of the cycle. In many insects wing twisting and the like are entirely under the control of direct wing muscles. However, in some of the better fliers, such as members of the orders Diptera and Hymenoptera, the basal wing articulations are designed such that the appropriate twisting of the wing occurs at the proper time during a wingbeat cycle. This is not to say that direct muscles do not exert some degree of influence, even in these insects. The propulsive action of the wings is quite efficient. According to Wigglesworth (1965). "the flying insect produces a polarized flow of air from front to rear during approximately 85% of the cycle." In typical forward flight, a wing traces out a figure "8" relative to the body at its base (Fig. 6–11A). Relative to a point past which an insect is flying, the wing traces out a pattern resembling a sinusoidal curve. Many insects can hover by changing the inclinations of the figure 8 relative to the body (Fig. 6–11B). Many of the especially good fliers (Diptera, Hymenoptera, and Lepidoptera) are able to fly backward (Fig. 6–11C), sideways, or rotate around the head or tip of their abdomens. Rotation and flight sideways is apparently produced by unequal activity of the wings on either side. Besides being able to vary the patterns of wingbeat, insects may vary the power of a wingstroke by varying the number of actively contracting muscle fibres.

The number of wingbeat cycles per second varies tremendously from one group of insects to another. Wigglesworth (1965) offers a number of representative values for wingbeat frequency from a variety of sources. These are presented in Table 6–1. Not only does wingbeat frequency vary in different insectan groups, but it also varies with age, sex, season, humidity, load (wingbeat frequency increases uniformly as wing size decreases), air resistance, air density, air composition, wing inertia (moment of inertia), muscle tension, and fatigue. However, variation in frequency of wingbeat is apparently not generally used by insects to control flight. Assuming a single series of muscle contractions for each wingbeat cycle, many frequency measurements seemed in the past to be fantastically high. Recording of the activities of the nerves associated with flight muscle showed that in certain insects the number of nervous impulses impinging on a wing muscle were much fewer than the number of muscle contractions and high wingbeat frequency observed. In other cases there was a one-to-one relationship between nerve impulses and muscle contractions. This phenomenon remained a puzzle until the elastic nature of the insect thorax and the action of resonating flight muscles were discovered.

In addition to the actions of direct and indirect muscles, it has been found that in many insects the elasticity of the thoracic skeleton plays a significant role in the movement of the wings. As the wings move in one direction, some of the energy of muscle contraction may be stored as elastic energy in the thoracic skeleton and antagonistic muscles and

Table 6–1
Representative Wingbeat Frequencies[a]

Insect	Wingbeats per Second [b]
Odonata	
Libellula	20
Aeshna	22, 28
Coleoptera	
Melolontha	46
Coccinella	75–91
Rhagonycha	69–87
Lepidoptera	
Pieris	9, 12
Colias	8
Saturnia	8
Macroglossa	72, 85
Acidalia	32
Papilio	5–9
Diptera	
Tipulids	48, 44–73
Aedes, male	587
Culex	278–307
Tabanus	96
Musca	190, 180–97, 330
Muscina	115–220
Forcipomyia	988–1047
Hymenoptera	
Apis	190, 108–23, 250
Bombus	130, 240
Vespa	110

[a] Data from Wigglesworth (1965) based on various sources.
[b] Two or more sets of numbers indicate data obtained by different investigators.

used in the movement of the wings in the opposite direction. This elastic characteristic can be seen in the locust *Schistocerca*. When the major flight muscles are relaxed, the wings are completely depressed. This is the stable position for them in relation to the elasticity of the thoracic skeleton. Any movement away from this stable position is opposed by the elastic forces.

In several insects, such as certain Diptera and Coleoptera, the wings have two stable positions, completely elevated and completely depressed. Movement of a wing away from one of these stable positions is opposed by the elasticity of the thorax to a point at which the direction of the elastic force reverses. At this point the wings are suddenly driven by the elastic energy into the opposite stable position. This has been called the "click mechanism." In Chapter 4 it was pointed out that insects possess both resonating and nonresonating muscles and that under appropriate mechanical conditions, the resonating type of muscle is capable of undergoing several successive contractions when stimulated with a single nervous impulse. The insect thorax with sufficient elasticity to produce a click mechanism is an example of such an appropriate mechanical situation. The resonating muscles involved are the mutually antagonistic dorsal longitudinal and dorsoventral flight muscles. As one set of these muscles is stimulated via nerve impulse to contract, a point is reached where the "click" occurs (i.e., where the elastic forces reverse). When this happens, the tension on the contracting muscles is suddenly released and the antagonistic pair is suddenly stretched. This sudden stretch stimulates the antagonistic pair to contract and the cycle then repeats itself, which produces several oscillations of the wing mechanism before another nervous impulse initiates the series all over again. These oscillatory contractions have been identified in many of the higher Diptera and Coleoptera, and others.

According to Chadwick (1953): "The airspeeds, or velocities of translation, possible of attainment by flying insects have been the subject of heated debate especially since about 1926. This question is difficult to investigate since any experimental interference raises doubt that the performance observed is a fair measure of the insect's native capabilities." Keeping Chadwick's point in mind, the reader is referred to Table 6–2.

The sensory cues pertinent to flight control undoubtedly come from a number of different sources, both from sensilla in the wings themselves and from sensilla located in other parts of the body. However, no sensory endings have ever been found in insect flight muscles. Wing sensilla fall into three main categories: (1) bristles and hairs, (2) campaniform sensilla, and (3) chordotonal sensilla. The bristles and hairs may be widely distributed over the wing's surface and respond to tactile stimulation. Thus they may respond to the movement of air over the wings during flight. However, Pringle (1957) points out that since the diameters of the fiber associated with these hair sensilla are so small, they may not be sufficient to ". . . mediate the rapid reflex actions necessary for the control of flight." The campaniform sensilla are arranged in definite patterns in the wings. Apparently different ones respond to different torques acting on the wings during flight, depending on their specific orientation. Since chordotonal sensilla are sensitive to changes in length, those in the wings probably respond to displacements of the cuticle at the wing-base resulting from various torques. Like the campaniform sensilla, they are arranged in definite patterns. In certain instances they may actually respond to changes in air pressure or even have an auditory function.

Sensilla in parts of the body other than the wings which collect information useful in flight control have been described in the head, neck, halteres, legs, and other locations. Sensilla in the head are commonly in the form of hair plates on the frontal region and mechanoreceptors (Johnston's organs) in the antennae. Both these groups of sensilla are sensitive to airstreams across the head. The compound eyes must not be overlooked as sense organs important to flight control. Dragonflies have hair plates on the neck which respond to changes in head position.

Probably any stimulus that in some way indicates danger would cause an insect to take flight as a means of escape. However, a number of specific stimuli that induce flight have been identified. The most widespread and important of these is loss of contact with the substrate. This may be easily demonstrated by suspending a cockroach or other insect from a string by means of a drop of paraffin. As long as the legs of the insect are in contact with a surface, there will be no flight movements, but if the insect is suddenly lifted into the air and surface contact lost, the wings will immediately commence to beat. This is the "tarsal reflex" since sensilla in the tarsal segments are involved. Other flight-initiating stimuli, such as a change in the relative positions of the pro- and mesothorax of the cockroach *Periplaneta*, have been

Table 6–2
Representative Measurements of the Velocity of Flight [a]

Insect	Velocity of flight (m/sec) [b]
Odonata	
Anax	8
Libellula	4–10
Neuroptera	
Chrysopa	0.6
Coleoptera	
Melolontha	2.2–3, 2.5
Lepidoptera	
Pieris	1.8–2.3, 2.5
Macroglossa	5
Sphingids	15
Diptera	
Tabanus	4, 14
Bombus	3, 3–5
Apis	2.5, 2.5–3.7

[a] Data from Wigglesworth (1965) based on various sources.

[b] Two or more sets of numbers indicate data obtained by different investigators.

described. It is interesting to note that some insects, as for example moths in the family Sphingidae (sphinx moths), are unable to fly until the wing muscles reach a temperature of 30°C. If the ambient temperature is 30°C or above, they have no trouble taking flight, but if it is below, these moths must rapidly flutter the wings until the necessary temperature is reached.

During flight, constant corrections must be made for any forces that tend to disrupt the stability of the insect. Instabilities can be resolved into tendencies of the insect to move in one or a combination of three planes: movement about the longitudinal-horizontal axis, or *roll*; movement about a transverse-horizontal axis, or *pitch*; and movement about a vertical axis, or *yaw*. An unstabilizing force or combination of such forces undoubtedly affects different sensilla to different degrees and in different ways and thus results in some change in the pattern of stimuli that impinge on the insect during stable flight. In response, the insect modifies its flight movements in such a manner that the impinging stimuli again are in the pattern that indicates stability. These stabilizing responses can most probably be explained by simple reflexes, but the presence of innate patterns of coordination of wing-beat rhythms and flight muscle actions are likely to be resident in the central nervous system.

The halteres of true flies are a specific example of stabilizing organs. These "modified hindwings" move in synchrony with the forewings. Groups of campaniform and chordotonal sensilla are arranged in specific patterns at the base of each haltere. Since the halteres are constantly vibrating during flight, they act like a gyroscope, which generates forces when there is any tendency to change direction. You will probably recall from basic physics the simple demonstration of gyroscopic action in which a rapidly rotating bicycle wheel held by two hands exerted resistive forces when you attempted to turn it away from its original plane of rotation. The forces generated by the vibrating halteres when subjected to turning forces are "sensed" by the campaniform and chordotonal sensilla in a manner similar to the way the resistive forces were "sensed" by the sensilla in your hands and arms. The halteres apparently monitor changes in all three of the planes described above.

Other sensilla that respond to unstabilizing forces during flight are the hairs on the neck of the dragonfly. During stable flight a characteristic distribution of pressure from the head is exerted on these sensilla and causes a "normal" pattern of signals. However, if a force tends to turn the insect in some direction, the pressure on these hairs becomes changed as a result of the inertia of the head, causing it to tend to resist the directional change. This change in pattern of stimulation of the hairs results in compensatory changes in the wing movements and hence correction for the unstabilizing forces. In addition to the neck hairs, the flight stability of the dragonfly is maintained by a dorsal light response by which the insect moves to keep the source of illumination dorsal to itself and an optomotor reaction to the general visual pattern (Wigglesworth, 1965). Johnston's organ in the antennae and the sensitive hair plates on the head may also detect changes in air flow and thus collect information pertinent to guidance as well as flight maintenance.

Selected Reference

Chadwick (1953); Hughes (1965); Nachtigall (1965); Pringle (1957, 1965); Wilson (1966).

7

Behavior

In the preceding five chapters we have discussed the various organ systems more or less individually. Now we are ready to consider insects as total organisms and examine the results of the very complex, integrated actions of these various systems in response to changes in the external and internal environments—what insects as living, integrated systems do.

The behavior of insects has been extensively investigated under natural and laboratory conditions, both approaches being valid, appropriate, and useful. The necessary first step in the study of any insect's behavior is to describe as accurately and completely as possible its behavioral patterns. This includes the time of appearance of specific behavior patterns in the insect's life cycle. For instance, many behavior patterns do not appear until an insect is sexually mature. The second step is to attempt to determine the specific external and internal conditions necessary for a given pattern to occur: in other words, to determine the specific stimuli that elicit observed behavioral phenomena. A major objective, of course, is eventually to relate behavior to central nervous system and endocrine mechanisms. There is a definite disciplinary interface between descriptive-experimental behavioral studies and neurophysiology, and there has been much activity in this interface area in the past twenty years or so. Another major objective that is considered after sufficient descriptive and/or experimental data are available for related species is the phylogeny, or evolutionary development of behavior, the rationale being that patterns of behavior have been subjected to natural selection in the same way as morphological traits.

One of the pitfalls encountered during the study of insect behavior has been the temptation to ascribe human purposiveness or goal seeking to many behavior patterns. This is commonly referred to as *anthropomorphism* and should be avoided. Anthropomorphic usage tends to obscure the fact that adaptive behavior is the result of natural selection and anticipation of a goal is unnecessary.

The approach of this chapter will be to consider first the basic kinds of behavior characteristic of insects and then to discuss the biological functions of the result of these types of behavior acting together in the insect. A number of excellent textbooks are available on animal behavior (see Selected References).

Kinds of Behavior

In very broad terms, insects exhibit two kinds of behavior, innate and learned. *Innate behavior* to a large extent consists of a more or less fixed response or series of responses to a given stimulus or pattern of stimuli. The "more or less" in the preceding sentence should be emphasized, since innate behavior is usually somewhat flexible and may be modified by experience (learning). Innate behavior is generally considered to be based upon the inherited properties of the nervous system. On the other hand, *learned behavior* is not inherited but is acquired through interaction with the environment during the life of the individual. Obviously, although specific patterns of learned behavior are not inherited, the potential for learning is. In many instances it is difficult to determine whether an observed behavior pattern is inherited or learned. There is no evidence of the ability to reason among insects, the observed behavior patterns being explainable on the basis of innate patterns and learning.

Innate Behavior

Some innate behavior patterns are comparatively simple, for example, the *reflexes*. Reflexes may involve only a part of the body, as in proboscis extension, or the whole body, as in the righting reflex. Reflexes can be grouped into two classes, *phasic* and *tonic*. Phasic reflexes are comparatively rapid and short-lived and are involved in rapid movements such as proboscis extension. Tonic reflexes are slow and long-lived and are involved with the maintenance of posture, body turgor, muscle tone, and equilibrium. Reflexes vary in complexity, many of the simpler ones being mediated by a single afferent impulse from a receptor to an interneuron and then along an efferent neuron to an effector (see Fig. 4–1). This theoretical simplification is often called a *reflex arc*. Probably all reflexes are more complex than this, involving many more neural connections. Individual segmental ganglia may show considerable reflex autonomy. For example, in the silkworm moth (*Bombyx mori*) the oviposition reflex, which results from the contact of the ovipositor with a surface, resides entirely in the last abdominal ganglion.

On the other hand, most insect reflexes occur in patterns or groups and involve several ganglia. These reflexes are generally thought to be under control by higher centers of coordination, which may furnish stimulatory or inhibitory signals. Chapter 6 contains a number of examples of complex coordinated reflex patterns. It should be noted that reflex patterns in insects tend to show much plasticity. For example, you may recall from Chapter 6 that an insect's walking pattern changes appropriately with the loss of one or more appendages.

More complex innate behavior includes the various orientation patterns. Orientation may be defined as "the capacity and activity of controlling location and attitude in space and time with the help of external and internal references, i.e. stimuli" (Jander, 1963). The reviews of Frankel and Gunn (1961), Markl and Lindauer (1965), Jander (1963), and Birukow (1966) are all useful in their consideration of this topic. The terminology used in describing the kinds of orientation is confusing. Frankel and Gunn (1961) offer a classification scheme for the various kinds of orientation and it will be followed in this text. They first recognize two broad kinds of orientation, primary and secondary. *Primary orientation* is the assumption and maintenance of the basic body position in space, that is the normal stance, either in a stationary or in a moving insect. For example, those reflexes which keep a flying insect on an even course are responsible for the primary orientation. *Secondary orientation* is superimposed upon primary orientation and has to do with the positioning of the insect in response to various stimuli. These stimuli may be external (e.g., light, humidity, temperature) or internal (e.g., the presence of a particular hormone). Secondary orientation may be of value to an insect by leading it toward potential prey or a potential host, or away from potential danger or an unsuitable environment, or toward a favorable one. In addition to the primary and secondary divisions, Frankel and Gunn (1961) also offer a classification based on the mechanisms of orientation, dividing them into *kineses, taxes,* and *transverse orientations.* We shall offer only a few examples here.

Kineses are essentially undirected locomotor reactions of an insect to stimuli, and hence there is no particular orientation of the long axis of the body relative to the source of a given stimulus. For example, in the tsetse fly, *Glossina*, there is an increase in activity in an arid atmosphere relative to activity in a humid one. This behavior would result in the insect "finding" and remaining in a humid atmosphere. The body louse, *Pediculus*, makes fewer and fewer directional changes as the temperature, humidity, and odor become increasingly intense as the louse moves closer to a potential host. This is of obvious value to the louse in host location.

Taxes are different from kineses in that they are characterized by directed responses of the insect relative to the stimulus source. Hence the long axis of the body takes on a definite position relative to the stimulus and moves toward (positive) or away from (negative) the stimulus source. When an insect is maintaining a course relative to a stimulus, any angular deviations from this course theoretically result in the initiation of a "turning tendency," that is, the tendency for the insect to correct for angular deviation by turning back to its original course. The magnitude of this turning tendency increases with increasing angular deviation. Hence, if the magnitude of this turning tendency is plotted as the ordinate and angular deviation as the abcissa, a sine curve is generated (Fig. 7–1).

Taxes are commonly classified according to the stimuli involved [e.g., phototaxis (light), geotaxis (gravity), anemotaxis (air currents), rheotaxis (water currents), and thigmotaxis (contact)].

Fig. 7–1

Relationship between magnitude of turning tendency and angular deviation. (Redrawn with modifications from Jander, 1963.)

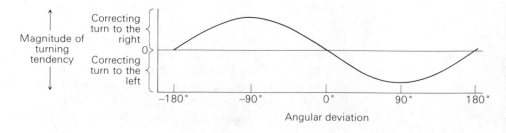

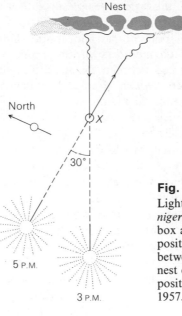

Fig. 7–2

Light-compass reaction in the ant *Lasius niger*. At point *X*, the ant is covered with a box at 3 P.M. and released at 5 P.M. (sun positions indicated). Note that the angle between the paths followed from and to the nest corresponds to the angular change in the position of the sun. (Redrawn from Goetsch, 1957.)

A good example of a taxis is klinotactic orientation. An insect orienting klinotactically swings all or part of the body back and forth across a stimulus field and moves toward or away from the region of maximum stimulation, depending upon whether it is attracted or repelled by the particular kind of stimulus. For example, when fly maggots have completed feeding and will soon enter the resting (pupal) stage, they move the head region back and forth until they are heading directly away from a source of light. You will recall from Chapter 6 that this is the position in which the light sensitive cells are the least affected by the light stimulus. These fly maggots are then photonegative.

Transverse orientations are those in which the body is oriented at a fixed angle relative to the direction of the stimulus. Locomotion may or may not be involved. Frankel and Gunn (1961) discuss two types of transverse orientations found in insects: the *light-compass reaction* and the *dorsal light reaction*. The light-compass reaction consists of an orientation such that locomotion occurs at a fixed angle relative to light rays. This orientation was first demonstrated in ants and subsequently in caterpillars, bees, and certain beetles and bugs. The light-compass reaction serves, for example, to aid an ant on a return trip to its nest. One simple but informative experiment (Fig. 7–2) demonstrating that ants use the sun in this manner involved placing an ant on the way back to its nest in a dark box for a period of time during which the sun's position in the sky would change significantly. Subsequently, the ant was released and its course was then the same as before relative to the sun, but incorrect in terms of its nest because the sun's position had changed. Similar orientations to gravity have also been described.

In the dorsal (or ventral) light reaction, both the long and transverse axes of the body are kept perpendicular to a directed source of light (Frankel and Gunn, 1961). Although symmetrical receptors (the eyes) are involved, this type of reaction is different from phototaxis because orientation is transverse to the light rays and not parallel to them. Examples of both dorsal and ventral light responses are found in certain aquatic members of the order Hemiptera. When the air bubbles trapped between the head and the antennae are removed, backswimmers, *Notonecta* and *Plea*, depend entirely upon a ventral light reaction for the maintenance of their normal swimming position while the converse is true for the water boatman, *Corixa*, and the creeping water bug, *Naucoris* (Rabe, 1953). Some caterpillars are shaded such that, provided they are oriented properly with respect to light, they are effectively camouflaged. Dorsal and ventral light responses come into play in these insects, ensuring "proper" orientation. The dorsal (ventral) light response is a good example of a means for maintaining primary orientation.

The most complex innate behavior patterns are the *fixed action patterns*. These patterns are characterized by being unlearned, species specific, and adaptive and differ from reflexes and orientation mechanisms in that they frequently require an internal readiness before they can occur and the eliciting stimulus (*releaser*) is commonly not required to act during the entire course of the pattern. The releasing stimulus may be internal in the form of hormone secretion or an excitatory signal from a higher neural center. The stimulus may also be external in the form of a physical factor—light, temperature, mechanical contact, and so on—or the behavior, appearance, odor, and so on, of another animal.

A variety of factors may affect the development of a state of internal readiness or motivation necessary for a given behavior pattern to occur. These may be either internal or external in origin (Markl and Lindauer, 1965).

Internal motivational factors are humoral or neural in nature and commonly depend on the developmental stage of the insect. For example, the relative concentrations of the juvenile hormone (neotenin) and the molting hormone (ecdysone) determine the spinning behavior of the caterpillars of certain moths (*Galleria* and others). A hormone secreted by the ovaries is necessary for the female grasshoppers (Acrididae) in the subfamily Truxalinae to sing (Markl and Lindauer, 1965, citing the work of Haskell, 1960). Internal receptors are commonly involved when a behavior pattern appears spontaneously without any discernible external stimulus. An excellent example of this has been found in the blow fly (*Phormia*) in which receptors on the foregut carry stimuli that inhibit food ingestion as long as the foregut is full. No inhibitory stimuli arise from these receptors when the foregut is empty and hence ingestatory behavior can occur.

In other instances of spontaneous behavior regular cycles of behavioral activity of internal origin (endogenous rhythms) determine when a particular pattern may be released. For example, crickets are active only during a certain period each day and this daily rhythm continues for several days when the insects are kept in constant darkness in the laboratory. We shall consider rhythmic behavior at greater length later in this chapter. The last internal motivational factor listed above was the stage of development. Obviously behavior patterns of immature insects are quite different from those of adults. The blood-ingesting behavior of the female mosquito has very little similarity to the behavior of filter-feeding aquatic larvae.

External factors—light, temperature, substances produced by or activities of other insects of the same species, and others—have definite effects on motivation in addition to serving as releasers. Light and temperature commonly determine whether an insect is active or inactive. Recall when we discussed the dorsal ocelli as stimulatory organs we gave examples in which insects failed to demonstrate certain light responses when the ocelli were blocked. Every insect has a temperature below which and one above which it becomes inactive. Within the range of these temperatures, specific temperatures may be necessary for particular activities. For example, the readiness of certain butterflies (e.g., *Pieris*) to copulate is determined by temperature.

Many instances are known where a substance secreted by one insect influences the behavior of another of the same species. The term *pheromone* is commonly used to denote such a substance. Pheromones may serve as releasers or as motivational factors or both. A good example of a pheromone affecting motivation is found in the honey bee. The queens produce a substance in the mandibular glands which is passed from worker to worker orally and which inhibits ovary development as well as the behavior involved with queen cell construction. Queen cells in which new queens develop are constructed when a hive has lost its queen and hence the effects of the substance of her mandibular glands. Wilson (1965) reviews the literature of chemical communication in social insects.

When an insect is in a state of internal readiness (motivated), it may begin to behave in a manner that increases the probability of exposure to a releasing stimulus. This has appropriately been called *appetitive behavior*. Probably the most common appetitive behavior is increased locomotion. Appetitive behavior does not reduce the motivation and is likely to continue until the insect comes into contact with the appropriate releasing stimulus. At this time the specific behavior pattern involved with the motivation is manifested. This occurrence has been called the *consummatory act* and reduces or lowers the motivation. Male cicadas that have been sexually aroused move about randomly (appetitive behavior), thus increasing the probability of encountering a female. This random movement will continue until contact is made with a female. At this time copulatory behavior (consummatory act) is released and sexual arousal reduced. There are probably a number of behavioral steps between the initial motivation and the consummatory act. It is generally felt that behavior is organized into a hierarchical sequence of patterns ultimately based on a hierarchical organization of nervous centers. From this point of view a given fixed action pattern (e.g., sexual behavior, food getting, etc.) then consists of a definite sequence of patterns, each with a specific releaser, the final pattern being the consummatory act that reduces the motivation.

Markl and Lindauer (1965) cite an excellent example from the work of Baerends, who studied the sequence of patterns involved in brood care in the digger wasp, *Ammophila adriaansei*:

In a definite sequence, three groups of activities follow one another: (1) digging of a nest, hunting for a caterpillar, and oviposition; (2) bringing more caterpillars and temporary closing of the nest; (3) fetching still more caterpillars that are required and the final closing of the nest. The process is complicated by the fact that several nests are simultaneously cared for. At a second nest, work can only proceed in the spaces between the three phases of supplying the first. Therefore, a characteristic pattern of activities appears in the supplying of the nests. The single parts of the process are also organized hierarchically. In the process of bringing in the first caterpillar this sequence is always as follows: the depositing of the caterpillar in the front of the entrance of the nest; digging up of the entrance; turning around at the open nest; and pulling in of the caterpillar. The sight of the nest situation releases the dropping of the caterpillar and scraping, the half-open nest site releases digging, the open nest releases the turning around, and the sight of the caterpillar, releases the pulling in.

Learned Behavior

Although there are a number of possible definitions of learning, we shall use that of Thorpe (1963): learning is "that process which manifests itself by adaptive changes in individual behavior as a result of experience." Thus learning involves the accumulation and storage (memory) of environmental information and the subsequent effects of this stored information on an animal's behavior. Extreme care has been necessary in calling a given pattern of behavior learned, since an innate pattern that appears at some point during the development of an animal may give the distinct impression of having been learned.

Alloway (1972) discusses four types of learning situations that have been described among insects: *classical conditioning, instrumental learning, "shock avoidance learning,"* and *olfactory conditioning.* The first two types have been described in several groups of animals. The other two seem to fall into the learning category.

Classical conditioning was originally discovered by the Russian scientist Pavlov. In this type of learning, two stimuli are involved. One of the stimuli (the unconditioned stimulus, UCS) elicits a response (the unconditioned response, UCR). The other, the conditioned stimulus (CS), prior to conditioning, does not elicit the UCR. However, when these two stimuli are repeatedly presented to an animal in very close succession (CS briefly preceding the UCS), the animal gradually begins to respond to the CS in the absence of the UCS.

Classical conditioning has purportedly been shown to occur in honey bees. In these insects, the proboscis extension reflex can supposedly be conditioned to respond to the essence, coumarin, in the absence of sugar. In this case, proboscis extension is the UCR, the sugar is the UCS, and the odor of coumarin the CS. However, Alloway (1972) explains that the experiments carried out to demonstrate the conditioning of the proboscis extension reflex were insufficiently controlled. Thus he concludes that classical conditioning has yet to be found among insects.

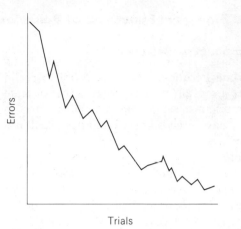

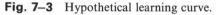

Trials

Fig. 7–3 Hypothetical learning curve.

In instrumental learning, a response or series of responses is induced or inhibited by the presentation of a stimulus or pattern of stimuli [reinforcer(s)]. If the presentation of a stimulus or stimulus pattern acts to increase the probability of the occurrence of a given response or responses, then this stimulus or stimulus pattern is a positive reinforcer. On the other hand, if removal of the stimulus or stimulus pattern raises the probability of the response, the stimulus or stimulus pattern is a negative reinforcer.

Maze learning is a good example of instrumental learning. A maze of some sort is placed between an insect and its nest, food, or some other strong positive reinforcer; or a negative reinforcer, such as an electric shock, is applied for an incorrect response. If an insect is capable of learning the maze, the number of errors (wrong turns) will decrease with the number of trials. For example, ants are capable of learning a rather complex maze placed between themselves and their nest (positive reinforcer). A hypothetical example of a learning curve that might be generated by an insect learning a maze is shown in Fig. 7–3.

Alloway (1972) describes the work of Horridge (1962a and b) who showed that a single leg of a decapitated cockroach suspended above a container of electrified saline solution can be trained to stay above a level where it would receive a shock ("shock-avoidance learning"). The experiments were so set up that a test animal received shocks according to the level of the test leg, while a control animal received the same number and intensity of shocks, but in a random fashion, without regard to leg position. Later investigators, Eisenstein and Cohen (1966), demonstrated that even a single isolated ganglion in the ventral nerve cord could mediate the shock-avoidance learning of a single leg.

Olfactory conditioning may also be a form of learning. An example of this phenomenon is a situation in which exposure to an ordinarily repellent odor during a certain period of development results in the loss of repellency of this odor to adults. This is the case in *Drosophila melanogaster,* the adults of which are normally repelled by the odor of peppermint oil. However, if the larvae have been reared in the presence of peppermint, they are strongly attracted to it as adults.

The Biological Functions of Behavior

Reproductive Behavior

Mate Location. The first problem pertinent to reproduction that faces a sexually mature insect is that of mate location. A wide variety of mechanisms that function in bringing together members of both sexes have been observed and described in insects. Basically they can be placed in one or more of three groups: visual, olfactory, and auditory.

The visual stimuli associated with mate location and the responses they elicit vary in complexity. The simplest involves a male insect approaching any moving object of appropriate size that happens to enter its visual field (e.g., *Pyrrhocoris* and *Gerris*, order Hemiptera; Markl and Lindauer, 1965). Other insects require more specific stimuli to induce approach. For example, male damselflies in the family Lestidae approach any insect that has transparent wings and flies in a manner similar to that of damselflies (Markl and Lindauer, 1965). The agrionid damselflies in the genus *Calopteryx* require even more specific stimuli, the males of different species being able to recognize members of their own species by the amount of light allowed to pass through the wings (Carthy, 1965). Once an insect has been induced to approach a possible mate, other stimuli are usually required to release further pursuit. These are generally olfactory, but they might be a characteristic behavior on the part of the pursued. For example, female damselflies in the family Lestidae permit themselves to be grasped behind the prothorax only by males of the same species (Markl and Lindauer, 1965). An insect that recognizes a member of its own species may or may not be able to distinguish the sex of the individual it pursues. Males of *Pyrrhocoris, Gerris,* and *Leptinotarsa,* the latter being a beetle in the family Chrysomelidae, recognize members of their own species by specific odors but will attempt to copulate with either sex (Markl and Lindauer, 1965). On the other hand, fruit flies are able to recognize the sex of an individual by its odor.

More complex visual stimuli involved in mate location include the use of luminescent organs in signaling for a mate. This is best exemplified by the fireflies in the beetle family Lampyridae. In some species both the males and females produce light and have a very complicated signaling system; in other species only the males produce the light signals. In either case the light-signaling systems are species–specific. In fact, several species have been discovered based on differences in their signaling systems. The mating behavior of *Photinus pyralis,* a well-known American species, has been described by McDermott and Mast (McElroy, 1964, quoting from Buck, 1948):

> At dusk the male and female emerge from the grass. The male flies about 50 centimeters above the ground and emits a single, short flash at regular intervals. The female climbs some slight eminence such as a blade of grass and perches there. She ordinarily does not fly at all and she never flashes spontaneously as does the male but only in response to a flash of light which is produced by the male. If a male flashes within a radius of 3 or 4 meters of the female she usually responds after a short interval by flashing. The male

then turns directly towards her in his course and soon glows again. Following this the female again responds by glowing and the male again apparently takes his bearings, turns and directs his course towards her. This exchange of signals is repeated usually not more than 5 or 10 times until the male reaches the female and mates with her.

In some tropical species of fireflies, several males congregate in a single tree and begin to flash synchronously. McElroy (1964) describes this phenomenon as follows:

In Burma and Siam and other eastern countries, all the fireflies on one tree may flash simultaneously, while on another tree some distance away this same synchronous flashing would be apparent, but out of step with those of the first tree. Observers have been particularly impressed by the display which is one of the interesting sights of the Far East. Unfortunately, very little experimental work has been done on this problem and every possible attempt to explain this phenomenon has been made. It is difficult to understand how one female could be so located in a tree so that it could trigger the flash of a large number of males that are uniformly distributed throughout the tree.

Certain plants produce stimuli that serve to attract insects and even release copulatory attempts. For example, some orchids in the genus *Ophrys* are so similar in appearance and odor to female wasps *(Gorytes mystaceus)* in the family Specidae (Hymenoptera) that males actually attempt to copulate with them. This conveniently results in pollination (Markl and Lindauer, 1965).

The use of olfactory cues in mate location is widespread among the insects. Sex attractants or lures are produced by highly specialized glands in males, females, or both sexes of a given species. These substances are, along with a number of others, examples of pheromones. Sex attractants are very potent substances as evidenced by the following statement: "Detected by the insect in fantastically minute amounts, these attractants are undoubtedly among the most potent physiologically active substances known today" (Jacobson, 1965). The majority of these sex-attracting substances are apparently species-specific; however, there are several known examples of nonspecificity.

The production by female insects of scents that are utilized by males in locating a mate has been observed and described for several species in a number of orders. In his very informative review of the literature pertaining to insect sex attractants, Jacobson (1965) lists 10 species of cockroaches (Orthoptera); 109 species of moths and butterflies (Lepidoptera); 17 species of beetles (Coleoptera); 17 species of bees, wasps, sawflies, and ants (Hymenoptera); 4 species of true flies (Diptera); and 2 termite species (Isoptera), in which the females produce sex attractants. It should be noted that these figures were based on a literature review up to 1965. Many more species have undoubtedly been added since then. At this point we can offer only a couple of examples of this phenomenon. In the Gypsy moth (*Porthetria dispar*) virgin females produce an attractant that is capable of luring males from up to 100 meters distant. The containers in which virgin females have been held apparently absorb some of the odorous substance since they remain attractive to males for 2 to 3 days following the removal of the females (Jacobson, 1965). Virgin *Periplaneta americana* (American cockroach) females produce an attractant that stimulates ". . . male alertness, antennal movement, searching locomotion, and vigorous wing flutter" (Jacobson, 1965). Filter paper that has been in contact with virgin females has the same effect.

As mentioned earlier, males in many species are also capable of producing sex attractants. As he did for the female producers, Jacobson (1965) presents a list for males which includes 3 species of cockroaches, 2 species of true bugs (Hemiptera), 40 species of moths and butterflies, 1 species and 1 family of beetles (Coleoptera), 1 bee, 3 true flies, 2 species of scorpionflies (Mecoptera), and 1 Neuropteran. The odorous substances produced by the males of some of these species serve both as attractants and excitants or excitants alone. Again we will offer only a couple of examples. The males of the greater wax moth, *Galleria mellonella,* secrete an odorous substance from glands on their wings which is very attractive to females. The odor is dispersed by the male vibrating his wings and dancing around. Bumblebee, *Bombus terrestris*, males produce an attractant in their mandibular glands that lures females. This substance has been extracted from their mandibular glands with pentane and identified as farnesol, a substance present in the flower oils of many plants. These flower oils may be the bee's source of farnesol (Jacobson, 1965).

Jacobson (1965) describes 3 species of beetles in the genus *Dendroctonus* family Scolytidae, which provide an interesting variation of the function of sex attractants. In these species a substance that is attractive to both sexes is produced by sexually mature, unmated females feeding on fresh Douglas fir phloem. Since both sexes are attracted by this odorous substance, it serves to bring them together, which eventually leads to courtship and copulation. *Trypodendron lineatum,* an ambrosia beetle, produces a substance that has a similar effect. Jacobson (1965) refers to all these substances as assembling scents.

For a large number of insects, acoustic signals (calling songs; Haskell, 1964) serve to bring together the members of the opposite sexes. This is especially true in the orders Orthoptera, Homoptera, and Diptera and somewhat true in Lepidoptera, Hemiptera, and Coleoptera. Calling songs may be quite elaborate and are usually species-specific. In general, a response on the part of a member of the opposite sex to one of these songs depends to a large extent on the state of internal readiness for courtship and copulation. In the majority of cases, specialized structures are involved in the production of sounds (see Chapter 5), but several insects produce sounds as a direct result of wing movements in flight. Examples of sounds produced in this fashion that serve to bring the sexes together have been found in members of the orders Diptera, Hymenoptera, and Orthoptera. In the order Diptera, mosquitoes are a particularly well-studied example. The flight sound of a mature female mosquito is very attractive to a male and releases copulatory behavior. The males react similarly when the frequency of the flight sound is produced by a tuning fork. In some insect groups the acoustical counterpart of the "assembling scent" has been found. Haskell (1964) cites the work of Alexander and Moore (1962) on the periodical cicada: ". . . in the *Magicicada* complex of species . . . is an extraordinary chorusing

behavior in which a calling or aggregating song, produced individually, is sung in chorus and is responsible for activating and assembling both males and females."

Courtship and Mating (Fig. 7–4). Following the location of a
mate, highly variable and often elaborate types of behavior occur
which ultimately lead to copulation. During this courtship behavior,
escape and attack responses are inhibited, at least to the extent that
copulation may occur. However, they may not be inhibited completely. For cxample, the female praying mantis very often devours
the head and part of the thorax of a male that is attempting to initiate
copulation. Roeder (1963) describes the courtship of the mantis as a
"sneak attack" rather than a courtship. However, the outcome
is quite favorable for the mantis since removal of the head and in

Fig. 7–4
Copulation and
oviposition in the
yellow dung fly,
Scatophaga stercoraria
(Diptera;
Anthomyiidae).
A. Copulation. Female
on substrate; copulating
male above; competitive
male partly out of
picture. B. Oviposition
with the male protecting
the female from other
males. This behaviour
prevents further
mating activity from
interfering with
oviposition. (Courtesy
of W. A. Foster.)

A

B

particular the subesophageal ganglion has the effect of increasing the vigor of copulation on the part of the male. Evidently, signals from the intact subesophageal ganglion inhibit an endogenous and autogenous nervous pattern in the rest of the body, which initiates vigorous copulation (Roeder, 1963).

Many of the cues that serve in mate location act further as releasers of courtship and copulation. In fact, it is often difficult or impossible to make a distinction between those cues which are involved in bringing the sexes together and those involved with release of court-ship and copulation. For example, the flight sound of the female mosquito not only attracts males but releases copulatory behavior. In many insects the sex attractant produced by one of the sexes also serves as an excitant (aphrodisiac), stimulating courtship or attempts at copulation. For instance, the sex attractant produced by female American cockroaches not only attracts males but induces a wing-raising display and attempts to copulate. In the absence of a female, males may attempt to copulate with one another if the female odor is present. This female sex attractant is also capable of eliciting courting behavior in males of other *Periplaneta* species and males of *Blatta orientalis* (oriental cockroach; Jacobson, 1965). In some insects, the sexually excited male or the female may produce a sub-stance that acts as an excitant for the opposite sex. Jacobson (1965) describes the following two examples, the first referring to *Lethocerus indicus*, a giant water bug, and the second to a family of beetles:

During sexual excitement the male is easily recognized by its odor, for its abdominal glands secrete a liquid with an odor reminiscent of cinnamon . . . This substance, produced in two white tubules 4 cm long and 2–3 mm thick, occurs to the extent of approxi-mately 0.02 ml per male and is used in south east Asia as a spice for greasy foods. The female does not secrete the substance, which is believed to act as an aphrodisiac to make her more receptive to the male.

Males of Malachiidae, a family of tiny tropical beetles, entice females first with a tarty nectar and then expose them to an aphrodisiac. The males possess tufts of fine hair growing out of their shells (in some species on the wing covers, in others on the head). These hairs are saturated with a glandular secretion that the females cannot resist. During the mating season, the male searches for a female; when he finds one he offers his tuft of hair, which the female then accepts and nibbles upon. In so doing, her antennae come in contact with microscopic pores in his shell, through which the aphrodisiac substance is excreted, thus putting her in a state of wild excitement.

Insects show a seemingly endless variety of sexual patterns. Some are quite simple, the male and female simply coming together and copulating with little or no courtship manuevers. On the other hand, many have quite elaborate and rather bizarre patterns of courtship. Two examples will serve to give you an idea of how elaborate some of these courtships are. The silverfish, *Lepisma saccharina,* guides the female to a spermatophore (a case or capsule containing spermatozoa) which he has previously deposited on the substrate by spinning a series of threads that restrict her movements to those which bring her closer and closer to making contact with it (Fig. 7–5). Jacobson (1965) cites the work of Bornemissza (1964) in the following description of courtship and copulation of two species of scorpionflies (*Harpobittacus,* order Mecoptera):

> Males of both species hunt for the soft-bodied insects on which they feed; females have never been observed hunting, capturing, or killing prey in the field. When the male holds its prey and begins to feed, two reddish-brown vesicles are everted on the abdomen between tergites 6–7 and 7–8 and begin to expand and contract, discharging a musty scent perceptible to humans. This scent attracts females to the vicinity of the male, moving upwind

Fig. 7–5
Courtship in the silverfish, *Lepisma saccharina.*
A. Approach. B. Male affixes threads to wall and floor and deposits spermatophore (only the main thread and one secondary thread are shown, although many secondary and irregular threads are produced). C. Female is guided to the spermatophore by the threads. (Redrawn from Sturm, 1956.)

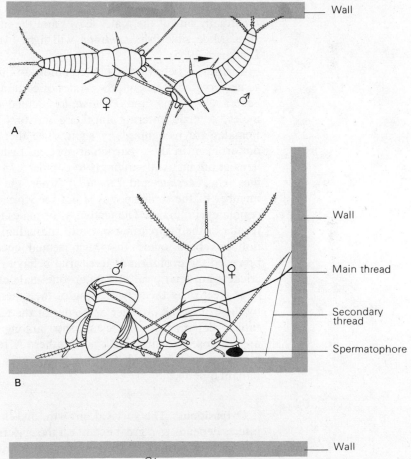

in his direction. As soon as the female is within reach, the male retracts his vesicles and brings the prey to his mouthparts. The female attempts to get hold of the prey but is prevented by the male, whose abdomen seeks out the tip of the female's abdomen and copulation takes place. Once in copula, the male voluntarily passes the prey with his hind legs over to the female.

Rivalry and Territoriality. Rivalry between males may arise over the courtship of the same female. This rivalry may result in direct physical aggression or displays of various sorts. These reactions to members of the same species (conspecifics) are different from those involved with the escape or defense, which are stimulated by threats of danger. When a male grasshopper that is serenading a female with a courtship song is suddenly in competition with another male they face one another and sing a characteristic "rivalry" song (Haskell, 1964). Eventually one of the males leaves and the other continues the courtship. Male silverfish, *Lepisma*, will fight over a female (Markl and Lindauer, 1965).

Territorial behavior is not common among insects, but there are some very definite examples. Males of certain dragonfly species in the genera *Calopteryx* and *Pachydiplax* defend territories against other males. A male entering another's territory is met with an attack. Females are recognized by sight since there is considerable sexual dimorphism in these particular species. Instead of aggression, these females are met with attempts to copulate. In other genera of dragonflies (e.g., *Aeshna* and *Libellula*), when the sex of an approaching member of the same species is not recognized, males will attempt to copulate with males. The tendency for males to avoid such encounters results in their spreading out into individual territories. The suggestion has been made that such sexual encounters between males preceded the evolution of territorial behavior (Manning, 1966). Male crickets are territorial and when one male enters anothers territory, he is greeted by the "rival" song of the other. This is followed either by the exit of the intruder or holder of the territory or by fighting. A rank order may become established among males whose territories are in close proximity to one another. A female entering a male's territory is also greeted with the "rival" song, and will either leave or remain quietly.

Oviposition. The survival, growth, and development of immature insects depends to a great extent on the eggs from which they emerged having been placed in an appropriate environment. The problem is especially acute in those insects which are very specific as to what they feed upon (e.g., a specific plant or host in the case of many parasitic insects). The responsibility for the "proper" choice of oviposition site rests almost entirely upon the parent insect. For example, adult mosquitoes that are ready to oviposit must do so in a place where the eggs are in water or will eventually be. Some insects merely drop their eggs wherever they may happen to be (e.g., some mayflies). However, more often they are quite specific as to their choice of oviposition site, locating it by means of a variety of different stimuli, depending upon the kind of insect. For example, the beetle *Hylotrupes* is attracted to the terpene odor of the wood in which it deposits its eggs. The parasitic wasp, *Nasonia,* locates the puparial cases of host blow flies by being attracted to the odor of the decaying flesh in which blow fly larvae and pupae are found. Insects ready to oviposit may respond to stimuli which prior to that time elicited no response whatsoever (e.g., *Pieris* butterflies show a definite preference for objects with a green color when they are ready to oviposit). On the other hand, when they are searching for food, they demonstrate no such preference. When an insect is highly selective in the choice of a site for oviposition, it does not mean that it has foreknowledge of the needs of its offspring. Oviposition in response to specific stimuli that are associated with an environmental situation in which the young can survive has, during the course of evolution, no doubt been selected for.

Brood Care. Once a parent insect has fulfilled its responsibility for placing the egg (or larva in some instances) in an appropriate environmental situation, it may simply leave it at that. However, some continue an association with the eggs and immature stages. This association is most highly developed in the social insects (ants, bees, wasps, and termites) in which the brood form a core around which all activity is centered and these brood are "reared" from egg to adult by adults. Social insects differ from social vertebrates in that in the latter a colony is composed of a number of mating pairs and offspring, while in the former a colony is usually the product of a single female or single male–female pair. We shall consider the various activities of social insects later in this chapter.

Many nonsocial insects do more for their offspring than merely deposit the egg. Earwig females (order Dermaptera) lay their eggs in burrows in the ground and guard them until they hatch. Female beetles in the genus *Omaspides* protect their brood from the ravages of invading ants. Some nonsocial insects go to the extent of preparing elaborate nests and stocking them with food for their young. For example, a female solitary wasp in the genus *Bembix* digs a nest in the soil with her legs and mandibles. She then captures an adult fly (or sometimes another insect) and brings it to the nest. Evans (1957) describes the prey capture:

> The capture and stinging of the prey occur with great rapidity. The female wasp proceeds slowly through the air or hovers over a source of flies, pouncing upon the flies either in flight or at rest; then she descends to the earth or to some other solid object, where she quickly bends her abdomen downward and forward, inserting the sting on the ventral side of the thorax or in the neck region of the fly . . . .

Following prey capture, the wasp carries the fly back to the nest and lays an egg on it. Thereafter, a number of flies are brought to the nest as the larva grows. Eventually, the larva spins a cocoon, pupates, and emerges from the ground as an adult. Depending on the species and the time of year, it may remain over the winter in the cocoon as a prepupa.

In this section we shall consider the biological sources of nutritive materials, how insects are able to locate them, and the control of feeding.

Insects may be very broadly, although somewhat loosely, grouped on the basis of the biological source of nutriment as follows: saprophagous, phytophagous, and zoophagous.

Saprophagous Insects.   Saprophagous insects are those which feed on dead and decaying organic materials: leaf litter, carrion, dung, and so on. Such materials were probably the basic food source for the most primitive insects, which lived among the dead and rotting materials on the forest floor. Many of today's insects also utilize these materials throughout all or a part of their lives and are extremely important in the progressive degradation of decaying organic material. Cockroaches are excellent examples of saprophagous insects which subsist on a wide variety of dead plant and animal matter. They are commonly described as scavengers. Other examples of saprophagous insects are dung beetles, carrion beetles, and many of the soil-inhabiting apterygotes (e.g., Collembola). Actually many insects that upon cursory examination appear to be saprophagous are found upon closer scrutiny to be entirely dependent upon the microorganisms associated with dead, decaying organic material. The fruit fly, *Drosophila*, is an example of such an insect. The larvae cannot be maintained on a sterile culture medium. Thus, although one associates them with decaying organic matter, they are actually phytophagous or more specifically microphagous (Brues, 1946).

Phytophagous Insects.   Phytophagous means literally "plant eating". Living plants are the source of nutriment for the greatest number of insects and the habits of many insects have evolved closely along with the plants. Every species of plant has a number of different insectan species which feed upon it. Members of the orders Orthoptera, Homoptera, Lepidoptera, Thysanoptera, Isoptera, and some other smaller orders are mostly phytophagous. In addition, large numbers of insects in the orders Coleoptera, Hymenoptera, and Diptera are also phytophagous. Although flowering plants are by far the most common plant hosts, there are many insects which feed on fungi, bacteria, algae, ferns, and so on.

There are numerous examples of relationships in which the plants provide food for an insect and the insect in turn facilitates cross pollination of the plants. Probably the best-known pollinators are the honeybees, which are valued highly for these activities, particularly by orchardists. Honeybees utilize a wide variety of flowering plants for nectar and pollen. However, other pollinators are much more specific in their relationship to the host plant. A classic example (Fig. 7–6) of such a relationship is found between the yucca plant (*Yucca* spp.) and the yucca moth (*Tegeticula* spp.) (Powell and Mackie, 1966). The adult female moths have specialized mouthparts which they use to gather the pollenia of the yucca plant into a ball. This ball is then scraped across the plant's stigma, accomplishing cross pollination. The dependence of the yucca plant on the pollinating activity of the moth is demonstrated by the failure of the plant to develop pods in the absence of the moth. The yucca plant benefits the moth in that it provides a home and food for the larvae, which remain within the plant until they emerge as adults.

Other examples of these "mutualistic" relationships (both associates benefit) are found among the many insects which maintain fungus gardens within their nests. Outstanding among these are the fungus-growing ants in the tribe Attini (family Formicidae). Weber (1966) provides a brief description of their activities in the summary of a paper dealing with these insects:

Fig. 7–6
Yucca moth and yucca plant. A. Female moth gathering pollen. B. Female scraping the pollen ball across the plant's stigma. C. Female ovipositing (several petals removed). D. Larva within mature pod (portion of pod removed). E. Pod with exit holes and constrictions caused by larvae. (Redrawn from Schneirla, 1953, after Riley.)

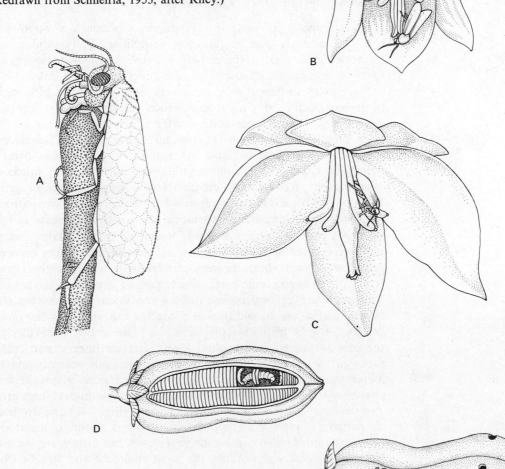

Fungus-growing ants (Attini) are in reality unique fungus-culturing insects. There are several hundred species in some dozen genera, of which *Acromyrmex* and *Atta* are the conspicuous leaf-cutters. The center of their activities is the fungus garden, which is also the site of the queen and brood. The garden, in most species, is made from fresh green leaves or other vegetable material. The ants forage for this, forming distinct trails to the vegetation that is being harvested. The cut leaves or other substrate are brought into the nest and prepared for the fungus. Fresh leaves and flowers are cut into pieces a millimeter or two in diameter; the ants form them into a pulpy mass by pinching them with the mandibles and adding saliva. Anal droplets are deposited on the pieces, which are then forced into place in the garden. Planting of the fungus is accomplished by an ant's picking up tufts of the adjacent mycelium and dotting the surface of the new substrate with it. The combination of salivary and anal secretions, together with the constant care given by the ants, facilitates the growth of the ant fungus only, despite constant possibilities for contamination. When the ants are removed, alien fungi and other organisms flourish.

Certain species of scale insects (family Coccidae), a wood wasp (*Sirex*), ambrosia beetles (families Scolytidae, Paltypodidae, and Lymexylomidae), and termites (*Macrotermes* and *Odontotermes*) also carry on mutualistic relationships with fungi (Batra and Batra, 1967).

Phytophagous insects may show very little preference with regard to the particular kind of plant upon which they feed, while others show some degree of preference and still others are very specific as to their host plants. Insects such as the migratory locust (*Schistocerca gregaria*) are *polyphagous* and will eat nearly every green plant in their path. On the other hand, the milkweed bug and many others are *monophagous*, feeding only on plants of a single species or genus. Most phytophagous insects lie somewhere between these two extremes (*oligophagous*). Considerable variation also exists with regard to what part of a plant is utilized as food. No major part of plant anatomy has been ignored by insects. There are those which feed in or on roots, stems, wood parts, buds, flowers, and fruits. However, a given insect is usually rather specific as to which part of a plant it feeds upon. Thus there are leaf rollers, leaf miners, rootworms, root borers, stem borers, and so on. In addition to providing nutriment for the phytophagous insects, plants usually serve as a place to live, particularly in the case of larval and nymphal forms. Certain insects, particularly those in the dipteran family Cecidomyiidae (gall midges) and hymenopteran family Cynipidae (gall wasps), induce plants to form characteristic growths in which the larvae of these insects feed, grow, and develop. These growths are called *galls* (Fig. 7–7) and are made up entirely of plant materials. Such organisms as fungi, nematodes, mites, and others also cause galls to develop, but insects are the most prominent cause and produce the most elaborate and specific ones.

Zoophagous Insects. Zoophagous insects are those which utilize living animals as a source of nutriment. They may be divided somewhat arbitrarily into two groups: *predators* and *parasites*. Predatory

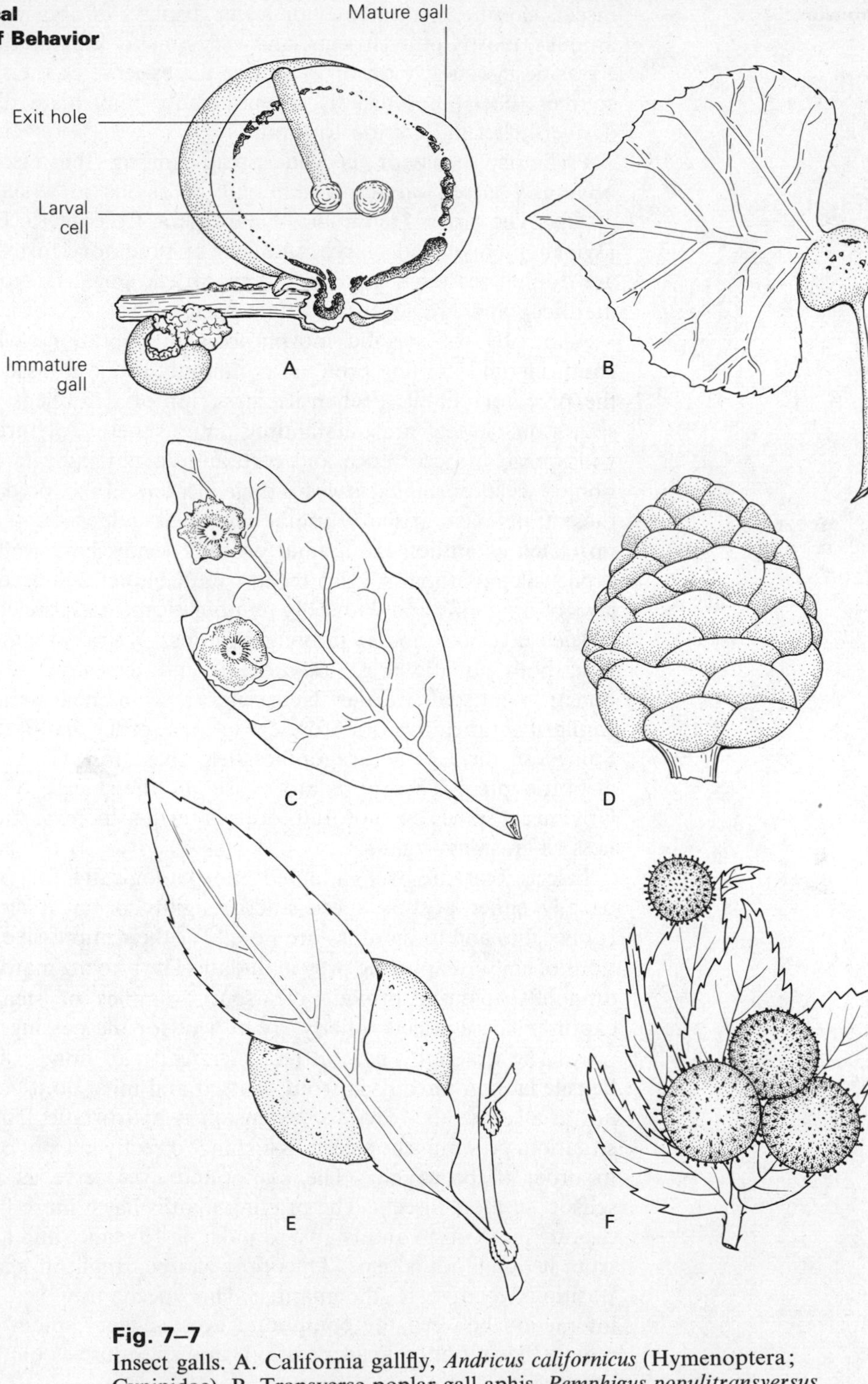

Fig. 7–7
Insect galls. A. California gallfly, *Andricus californicus* (Hymenoptera;
Cynipidae). B. Transverse poplar gall aphis, *Pemphigus populitransversus*
(Homoptera) on stem of poplar leaves. C. *Andricus pattersonae* (Hymenoptera;
Cynipidae) on leaf of blue oak. D. Pine-cone willow gall caused by *Rhabdophaga
strobiloides* (Diptera; Cecidomyiidae). E. Willow apple gall caused by the sawfly
Pontania pomum (Hymenoptera; Tenthredinidae). F. Spiny rose gall caused by
Rhodites puslulatoides (Hymenoptera; Cynipidae). (A–C redrawn from
Essig, 1958; D–F redrawn from CCM: General Biological, Inc., keycard.)

insects capture, with greater or lesser displays of ability, other living animals, mostly other insects, and very quickly kill and devour them. Parasitic insects live on or in and at the expense of their living food sources (hosts) but usually do not cause immediate death; many derive shelter in addition to food.

Predatory behavior is widespread among the insects, having obviously arisen on many different occasions in widely divergent groups. The orders Hemiptera, Neuroptera, Coleoptera, Diptera, and Hymenoptera include large numbers of predatory forms and many are found scattered among other orders [e.g., Orthoptera (e.g., mantids), and Trichoptera (certain species)].

A number of specific morphological adaptations which aid in capturing and holding prey are commonly found in association with the predatory habit. Prehensile jaws, raptorial forelegs, and strong spines on the legs are outstanding among these. All three traits are widespread in occurrence and certainly do not indicate close evolutionary relationships between their bearers. Like predatism itself, these traits have arisen a number of times independently. Insects as unrelated as antlion larvae and ground beetles have well-developed, prehensile mouthparts. Also recall from Chapter 2 the grasping labial jaws of dragonfly and damselfly nymphs. Raptorial forelegs are found in such diverse forms as praying mantids; several predatory Hemiptera, both aquatic (e.g., *Belostoma*) and terrestrial (e.g., ambush bugs); mantispids (order Neuroptera; so named because of the similarity of their raptorial forelegs to those of the mantis); and others. Spines on the legs of dragonflies help these insects catch and hold airborne prey. These predators use all their legs to grasp prey. Obviously, spines on raptorial forelegs further increase their effectiveness as grasping organs.

Insects that are well adapted morphologically for prey capture usually either possess speed and/or agility or great stealth. Adult robber flies and dragonflies are noted for their impressive demonstrations of agility, capturing prey in midair. The praying mantis and many dragonfly nymphs provide clear-cut examples of stealth in prey capture. These insects quietly lay in wait for the passing of potential prey. The dragonfly nymph (e.g., *Aeshna*) can only extend its prehensile labium directly in front of itself and must do it completely or not at all since the force for extension is hydrostatic. This being the situation, prey must be a certain distance directly in front of the nymph in order to be caught. The compound eyes serve as the ranging sensors in these insects. The praying mantis has a more flexible prey-capture mechanism and is able to judge the distance and angle of prey from its longitudinal axis. Thus prey can be caught in many different positions relative to the mantis. This mechanism is based on an interaction between the compound eyes, sensory spines in the neck region that monitor head position, and the muscles of the raptorial forelegs.

Certain insects, such as antlion larvae and the larvae of some species of caddisflies, literally trap their prey. The antlion forms a shallow cone-shaped pit in a sandy area and buries itself at the bottom just beneath the surface. When an ant or other small insect passes near enough to the pit to cause grains of sand to roll down the sides, the antlion responds by creating a miniature landslide, bringing the prey within reach of its well-developed prehensile jaws, through which it sucks out the digested insides of the prey. The aquatic larvae of certain species of caddisflies construct silken nets in which they are able to catch small organisms.

Insect larvae that lack strong morphological or behavioral abilities with regard to capturing prey usually rely upon the parent insect placing them in proximity to prey species or depend upon sluggish, easily captured prey. Many examples of parent insects locating prey for their offspring are found among the Hymenoptera (e.g., the case of the female *Bembix* wasp described in the preceding section). In the higher Diptera, the larvae have poorly developed mouthparts and move about very slowly. Thus the predaceous ones must depend on weaker, more sluggish prey, such as aphids, scale insects, mites, other Dipteran larvae, spider and insect eggs, and so on.

By far the greatest number of prey animals eaten by insects are other insects, particularly the phytophagous ones. This is not at all surprising when one recalls that phytophagous insects are the most abundant. In several instances small crustaceans (especially aquatic), mites, and occasionally spiders fall prey to insect predators. Certain members of several families of beetles (Carabidae, Silphidae, Drilidae, and others) and true flies (Phoridae, Sarcophagidae, and other muscoid families) feed upon snails. Protozoans compose a significant proportion of the diets of aquatic insects, particularly in their earlier developmental stages. With regard to the specificity of prey choice by predatory insects, a fairly safe generalization is that the members of the more specialized orders (e.g., solitary wasps) tend to be quite restrictive in prey choice, while members of the more generalized, primitive orders exhibit a wider range of prey choice.

Parasitic behavior, like predatory behavior, has arisen independently on several occasions among insects. Thus parasitic insects, having no particular common origin and displaying great variability in the details of their parasitic habits, present an exceedingly complex group to discuss as a unit. They have been grouped in a variety of ways, none of which is entirely satisfactory. The classification of parasites herein presented represents a compromise of ideas from a number of sources. In this classification parasitic insects are divided into the following broad areas: *ectoparasites, endoparasites,* and *nest-invaders* or *social parasites.*

Ectoparasitic insects are those which although they derive all or a part of their sustenance from the host, live entirely on its external surfaces. These insects will be discussed on the basis of their hosts, vertebrates or insects.

Most insects that are ectoparasitic on vertebrates are found in the orders Anoplura (sucking lice), Mallophaga (chewing lice), Hemiptera (true bugs), Diptera (true flies, Fig. 7–8), and Siphonaptera (fleas). The vast majority of these, save most Mallophaga, are blood feeders (hematophagous). Among the members of this group, some remain on the host throughout their life cycle, some live on the host only during a particular stage of the life cycle, and others visit the host during a particular stage of the life cycle and are otherwise free-living. Those parasitic insects which remain on the host throughout their lives

Fig. 7–8

Blood feeding by *Phlebotomus longipes* (Diptera; Psychodidae) on a human host. This species is a vector of *Leishmania tropica*, a protozoan parasite of man in Ethiopia. (Length of fly approximately 3.0 millimeters.) (Courtesy of W. A. Foster.)

(e.g., the lice) tend to be highly host specific and have had a close evolutionary relationship with their host species and are most likely to be entirely dependent on it. Fleas provide good examples of parasites that spend only one stage of their life cycle on the host. Flea larvae are free-living and feed on the organic debris that accumulates in animal nests and dwellings; the adults tend to remain on the host, taking intermittent blood meals. However, adult fleas are quite active and do move from host to host and do not exhibit such distinct host specificity as the lice. Adult mosquitoes, no-see-ums (Diptera; family Ceratopogonidae), other hematophagous Diptera, a few Dipteran larvae, and nymphs and adults of the blood-sucking Hemiptera (Cimicidae, bed bugs; Reduviidae, conenose bugs) visit host vertebrates only to take a blood meal and exhibit varying degrees of host specificity.

It can safely be said that most ectoparasitic insects tend to feed on a single or very few host vertebrates and under most circumstances cause little or no harm to their hosts. This is not meant to imply that they never harm the host, as they obviously do when they occur in large numbers or carry disease-causing organisms. Most ectoparasitic insects that have other insects as hosts spend their larval stages on the host and generally destroy it. This situation is, of course, close to predation but differs in that the host is usually somewhat larger than the parasite, at least at the beginning of the relationship. Some parasitic wasps in the hymenopteran family Ichneumonidae behave in this fashion. Insects that fall into this category are commonly grouped along with related endoparasitic insects under the heading *entomophagous* ("insect eating"). It is quite often difficult to tell where ectoparasitism ends and endoparasitism begins in entomophagous insects, since some may feed both externally and internally during their development. Other insect (adult) ectoparasites take intermittent blood meals from host insects in the same way those mentioned earlier did from vertebrate hosts. In fact, many members of the same orders and families are involved, [e.g., mosquitoes and ceratopogonids (no-see-ums)]. Various species in the latter family have been observed feeding on caterpillars and attached to and feeding in the vicinity of wing veins of lacewings, dragonflies, and so on. As was the case with those which had vertebrate hosts, insects with insect hosts probably individually cause little damage as a result of their feeding.

Endoparasitic insects live and feed within the host, usually during the larval stages, and become free-living adults. However, they do not always leave the host in the adult stage. For instance, most female Strepsiptera (twisted-wing parasites) spend their entire life within their host insect. Some insects are endoparasites of vertebrates. Almost all these are dipteran larvae (maggots) in several families, which develop, for example, in the alimentary canal, nasal cavities, and open sores. These larvae attack a wide variety of vertebrates, including man, and like the ectoparasites of vertebrates, do not individually cause much damage to the host. An animal infested with fly larvae is said to be suffering from *myiasis*.

By far the greatest number of endoparasitic insects attack other insects, and the majority of these are members of the order Hymenoptera in the superfamilies Ichneumonoidea, Chalicidoidea, and Proctotrupoidea, although a few occur scattered in other hymenopteran groups. A number of dipteran families also contain entomophagous parasites (e.g., Tachinidae and Sarcophagidae). These flies attack a wide variety of insects, including grasshoppers, caterpillars, true bugs, beetle grubs and adults, and hymenopterous larvae. The beetle family Rhipiphoridae is composed entirely of parasitic members, which attack such insects as hymenopteran larvae and cockroaches. Most endoparasites exhibit a high degree of host specificity, parasitizing a single or very few species of insects. In most instances only a single parasite can exist in a host. However, there are instances where several and rarely several hundred or a thousand can develop in a single host. It is a wonder that the hosts survive for any length of time after invasion by one of these parasites, but they do. The parasites usually feed on nonvital tissues until they are ready to pupate, and at this point they devour the insides of the host. Some of these endoparasites attack other endoparasites. Thus one may find a caterpillar harboring a larva of a tachinid fly, which in turn harbors a chalcid wasp larva. This phenomenon is referred to as *hyperparasitism*. Many entomophagous insects, especially the hymenopteran parasites, have been utilized in attempts to control various pest insect species.

The parasitic activities of certain insects fall best under the headings "social parasites" and "nest invaders." Brues (1946) describes the social parasitic behavior of several species of wasps in the genus *Vespula*, which parasitize other members of the same genus, as follows:

All the parasitic species have lost the worker, or infertile female caste and consist only of males and fertile females, the latter corresponding to the "queens" in the host species. The parasitic females enter the nests built and maintained by their hosts where they lay their own eggs in the paper cells already provided. Their young are then fed and reared in cells, intermingled with those containing the brood of the host, all of the feeding being done by workers of the host species.

Several other social Hymenoptera display this kind of parasitism. Certain beetles in the family Meloidae invade the nests of solitary bees. The beetle larvae destroy the bee's eggs and consume the food stored in the cells of the nest (Brues, 1946). Several other insects are found living and feeding within the nests of social insects. These forms are commonly called *inquilines*.

Location of Food Sources. Dethier (1966) has described feeding behavior as "a complex and interacting sequence of responses to a variety of stimuli culminating in ingestion to repletion." He subsequently outlines the basic sequence of events involved in feeding: ". . . locomotion bringing the insect to its food, cessation of locomotion on arrival, biting or its equivalent (probing, sucking, etc.), continued feeding, termination of feeding." In this section we shall consider some of the factors involved in food location and initiation of feeding.

Much research has been carried out on the stimuli involved in food location by different insects in all the general groups discussed above. Several examples of these stimuli and the insects that utilize them follow. Visual stimuli, giving information regarding color, form, movement, and so on, are widely utilized by insects. For example, you will recall the function of color, form, and movement of flowers in flower location by honeybees (see Chapter 5). Many butterflies also respond to these stimuli. As mentioned in the preceding section, dragonfly larvae and adults and preying mantids rely heavily on visual stimuli in prey capture. Visual stimuli are important in host location in mosquitoes, tsetse flies, and other hematophagous flies and likewise in insects that are parasites of other insects. Visual and olfactory stimuli are probably the main ones utilized in food location by insects in general.

Saprophagous insects, such as the dung-feeding *Geotrupes* (dung beetle) and *Lucilia* (a blow fly; family Calliphoridae), are attracted by the volatile substances skatol and ammonia, which are found in their food (Carthy, 1965). These substances are referred to as *token stimuli* since they serve as indicators of substances with nutritive value but have none themselves. Such token stimuli are commonly involved with the attraction of phytophagous insects to their host plants. For example, the butterfly *Papilio polyxenes* is attracted to several of the volatile oils which impart an odor to their characteristic food plants, members of the Umbelliferae (Carthy, 1965). However, these oils have no food value for the butterflies (Dethier, 1966). Olfactory stimuli play a critical role in host location by *Aedes aegypti* females. Carbon dioxide, steroids, amino acids, and other volatile substances characteristic of vertebrates are involved (Markl and Lindauer, 1965).

In addition to olfactory cues, tactile, thermal, and hygrostimuli play an important role in host location in the ectoparasites of warm-blooded vertebrates. For example, human body lice, which live between the skin and clothing of their host, show a marked preference for rough-textured materials (Carthy, 1965). They show an increased frequency of turning on smooth surfaces, and this frequency is inversely proportional to the stimulus strength (orthokinesis). Lice also demonstrate a preference for the 26.4 to 29.9°C range in a temperature gradient and 76% relative humidity in a humidity gradient.

Backswimmers and whirligig beetles respond to the vibrations produced by the swimming or struggling of potential prey. The backswimmer locates the prey by orienting itself such that it receives equal stimulation of the sensory hairs on its oarlike legs (Carthy, 1965).

Some insects, in particular the larvae of ladybird beetles, catch their prey, aphids, by crawling along the veins of a leaf and periodically stopping and "scanning" the vicinity by turning their body to and fro about the temporarily secured abdomen (Carthy, 1965). When they are successful in procuring an aphid, they demonstrate an increase in the frequency of turns, a response that keeps them in the immediate locality. This is a good example of a klinokinesis.

There are several instances where specific stimuli are involved in the induction of feeding. These stimuli are often referred to as *phago-stimulants*. They may be purely token stimuli, having no nutritional value, or they may be substances that have definite nutritive value. Many of the volatile oils characteristic of certain species of plants are clearly phagostimulants. For example, the cabbage aphid, *Brevicoryne brassicae*, is induced to feed even on an abnormal host when the substance sinigrin, a material extracted from Cruciferae (mustard family) and other plants, is present (Dethier, 1966). Sucrose, a common sugar in the food of many insects, is an effective phagostimulant in the majority of insects (Dethier, 1966). Amino acids, glucose, certain proteins, ascorbic acid, and several others have also been shown to be feeding stimulants. In certain aphids an inverse correlation exists between the numbers of winged forms and the amino acid content of the leaves upon which they are feeding (Carthy, 1965). Obviously a low number of winged forms would tend to keep a population of aphids in the immediate vicinity of an adequate food source.

The most detailed and extensive studies of the initiation and control of feeding have been carried out on the black blow fly, *Phormia regina*. Dethier (1966) explains that:

> Stimulation of chemoreceptors on the tarsi triggers proboscis extension; extension places the chemosensory aboral hairs of the labellum in contact with the food; stimulation of these hairs results in spreading of the labellar lobes which places oral taste papillae in contact with the food; stimulation of the papillae triggers and drives ingestion . . . . Once feeding had commenced continuous sensory input is supplied principally by the inter-pseudotracheal papillae (and to a lesser extent by tarsal and labellar hairs). Without this input feeding is neither begun nor maintained.

The termination of feeding occurs when the sense organs become adapted.

Since insects have a high reproductive capability but population sizes remain fairly constant, it is obvious that there are many factors (e.g., predators) which limit numbers. Thus throughout their evolutionary history there has been and still is a significant selective advantage in favor of the development of protective mechanisms. In this section we will discuss the behavior associated with protection from two standpoints: escape and defense.

Escape.   Many insects will exhibit some sort of escape reaction when threatened by a potential predator or other enemy. The releasing stimuli may be any of several (e.g., visual, olfactory, mechanical). Even those insects which have protective means other than escape will attempt to flee if the stimulus is great enough. A coupling of defensive behavior with escape is shown by some. For example, Eisner and Meinwald (1966) point out that the beetle *Chlaenius* (family Carabidae), which discharges a secretion that is repellent to ants, remains invulnerable to them for 8 to 13 minutes following a discharge. They explain that the beetle can cover an estimated 100 meters during this period of time!

Particularly interesting examples of insect escape reactions are those which are elicited in certain moths (especially in the family Noctuidae) in response to ultrasonic impulses (Roeder, 1963, 1965, 1966). The tympanic organs of hearing have already been described (see Chapter 5). Insectivorous bats locate nocturnally airborne insects by emitting ultrasonic impulses and locating the sources of echoes. Moths detect these impulses beyond the range of the bats' echo-locating sensitivity and evasive behavior is released. The various patterns of evasive behavior were studied by the use of an electronic sound source that produced sound in the range of the bats "chirps." The sound was emitted from a loudspeaker mounted on an upright pole. The paths of moths arriving in the vicinity of the sound source were recorded photographically by means of a stroboscopic (regularly blinking) flash that illuminated the moths at regular, very closely spaced intervals of time. Each time an approaching moth was illuminated the reflected light, indicating the position of the moth, was recorded on the same photographic plate. With this technique a number of different evasive tactics on the part of moths were detected. These included turning and flying directly away from the sound source, oblique turns upward, "power dives," "passive drops," complex "looping dives," and zigzag movements. Some individuals landed on the ground and remained motionless for a period of time. Similar responses to bats were observed when moths were tossed into the air, the recording technique being the same as that just described.

Defense.   Insects display a fantastic amount of variation in their form, coloration, and patterns of coloration. Instances in which one or all of these characteristics play a definite role in the protection of the insect are legion. The close resemblance of the wings and other body parts of many insects to leaves, twigs, thorns, and other plant parts; insects with "eyespots" or bright colors on the wings which they suddenly display to a threatening predator; and those species of insects which closely resemble species that vertebrate predators learn to avoid because of results such as poisonous stings, irritating sprays, and disagreeable taste; and insects which in their normal surroundings blend in so well that they become nearly invisible to predators (and collectors) are but a few examples. It should be clear that an insect must behave in a very specific and appropriate way if it is to gain any protective benefit from its form and coloration. Thus an insect that closely resembles a leaf will tend to orient itself on a stem in an appropriate manner, one with "eyespots" or other brightly colored markings will display these characteristics when threatened by a potential predator, and one that resembles other species may behave in similar ways. Insects that blend with their surroundings tend to remain akinetic (motionless) when threatened. Readers interested in this topic should consult the works of Cott (1940) and Wickler (1968).

A number of skeletal structures, although not always used for defense, serve quite admirably in this capacity. The raptorial forelegs of a praying mantis, although functioning primarily in prey capture, are used to strike out at an attacker. The stingers of many ants, bees, and wasps also serve such a dual purpose. Many, if not all, of the sound-producing orthopterans produce sounds of warning and aggression. Certainly, any skeletal structure of sufficient strength and/or hardness can serve in a passive defense capacity.

A large number of insects produce nonskeletal barriers of various types. Any of the nests, cocoons, cases, and so on, no doubt function to a greater or lesser extent in defense. The cast skins and fecal material of the larvae of the beetles in the genus *Cassida* (family Chrysomelidae) remain on a "fork" of posterior abdominal appendages that are highly maneuverable. This waste material dries and forms a sturdy "fecal shield" which is used by the larva as a protective device and is said to be highly effective in blocking the bites of ants (Eisner, van Tassell, and Carrel, 1967).

It would seem that the poison-gland-containing hairs characteristic of many lepidopterous larvae and some pupae and adults would also serve in a defensive capacity.

The primary defense mechanism of many insects is chemical. A large variety of noxious substances are secreted by the repugnatorial or odoriferous glands associated with the integument of a variety of insects. These substances are described by Eisner and Meinwald (1966) as representing "the means by which predators and other potential enemies are 'told' to desist or withdraw." The glands producing repugnatorial substances are so variable in number, location, and morphology that they are considered to have arisen independently a number of times during the course of evolution. Repugnatorial material is discharged from the glands, or more specifically the storage reservoirs associated with the glands, by a variety of mechanisms (Eisner and Meinwald, 1966). Basically they fall into three categories: (1) the secretion oozes onto the integumental surface, (2) the gland is evaginated and the secretion allowed to volatilize, and (3) the secretion is forcibly discharged.

Caterpillars of the species *Papilio machaon* possess glands behind the head, *osmeteria*, which they evert when appropriately stimulated and which produce a secretion effective against ants (Eisner and Meinwald, 1965).

The forcible discharge of material is effected in a variety of ways. Contraction of muscles associated with the glands, hydrostatic pressure of the hemolymph, and air pressure are among them. Examples of insects which are capable of forcible discharge of odoriferous material are particular species of cockroaches, stink bugs, earwigs, stick insects, caterpillars, carabid and tenebrionid beetles, and others. Some of these are able to produce a spray that reaches several feet and many have a very accurate aim. A particularly impressive discharge mechanism is found in the bombardier beetles (*Brachinus* spp.). These beetles eject a hot spray from glands in the posterior part of the abdomen containing quinones with a distinctly audible "explosion," which is the result of an exergonic chemical reaction in which oxygen is liberated. The beetle's raised posterior at the time of the explosion adds to the impressiveness of this display. The answer to the question as to why the defensive secretions liberated by insects do not harm the insects themselves is thought to be a cellular mechanism which allows two or more harmless precursors to be mixed in tubes or chambers essentially isolated from the rest of the cell. This principle operates at the multicellular level in the repugnatorial glands of the bombardier beetle (Fig. 7–9). Each gland, which is two-chambered, contains phenolic precursors of quinones and hydrogen peroxide in an inner chamber and the enzyme catalase in the outer chamber. When these substances are allowed to come into contact with one another, oxygen is liberated as a result of the mixing of catalase and hydrogen peroxide and the phenols are oxidized to quinones, producing the reaction described above (Eisner and Meinwald, 1966).

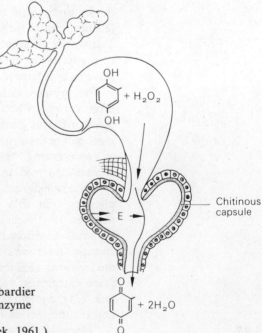

Fig. 7–9
Defensive gland of a bombardier beetle, *Brachinus* sp.; E. enzyme catalase. (Redrawn from Schildknecht and Holoubek, 1961.)

The papers by Roth and Eisner (1962) and Eisner and Meinwald (1966) contain tables listing the many specific chemicals and the arthropods in which they have been found. Besides functioning in defense, some of these chemicals are thought to have some antimicrobial effects. Some also act as "alarm" pheromones. For example, soldiers of certain termites discharge a spray that not only works against nest invaders but alerts other soldiers. Eisner (1970) is a good source of further information on the chemical defenses of arthropods.

Some insects are behavioral mimics of those which produce defensive secretions. For example, *Eleodes longicollis*, a species of tenebrionid beetle that possesses defensive glands, stands with its posterior well raised and discharges material from the glands. A second species, *Megasida obliterata*, mimics the stance of the first but lacks the defensive glands.

Not all defensive chemicals are directly associated with glands in the integument. Some are found in the hemolymph and impart a bad taste or render the insects indigestible. For example, the Monarch butterfly, *Danaus plexippus*, possesses such substances, called *cardenolides,* in its body tissues (Reichstein et al., 1968). They are derived from the characteristic food plants of this species. The Viceroy butterfly, *Limenitis archippus*, is a well-known mimic of the monarch and enjoys avoidance by "experienced" vertebrate predators even though it does not possess any distasteful substances.

Migration and Dispersal

Migration and dispersal are currently a matter of considerable controversy. The reviews by Johnson (1965, 1966, 1969) and Schneider (1962) are particularly pertinent and should give some insight into the nature of this controversy. We shall merely mention some of the highlights of this subject here.

Dispersal has been defined in a number of ways. For our purposes it can be looked upon as the tendency of a population or species to spread out spatially as the result of several mechanisms, probably the most important of which is migration. Other probable mechanisms include dispersal associated with the search for a mate or food, phoresy (a relationship in which one organism attaches to another and is nonparasitic but gains a means of dispersal), and responses to gradients of certain environmental factors (e.g., temperature, moisture, and CO_2). Many insects are carried for varying distances passively by the wind or by water currents. Johnson (1966) points out: "It is extremely difficult to measure how much dispersion occurs accidentally, incidentally, or adaptively."

Migration is a difficult term to define. Currently it is considered to be an active mass movement adapted to displace populations. It is characterized behaviorally "as an accentuation of locomotor function with a depression of vegetative function" (Johnson, 1966). In other words, migrating insects are persistent in their movement and do not typically respond to those stimuli associated with food, potential mate, and so on. Migration is accomplished mainly by flight, and the direction of displacement for many is determined by the wind. In migrant species the female sex always migrates and may or may not be accompanied by males. Migrant females are generally sexually immature, but there are definite exceptions to this.

Migrations of insects have been studied by a variety of techniques, including nets; suction traps; aerial balloons; aircraft; photography; marking with dyes, paint, radioisotopes; and genetic markers.

Johnson (1966) describes the following three types of migration based on the life span of the adult: (1) short-lived adults that emigrate and die within a season, (2) short-lived adults that emigrate and return, and (3) long-lived adults that hibernate or aestivate. Members of the first group usually leave the breeding site, oviposit elsewhere, and die. Examples of this type include locusts, termites, aphids, thrips, and butterflies. The distance covered is variable; in the locusts and butterflies it may be several hundred miles, in termites only a few feet or yards. The flights of many of these are windborne. Relatively short-lived adults, which emigrate and return, depart from the breeding site to feeding sites; the eggs mature at these sites and then the females fly back to the vicinity of their original breeding site and oviposit. This emigration and return may be repeated by the same individual. Many dragonflies behave in this fashion. Those insects in the third category fly to hibernation or aestivation sites and return to the original breeding site the following season. The monarch butterfly, *Danaus plexippus*, many Noctuid moths, and several beetles fall into this category.

The result of migration is the transference of a group of insects from an old to a new site. This is of particular importance when the "old" site was a temporary one in terms of its value to the survival of the insects. Temporary sites would include pools of water which tend to dry out, seasonal food plants, and so on. Schneider (1962) points out: "In the course of evolution, a low level of dispersive movement has been associated with the colonisation of permanent habitats and a high level has been closely correlated with the adoption of a temporary one." He further explains that this idea also applies to species in which the adults and larvae feed in quite different habitats (e.g., mosquitoes) and to those which hibernate or aestivate in habitats different from the breeding sites.

Johnson (1966) describes two current hypotheses that have been advanced to explain the causes of migrations. The first one suggests that insects respond to the onset of adverse conditions by flying away. The second explains migration on the basis of endocrine changes correlated with particular environmental effects such as crowding, food deficiency, and short days. For example, crowding in aphids results in the production of winged (alate) forms instead of the nonmigrant wingless (apterous) forms, and this seems to be associated in some way with activity of the corpus allatum. At present available evidence tends to favor the second hypothesis, but there appears to be no good reason why both could not be "correct" for different species or the same species under different conditions.

Gregarious, Subsocial, and Social Behavior

Seemingly all levels of intraspecific interaction, from completely solitary behavior to highly complex coordinated societies, are found among insects. However, comparatively few insect species are truly social, a few thousand perhaps, and are included in only two orders, Isoptera (termites) and Hymenoptera (ants, bees, and wasps). The societies of these insects are characterized by (1) being essentially permanent establishments (with certain exceptions); (2) mutual cooperation, the activities of each individual being subservient to the needs of the colony; (3) offspring that are a product of a single female (queen) or very few females but are tended by many; (4) the queen surviving over a period of several generations of her offspring; and (5) the individual members being entirely dependent upon the colony for their existence and possessing a definite attraction for one another. The evolution of mutual cooperation and an integration of activities has of course required the development of special means of communication among the individuals of a colony of social insects. The specific mechanisms of communication among social insects has been studied in the greatest detail in the honey bee *Apis mellifera*. This topic is discussed at considerable length by von Frisch (1950, 1967). Information regarding communication in social insects and some of the characteristics of their various societies will be discussed under the appropriate orders. Wilson (1971) gives an up-to-date, comprehensive treatment of insect societies.

Since social behavior centers on the care of the brood, such behavior on the part of other insects may be considered to be subsocial. A number of examples of this kind of behavior have been given in the section on "Reproductive Behavior." Other examples include the webspinners (order Embioptera) and *Geotrupes* spp. (dung beetles). Webspinners live in a series of chambers constructed from silk spun from glands in the forelegs of adults. The dung beetle male and female dig a burrow in the soil and store dung for the larva, which is tended for most of its life by the female.

Many aggregations of insects are not social or subsocial but have occurred as a result of a stimulus or stimuli attractive to the individuals of the aggregation. These commonly attractive stimuli include a particular temperature, light condition, food, particular chemicals, and humidity. In addition, there may be a degree of mutual attraction independent of external conditions. Examples of such aggregations are the hibernating groups of ladybird beetles, commonly found in human habitations during the winter months in temperate areas, and cockroaches, which are commonly found in feeding aggregations. These groupings, of course, lack most, if not all, of the traits that characterize subsocial and social behavior. This is not to imply that some mutual advantage is not afforded to the individuals of these aggregations. For instance, insects that produce defensive secretions and/or possess warning coloration may very definitely derive a certain amount of protection by amassing their defensive capabilities.

Periodicity in Behavior

Many of the activities of insects (e.g., locomotion, oviposition) show a recurrence at regular intervals. The term *periodicity* is appropriately applied to such recurrences of behavior. The patterns of recurrence of particular activities may be every 24 hours (diel), lunar, annual, or even longer. A diel interval is divided into a light period, a dawn and dusk period, and a dark period. When maximum behavioral activity occurs during the light, dawn and/or dusk, and dark periods the terms *diurnal*, *crepuscular*, and *nocturnal* are used, respectively. Periodicity in behavior is influenced by components from two basic sources (Corbet, 1966): endogenous, and exogenous. The endogenous components arise from within the insect and are ultimately related to its hereditary endowments. These include, in particular, rhythms, but the stage in a given physiological–behavioral cycle (e.g., gonotrophic cycle), developmental stage, and other endogenous factors may also play a role (Corbet, 1966). Exogenous components arise from the external environment.

Rhythms are periodicities that are controlled by an "innate time-measuring sense" or a "biological clock." They characteristically continue to occur when external conditions (temperature, light, and so on) are kept constant. Most rhythms which have been identified have a period of approximately 24 hours and are referred to as *circadian*. Several examples of circadian rhythms have been identified in insects. For example, cockroaches are nocturnally active; *Drosophila* emerge from the pupal stage at dawn; *Tettigonia* and other orthopterans sing at a particular time of day or night. A number of noncircadian rhythms are known [e.g., the lunar emergence of certain Chironomid flies and mayflies and the annual emergence of a species of dermestid beetles (Corbet, 1966)]. A particularly well-known example of noncircadian rhythmicity is found in the species of periodical cicadas (*Magicicada* spp.), the various broods of which emerge en masse at 13- or 17-year intervals.

A number of factors exert an influence on rhythmic patterns. Particularly significant are the external time cues (*Zeitgebers*), which set the phase of the rhythmic patterns of the individuals in a population. In a population under artificially constant environmental conditions (e.g., constant illumination) the rhythmic activities of the individuals may be out of phase with one another, but with the introduction of an appropriate time cue, they are set in phase and the activities of the individuals become synchronous. For example, mosquitoes (e.g., *Aedes aegypti*) held under constant illumination oviposit arrhythmically, but, when exposed to a dark period, even a very brief one, oviposit synchronously at 24-hour intervals thereafter (Corbet, 1966). This process of phase setting has been referred to as *entrainment* (Carthy, 1965). Under laboratory conditions, the phase of a rhythmic activity can be shifted about at the investigator's will. Among the exogenous factors that may act as time cues are length of the light or dark phase of a photoperiod ("a cycle consisting of a period of illumination followed by a period of relative darkness," Beck, 1968), light intensity, temperature, time at which food is available, and others. Length of the light or dark phases in a photoperiod have been shown to induce a state of hibernation called *diapause* in many insects (Beck, 1968). Temperature changes also exert an effect on diapause. In temperate zones, diapause may be looked upon as an adaptation to the march of seasons, an insect overwintering in this state. In tropical regions, diapause is considered to be an adaptation that favors survival during dry seasons.

Exogenous factors are known to exert an influence on rhythms. For example, in the dragonfly *Anax imperator*:

> Emergence (probably rhythmic) usually occurs abruptly soon after sunset; if, however, the air temperature falls below 10°C., after larvae have left the water but before ecdysis has begun, some of them will return to the water and emerge the next morning after temperatures have risen again. For these individuals the diel periodicity of emergence has been changed by a short-term response to unfavorable exogenous factors, but the phase-setting of their rhythms has remained unaltered (as evidenced by their attempting to emerge at the normal time).—Corbet, 1966.

Some investigators have maintained that rhythms are under the control of undetected exogenous factors rather than endogenous biological clocks. However, evidence favors the endogenous source, particularly the fact that the phase of a rhythm does not change even if the animal is transported to a place on the earth far from the locality of entrainment. In addition, two insects can be maintained on different light regimes in the same room, all other factors being equivalent (Carthy, 1965). Since rhythmic activities are found throughout the spectrum of living organisms (plant and animal), the general feeling is that the ultimate clock mechanism lies within cells, although its exact nature still remains a mystery.

Many external environmental parameters show a 24-hour periodicity, which may result in corresponding periodicity in insects. Among these are light intensity, moisture, temperature, wind velocity, and various possible biotic factors. An example of how one of these parameters may cause periodic behavior is in the swarming of many species of mosquitoes, which is determined by light intensity. These insects continually possess the "drive" to swarm, but do so only under appropriate conditions of light intensity and can be induced to swarm for hours if the light intensity is held artificially at the appropriate level. Under natural conditions, the appropriate light intensity may be periodic (e.g., that which occurs during crepuscular periods). In this situation the mosquitoes would swarm "periodically."

Selected References

GENERAL

Davis (1966); Dethier and Stellar (1964); Fraenkel and Gunn (1961); Hinde (1969); Johnsgard (1967); Maier and Schneirla (1935); Marler and Hamilton (1966); McGill (1965); Roe and Simpson (1958); Thorpe (1963); Wilson (1970).

INSECT BEHAVIOR

Alexander and Moore (1962); Alloway (1972); Askew (1971); Beck (1968); Birukow (1966); Blum (1969); Brues (1946); Buck (1948); Carthy (1958, 1965); Corbet (1966); Danilevsky, Goryshin, and Tyshchenko (1970); Dethier (1966); Eisenstein and Cohen (1966); Eisner (1970); Eisner and Meinwald (1966); Evans (1968); von Frisch (1950, 1967); Horridge (1962a, b); Jacobson (1965); Jander (1963); Johnson (1965, 1966, 1969); Lloyd (1971); Manning (1966); Markl and Lindauer (1965); McElroy (1964); Powell and Mackie (1966); Roeder (1963, 1965); Roth and Eisner (1962); Schneider (1962); Schneirla (1953); Wilson (1965, 1971).

Reproduction
and Morphogenesis

In this chapter we shall consider the systems and processes involved
with reproduction and will then discuss the major events, which begin
with the newly fertilized egg (the *zygote*) and terminate with the death
of the insect. The last subject dealt with is *diapause,* a phenomenon
with great ecological significance and hence one that facilitates a
smooth transition into Chapter 9, which deals with insect ecology.

Reproductive System and Gametogenesis

In this section we shall discuss the structure and function of the
male and female reproductive systems. Reproduction in most insects
is bisexual. The male reproductive system functions in the production,
storage, and delivery of spermatozoa; the female system produces and
stores eggs, stores spermatozoa, is the site of fertilization, and
deposits eggs or larvae when appropriate. As will be seen, the male
and female systems are similar in many respects.

The reviews of Davey (1965), DeWilde (1964a, b), Highnam
(1964), Engelmann (1970), and Wigglesworth (1965) are all quite useful
as sources of information regarding reproduction.

Male Reproductive System, Spermatogenesis,
and Spermatozoa

The male reproductive system (Fig. 8–1) is located in the posterior
portion of the abdomen and typically consists of paired gonads
(*testes*) connected by various ducts, which ultimately open into the
intromittent organ, the *penis* or *aedeagus. Accessory glands* of various
sorts are usually associated with the ducts. Although fundamentally
similar, there are many variations of this system among the different
species of insects. The following description is of a generalized male
reproductive system.

Testes. Although the testes are usually bilateral, paired structures, they have undergone a medial fusion in some insects (e.g.,
Lepidoptera). Basically each testis is composed of a varying number
of *testicular follicles* (sperm tubes), which are usually encased by a
layer of connective tissue. Each follicle is in turn enclosed by a layer
of epithelial cells which are thought to serve a trophic (nutrient)

function, absorbing nutrients from the hemolymph and making them available to the germ cells within.

Ducts. Tiny ducts, the *vas efferens*, lead from each follicle to a common lateral duct, the *vas deferens*, and finally the vas deferens from each testis join to form the *ejaculatory duct* which ends in the penis or aedeagus at the *gonopore*. The vas deferens are of ectodermal origin and are hence lined with cuticle. In some insects (Protura, Ephemeroptera, and some Dermaptera) each vas deferens has its own opening to the exterior (Snodgrass, 1935). Both the vas deferens and ejaculatory duct are invested with a layer of muscles which are involved in the propulsion of semen. Each vas deferens may be thrown into a series of convolutions forming an *epididymis* or have a dilated portion (i.e., *seminal vesicle*) in which the spermatozoa are stored in a quiescent state.

Accessory Glands. Various glandular structures are associated with the vas deferens or ejaculatory duct. These accessory glands and their ducts are either of mesodermal or of ectodermal origin and usually occur as a single pair (Snodgrass, 1935). However, in some insects they may be several in number and form a cluster (e.g., in male cockroaches, the cluster being referred to as the mushroom body). In addition, in some insects portions of the vas deferens or ejaculatory duct may also have glandular functions (Wigglesworth, 1965). One of the known functions of male accessory glands is the secretion of seminal fluid.

Spermatogenesis. The testicular follicles contain the *germ cells* and are hence the sites of the reduction divisions (meiosis) which eventually give rise to *spermatozoa*, the entire process being referred

Fig. 8–1
Generalized male reproductive system. A. Principal male organs. B. Detailed structure of a testis. C. Section of a testis. (Redrawn from Snodgrass, 1935.)

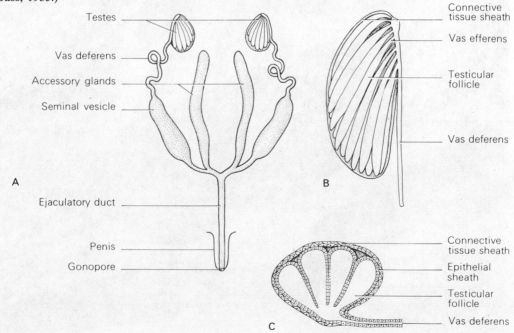

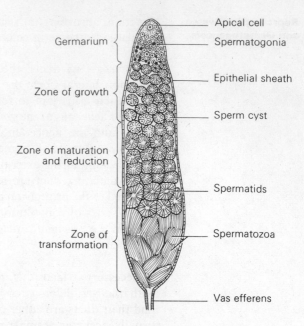

Germarium {
Apical cell

Spermatogonia

Zone of growth
Epithelial sheath

Sperm cyst

Zone of maturation
and reduction

Spermatids

Zone of
transformation
Spermatozoa

Vas efferens

Fig. 8–2
Section of a testicular follicle (diagrammatic).
(Redrawn from Snodgrass, 1935.)

Fig. 8–3
Photomicrograph of a sagittal section of a testis in a male mosquito,
Aedes triseriatus. 1, Germarium; 2, zone of growth; 3, zone of maturation
and reduction; 4, zone of transformation; 5, spermatozoa; 6, epithelial
sheath; 7, vas efferens. (Scale line = 50 microns.)

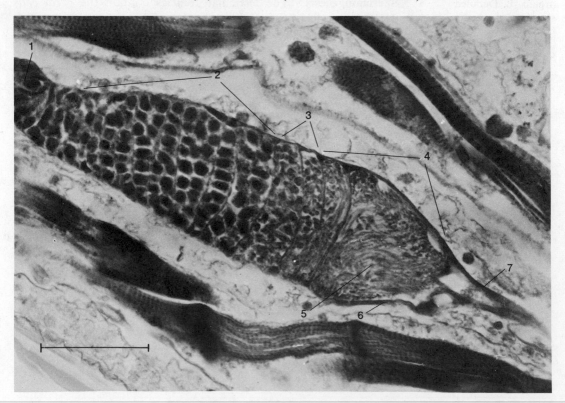

to as *spermatogenesis*. This process usually occurs during the last larval instar or pupal stage and in some species continues in the adult
stage. In most insects each follicle contains a large apical cell, while
some others have an apical complex of cells which apparently serve
a trophic function, providing nutrients for the developing *spermatogonia*. Each follicle is divided apically to basally into zones that
represent the different stages of spermatogenesis (Figs. 8–2 and 8–3)
(Wigglesworth, 1965; DeWilde, 1964a). Apically they begin with the
germarium or *zone of spermatogonia*, which is comprised of the germ
cells (spermatogonia) and somatic mesodermal cells. In the next
region, the *zone of growth* or *zone of spermatocytes*, the spermatogonia
undergo several mitotic divisions, forming *spermatocytes* which
become encysted in somatic cells. These diploid cells undergo the
reduction divisions of meiosis and produce haploid daughter cells
in the next region, the *zone of maturation and reduction*. With the first
and second reduction divisions the spermatocytes become *primary*
and *secondary spermatids*, respectively. In the basal *zone of transformation*, the secondary spermatids become transformed into flagellated
spermatozoa. When the cysts in which they have remained encased
throughout spermatogenesis rupture, the spermatozoa enter the
vas efferens and vas deferens. Commonly, when the spermatozoa
are released into the ducts they remain in bundles (*spermatodesms*)
held together by gelatinous material (Wigglesworth, 1965). The
spermatozoa of most insects studied are filamentous with poorly
developed "head" regions. Breland et al. (1968) present a diagram
that represents a more or less generalized, hypothetical insect spermatozoon (Fig. 8–4).

The movement of spermatozoa within the male reproductive
system are not due to their inherent mobility but to contractions
of the muscles associated with the vas deferens and ejaculatory duct.
This is, in effect, a mechanism that subserves the conservation of the
energy of the spermatozoa (Hinton, 1964).

Spermatophores. In some insects the spermatozoa are produced
and released in specialized packets held together by proteinous
secretions of the accessory glands. These packets are called *spermatophores* and often assume rather distinct forms (Fig. 8–5). Spermatophores are common in the lower orders (e.g., the apterygotes) and
rare or absent in some of the higher orders, such as Hymenoptera.

Fig. 8–4
Generalized insect spermatozoon. (Redrawn from Breland, Eddleman,
and Biesele, 1968.)

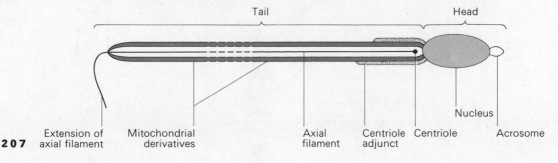

Tail   Head

Extension of   Mitochondrial   Axial   Centriole Centriole   Acrosome
axial filament   derivatives   filament   adjunct   Nucleus

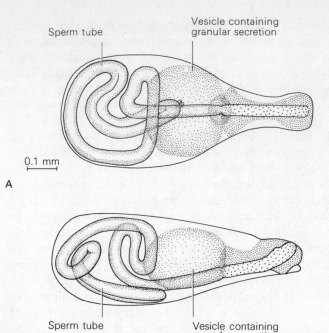

Sperm tube

Vesicle containing
granular secretion

Fig. 8–5
Spermatophore from a male silverfish,
Lepisma saccharina. (Redrawn from
Sturm, 1956.)

0.1 mm

A

Sperm tube

Vesicle containing
granular secretion

B

Female Reproductive System, Oogenesis, and Ova

Like the male reproductive system, the female system (Fig. 8–6)
is located in the posterior part of the abdomen and typically consists
of paired gonads (*ovaries*) connected by a series of tubes to the *vagina,*
which when present, opens to the exterior and receives the male
intromittent organ during copulation. Likewise, there are a variety
of *accessory glands* present, and although the female reproductive
system is basically similar between the different species, there is

Fig. 8–6
Generalized female
reproductive system.
A. Principal female
organs. B. Single
ovariole. (Redrawn
from Snodgrass, 1935.)

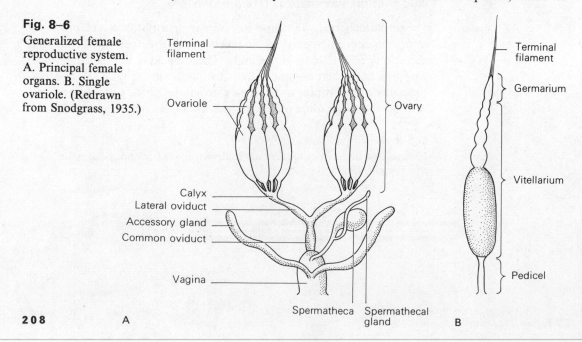

Terminal
filament

Ovariole

Ovary

Calyx
Lateral oviduct
Accessory gland
Common oviduct

Vagina

Spermatheca   Spermathecal
gland

Terminal
filament

Germarium

Vitellarium

Pedicel

A   B

considerable variation. The system described below will be a generalized one with comments on some variations.

Ovaries.   The ovaries are bilaterally located, mesodermal organs which produce eggs. They are composed of a number of functional units, *ovarioles*, which are invested in a layer of epithelial cells. This layer in turn is musculated and has tracheae associated with it.

The number of ovarioles per ovary varies greatly, from 1 in the tsetse flies, *Glossina* spp., and some aphids, to over 2,000 in the queens of certain termite species. A terminal thread from the cephalad portion of each ovariole joins those of its neighbors, forming a *terminal filament* which attaches to the dorsal diaphragm (Davey, 1965).

Ducts.   At the base of each ovariole is a small duct or *pedicel,* which joins those of the other ovarioles in a bulbous *calyx*, which in turn opens into the *lateral oviduct*. The lateral oviducts, which are, like the ovaries, of mesodermal origin, join to form the common oviduct. The common oviduct serves as a communicating tube between the lateral oviducts and the *bursa copulatrix* or *vagina*, which opens to the outside. The bursa copulatrix (Fig. 8–7) is a saclike expansion of the vaginal region. The common oviduct and bursa or vagina are of ectodermal origin and hence are lined with cuticle. There is usually a single outpocketing from the bursa, vagina, or common oviduct in which spermatozoa which have been introduced as a result of copulation are stored prior to fertilization. This outpocketing is called the *spermatheca*, and in some insects (e.g., certain flies) it is a paired structure.

Accessory Glands.   There are generally one or two pairs of these structures, which usually open into the apical portion of the bursa copulatrix. These glands are quite variable in structure and function and are normally involved with the secretion of materials that are adhesive (*colleterial glands*) and serve to cement eggs to the substratum, or materials that coat the surfaces of eggs or hold them together in masses. In cockroaches, secretions of the accessory glands

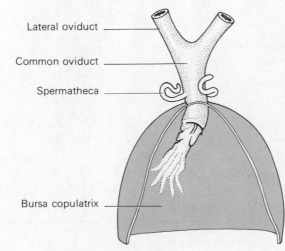

Lateral oviduct

Common oviduct

Spermatheca

Bursa copulatrix

Fig. 8–7
Ducts of a female triatomid bug, *Rhodnius* sp. (Hemiptera; Reduviidae), showing opened bursa copulatrix and entrance to the common oviduct (ventral view). (Redrawn from Davey, 1965.)

form a capsule or *ootheca* around eggs that have accumulated in the bursa copulatrix.

Oogenesis. Under the heading oogenesis we include all those processes which ultimately lead to the development of a mature ovum, capable of being fertilized. Oogenesis may be complete prior to or during the imaginal stage. Like the testicular follicles of the male, the ovarioles are divided into zones which contain germ cells or oocytes in various stages of development and maturation (Figs. 8–6B and 8–8). Basically there are two such zones, the apical *germarium* and the basal *vitellarium* both of which are invested in an outer layer of cells, the *ovariole sheath*. The germarium contains the primary female germ cells, the *oogonia*, which undergo successive mitotic divisions becoming *primary oocytes*. The germarium also contains prefollicular tissue, which comes to form the *follicular epithelium* in the vitellarium. The vitellarium is comprised of oocytes which are undergoing the deposition of nutrients, a process referred to as *vitellogenesis* (yolk deposition, Fig. 8–9). Each developing oocyte is surrounded by a follicular epithelium, and the oocyte and its associated epithelial layer comprise a *follicle*. The oocytes in the vitellarium have progressively more yolk in an apical to basal sequence, the most mature basal oocyte being separated from the lumen of the pedicel by an epithelial plug which ruptures when the oocyte is ready to leave the ovariole and proceed into the lateral oviduct. This process of exiting from the ovariole is called *ovulation*. Following ovulation, the follicular epithelial cells remain behind and eventually degenerate.

Fig. 8–8

Ovariole types. A. Polytrophic. B. Telotrophic. C. Panoistic.

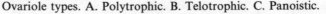
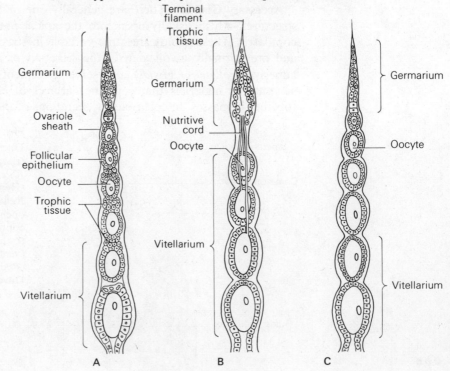

There are three types of ovarioles, based on the method by which yolk deposition occurs (Fig. 8–8): *polytrophic*, *telotrophic*, and *panoistic*. Polytrophic ovarioles have nutritive cells or *trophocytes* associated with each developing oocyte. These trophocytes are within the layer of cells, the follicular epithelium, which surrounds each oocyte. A follicular plug exists between each oocyte with its accompanying trophocytes. Polytrophic ovarioles have been described in Neuroptera, Lepidoptera, some Coleoptera, Diptera, and Hymenoptera. Telotrophic ovarioles are different from polytrophic ones in that the trophocytes are not directly associated with each developing oocyte but are all located in the apical region of the ovariole and are connected to the various oocytes by means of "nutritive cords." Otherwise, they are similar to the polytrophic types, each oocyte being surrounded by a follicular epithelium and being separated from the others by follicular plugs. Telotrophic ovarioles are characteristic of Hemiptera and some Coleoptera. Polytrophic and telotrophic ovarioles are sometimes collectively referred to as *meroistic ovarioles*.

The third type of ovariole listed, panoistic, has no trophocytes. However, each developing oocyte is surrounded by the follicular epithelium, and follicular plugs exist between adjacent oocytes. Panoistic ovarioles are found in the apterygotes, Orthoptera, Isoptera, Odonata, Plecoptera (stoneflies), Siphonaptera, and some Coleoptera. Recently, a fourth type of ovariole, *adenotrophic,* has been found in one species of beetle, *Steraspis speciosa*. In this type of ovariole, the germarium and vitellarium are very short, there are no nutritive cords or trophocytes, and there is a long region of glandular tissue proximal to the vitellarium (Engelmann, 1970).

In the vast majority of insects oogenesis occurs in the last larval instar, the pupa, and the adult stages. However, in some species immature stages are capable of producing mature oocytes and in some embryogenesis may commence and run its full course in the immature parent. This is the phenomenon of *pedogenesis* and has been described in a number of insect species. Among these are *Micromalthus debilis*, a strepsipteran, and cecidomyiid flies in the genera *Miastor*, *Oligarces*, and *Tanytarsus* (DeWilde, 1964a).

Eggs. Mature insect eggs are typically elongate and oval in longitudinal section (Fig. 8–10), although some assume other forms (Fig. 8–11). In most instances the largest portion of an egg is filled with *yolk* or *deutoplasm* and the cytoplasm and nucleus occupy only a small portion. The yolk contains carbohydrates, and protein and lipid bodies, the protein bodies being the most abundant. Cytoplasm is located around the nucleus (*nuclear cytoplasm*) and around the periphery of the yolk (*periplasm* or *cortical cytoplasm*). The egg may be encased in two membranes: the *vitelline membrane,* which apparently represents the cell wall of the egg, and the *chorion*, or eggshell, which is secreted by the follicle cells. Several insectan species lack a chorion; when present, it is nonchitinous, may be composed of two to several layers, varies considerably in thickness in different species, and may be smooth or sculptured in a variety of ways.

Since insect eggs are usually deposited outside the parent female, they face the same problems (e.g., water conservation) which plague

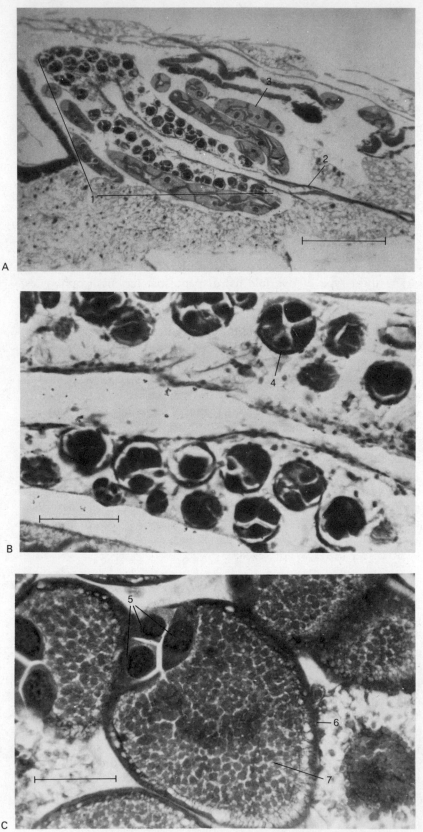

Fig. 8–9
Photomicrographs of
vitellogenesis in the
mosquito *Culex
nigripalpus*.
A. Sagittal section of
ovary in an unfed
female. B. Same as A,
higher magnification.
C. Oocyte, 48 hours
following a blood meal.
D. Oocyte, 96 hours
following a blood meal.
1, Ovary; 2, lateral
oviduct; 3, Malpighian
tubule; 4, follicle;
5, nurse cells; 6,
6, follicular spithelium;
7, yolk; 8, chorion.
(Scale lines: A, 250
microns, B–D, 50
microns.) (Tissues
prepared by Elaine
Cody.)

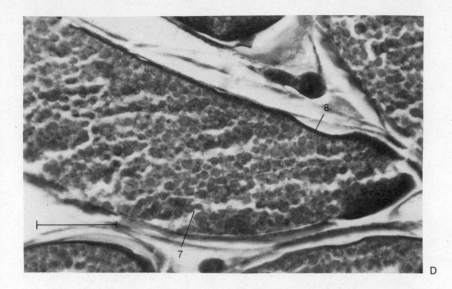

D

any terrestrial organism. The chorion serves the same functions as the cuticle of later stages. It serves as a protective coating, a barrier against water loss, and is intimately associated with the ventilation of the egg. Coatings secreted by the accessory glands which cover eggs may aid in water conservation in addition to serving to hold eggs together or glue them to the substratum. Some eggs laid in moist situations are capable of absorbing water from their surroundings. In insects in which the chorion is thin, ventillation may occur over the entire surface. In others the chorion may be lined with a porous, gas-filled layer which communicates with the outside of the egg by means of channels or *aeropyles* (Davey, 1965). Some eggs possess specialized structures which act as physical gills or plastrons, allowing them to obtain oxygen from water. Hinton (1969) reviews and synthesizes the major literature pertinent to respiratory systems of insect eggs. Spermatozoa gain entrance to an egg by means of one to several special channels or *micropyles,* which are perforations of the chorion and are located at various places on the eggs of different species.

Control of Oogenesis. It has been established for a number of species that the development of eggs, vitellogenesis in particular, is controlled by a secretion of the corpora allata. Some of the kinds of experiments that have enabled investigators to arrive at this conclusion

Fig. 8–10
Sagittal section of a typical insect egg. (Redrawn with modifications from Hagen, 1951.)

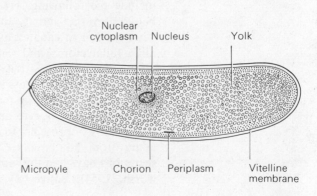

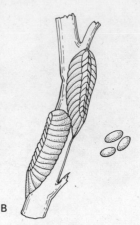

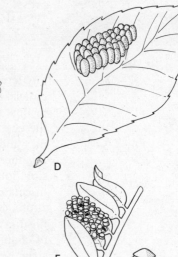

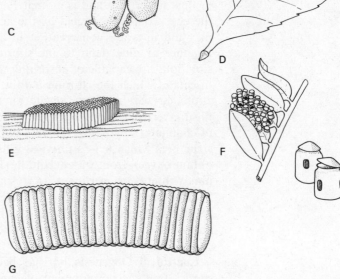

Fig. 8–11
Representative insect eggs.
A. California green lacewing,
Chrysopa californica. B. Egg
cases (oothecae) and eggs of a
praying mantid. C. Eggs and egg
masses of the mayfly,
Ephemerella rotunda. D. Eggs of
the convergent ladybird beetle.
E. Egg raft of the mosquito
Culex pipiens. F. Eggs and egg
masses of the stink bug,
Pentatoma lignata. G. Ootheca
of the German cockroach,
Blattella germanica. (Redrawn
from CCM: General Biological
Inc., key card.)

include the following: (1) histological observation of a correlation
between ovarian activity and activity in the corpora allata, (2) ovarian
activity being attenuated by the microsurgical removal of the corpora
allata (allatectomy) and restored by implantation, and (3) allatectomy,
attenuating ovarian activity and hemolymph transfusion from a donor
with active corpora allata restoring this activity. The corpus allatum
hormone may affect egg maturation by stimulating the incorporation
of yolk into the oocyte and simultaneously regulating metabolism,
particularly of proteins. In certain species (e.g., *Rhodnius* sp.) corpora
allata from adults have been transplanted into larvae and their
secretions have had the same effect as the juvenile hormone (Wigglesworth, 1954); conversely, larval corpora allata stimulate ovarian

activity (gonadotropic effect) in adults. This has lead many to hold the view that juvenile hormone and the gonadotropic hormone may be one and the same.

Recently, certain substances (e.g., farnesol and farnesyl methyl ether) have been found which mimic the effects of the corpus allatum hormone, acting as "juvenile" hormone and having a gonadotropic action. In some species of insects allatectomy does not prevent egg maturation, but there is evidence that the corpus allatum hormone may be present in the hemolymph from earlier stages. In several species, secretions from specialized neurosecretory cells in the pars intercerebralis in the brain are necessary for the activity of the corpora allata. Other factors that have been found to influence the activity of the corpora allata in different species include mating, light (photoperiod), chemical stimulation (pheromones from males), various nutritional factors, and the presence or absence of eggs in the brood chamber. There is, no doubt, considerable interaction of several factors operating in the control of oogenesis in any given insect species, and the level of understanding is not yet to the point where broad generalizations can be made. The reviews of Davey (1965), DeWilde (1964b), Engelmann (1968), and Wigglesworth (1965) should be consulted for extensive bibliographies and more detailed expositions of this topic.

Seminal Transfer, Fertilization, and Sex Determination

Seminal Transfer

Internal fertilization is looked upon as one of the several prerequisites that needed to be satisfied before animals could radiate into the terrestrial environment (Davey, 1965). Internal fertilization in insects is generally brought about by the act of copulation, during which time the semen (spermatozoa plus various glandular secretions) produced in the male reproductive system is transferred to an appropriate site in the female reproductive system. The external structures and behavioral patterns which govern their use were discussed in Chapters 2 and 7, respectively. Suffice it to say at this point that both these aspects show considerable variation from species to species, although their common underlying "goal," seminal transfer, is the same. Seminal transfer may involve the passage of either free semen or in many insects one or more spermatophores from the male to the female. The involvement of a spermatophore is considered to be the more primitive situation (Hinton, 1964).

Free semen is usually deposited in the bursa copulatrix or vagina, but in some species it may be deposited in the common oviduct, lateral oviducts, or even directly into the spermatheca (DeWilde, 1964a). Male dragonflies and damselflies deposit semen in the specialized organ on the venter of the second abdominal sternite. A portion of that organ is then placed in the female's vagina and seminal transfer accomplished. Many species in the superfamily Cimicoidea (e.g., bed bugs; family Cimicidae) have a rather bizarre method of seminal transfer. The males of these species inseminate the females by perforating the integument at a specialized site in their abdomen and ejaculating semen directly into the hemocoel. This method of semen introduction has been called *hemocoelic* or *traumatic insemination*. The seminal fluid and many of the spermatozoa are digested by the female, while some of the spermatozoa reach the ovaries. The adaptive value of this method of insemination is thought to be that it provides the female with additional nutritive substance (Hinton, 1964).

Spermatophores are usually deposited by the male somewhere in the female reproductive system: the bursa copulatrix, vagina, or rarely the spermatheca. However, in the apterygotes (e.g., Thysanura), the male deposits a spermatophore on the substrate and the female picks it up and deposits it within herself. Once a spermatophore is in the female reproductive system, various mechanisms may account for the release of the spermatozoa. In most insects the spermatozoa are either forced out by pressure applied by the female or by the mechanical perforation of the spermatophore. In the house cricket an osmotic-pressure mechanism is involved where there is a significant difference between the osmotic pressure of a block of protein or pressure body within the spermatophore and the osmotic pressure of the evacuating fluid, which surrounds the spermatophore at the time of ejaculation. This pressure difference results in the absorption of evacuating fluid by the protein body, which swells up and forces the exit of the spermatozoa (Hinton, 1964). In some insects the spermatophore may be digested away, causing the release of the spermatozoa. After the spermatozoa have been freed from the spermatophore, it is in most cases probably digested and absorbed by the female and thus may have some nutritional significance.

After release from the spermatophore or deposition in the form of free semen, the spermatozoa ultimately appear in the spermatheca. This is apparently due to muscular contractions of the female ducts but may also have something to do with sperm motility. In the honey bee the spermathecal duct has a "pumping" structure, which seemingly transports the sperm to the receptacle following copulation and subsequently releases them when it is time for an egg to be fertilized.

Fertilization

The processes involved in fertilization may be resolved into three parts: (1) release of spermatozoa from the spermatheca, (2) entry of the egg by spermatozoa, and (3) formation and fusion of the male and female pronuclei.

As mentioned earlier, spermatozoa are stored in the spermatheca of the female until it is time for fertilization. The females of many insect species (e.g., *Apis, Glossina, Rhodnius,* and *Triatoma*; Wigglesworth, 1965) mate only at a single time during their lives with one or more males, and the spermatozoa introduced at that time are stored and used to fertilize their eggs for the rest of the reproductive period. Spermatozoa stored in this fashion may survive for several months. Other insect species are inseminated periodically throughout their reproductive lives, and in these species the storage of spermatozoa in the spermatheca may only be for a short period of time. The mechanisms involved with the release of spermatozoa from the spermatheca are not clearly understood. Stimulation of sensory hairs by the passage of eggs in the oviducts, heomcoelic pressure forcing the exit of the sperm, and the inherent mobility of spermatozoa which have been "activated" by some secretion have all been advanced as mechanisms in various species.

Following ovulation, the egg is oriented in the reproductive passage in such a way that the micropylar region is in rough proximity to the site of sperm release. The sperm then enter the egg through the micropyle, having demonstrated what is considered by some to be a chemotactic response relative to the micropylar region of the egg (Wigglesworth, 1965). In the vast majority of insects more than one spermatozoan enters the egg but usually only one fuses with the egg pronucleus. Excess sperm usually degenerate without disrupting the development of the zygote.

Shortly following the entry of sperm into the egg, the egg nucleus undergoes the reduction divisions of meiosis, forming the female pronucleus (see Fig. 8–13A). The spermatozoan that will fuse with this female pronucleus loses its tail, becoming the male pronucleus. The two pronuclei fuse, forming the zygote, which signals the commencement of morphogenesis.

Sex Determination and Parthenogenesis

Nearly all insects are bisexual and the sex of a given individual is determined at fertilization. The majority of species that are not bisexual are *parthenogenic* (i.e., produce individuals from unfertilized eggs). Hermaphroditic species, those in which individuals possess both male and female germ tissues (e.g., certain species of scale insects), are extremely rare. Sex determination in bisexual insects is considered to depend upon a balance between genes for maleness and genes for femaleness. This balance is in most forms tipped in the direction of one sex or the other by a sex-chromosome mechanism in which one sex possesses two X (sex) chromosomes (i.e., the homogametic sex, XX) and the other a single X chromosome (XO) or a single X chromosome plus a differently appearing Y chromosome (XY). Individuals possessing the XO or XY configurations comprise the heterogametic sex. In most insectan groups the males are heterogametic and the females homogametic. However, the reverse is true in the Lepidoptera and Trichoptera. In some insects (e.g., Hymenoptera, Thysanoptera, and certain Homopterous insects) the male sex is determined by being haploid. White (1964) discusses sex determination in insects in some detail.

In some instances environmental factors have been shown to exert an influence on sex determination. For example, in the butterfly *Talaeporia* sp., more males than females are produced at high temperatures; the converse is true at low temperatures (Davey, 1965).

Since extirpation of gonads or implantation of gonads into a previously castrated individual produce no effects on individuals, it is generally considered that secondary sexual characters are not determined by any humoral secretions associated with the gonads (Wigglesworth, 1965). Apparently every cell in the insect body is involved in sex determination, as is particularly well evidenced by the occasional occurrence of *gynandromorphs*. These individuals are literally sexual mosaics, some parts of the body possessing typically male traits and other parts typically female traits (Fig. 8–12). This phenomenon is explained by differences in the sex chromosomes in the cells comprising the various tissues (i.e., some cells are "male" and others are "female"). These differences in sex chromosomes in cells of the same individual are known to occur by a number of mechanisms. One such mechanism is the loss of one X chromosome in the cleavage cells of a female (XX becoming XO, e.g., in *Drosophila*). These cells thereafter give rise to male traits in whatever tissue they happen to form (Wigglesworth, 1965). Another mechanism occurs in the honey bee, in which the fusion nucleus and an extra sperm that has entered the egg both undergo cleavage. The cells that result from cleavage of the fusion nucleus (diploid) give rise to female traits, and those from the sperm nucleus (haploid) produce male traits (DeWilde, 1964a). In addition to gynandromorphs, individuals called *intersexes* are sometimes produced. These are not mosaics but have characteristics that place them somewhere between male and female. This phenomenon is due to an imbalance of the sex-determining genes and in some instances is a function of the environment, particularly with regard to temperature.

As mentioned previously, parthenogenesis is the production of individuals from unfertilized eggs. Some species reproduce solely by parthenogenesis (obligate), and males of these species are not known to exist (e.g., certain stick insects and certain aphids; Davey, 1965). In other species, parthenogenetic reproduction is facultative (i.e., eggs can develop in the absence of fertilization, at least under certain circumstances). Parthenogenetic offspring may be solely male (arrhenotoky), solely females (thelytoky), or a mixture of males and females (amphitoky). In the first type just listed, males and females develop from unfertilized and fertilized eggs, respectively (e.g., honey bees); in the second type, no fertilization occurs at all and males are rare or nonexistent (e.g., a few members of several orders). An example of an insect with amphitokous parthenogenesis is the wasp *Habrobracon* sp. (Wigglesworth, 1965). In this species males are haploid and females are diploid. Uniparental females develop when eggs arise from tetraploid cells among the diploid cells of an ovary. These tetraploid eggs become diploid when meiotic division occurs and hence produce females, even though they are unfertilized. White (1964) suggests that the capacity for parthenogenesis is universal, since in nearly all bisexual insect species studied thus far, a few eggs laid by a virgin female complete embryogenesis and hatch. White terms this phenomenon *tychoparthenogenesis*. In some instances parthenogenetic generations alternate with bisexual generations. Such alternation of generations occurs, for example, in species of Cecidomyid flies (*Miastor* sp. and *Oligarces* sp.; Wigglesworth, 1965) and aphids.

Pre-eclosion Morphogenesis

Morphogenesis may be defined as all the morphological changes that occur during the development of an animal (insect) to its mature condition (Clements, 1963). Under the heading "pre-eclosion morpho-

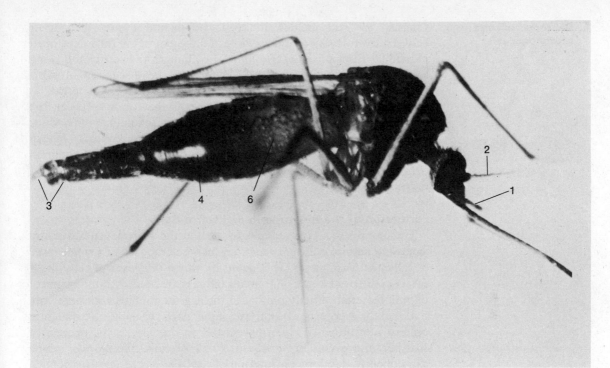

A

Fig. 8–12

Example of gynandromorphy in a mosquito, *Culex salinarius*. A. Wild-caught gynandromorph; the antennae and palpi are characteristic of a female; the terminalia are characteristic of a male. B. Normal female. Both specimens have taken a blood meal, a female trait, 48 hours earlier. 1, Palp; 2, antenna; 3, terminalia; 4, blood in midgut; 5, developing ovary; 6, bubbles in ventral diverticulum (see Fig. 3–2). (Courtesy of J. D. Edman.)

B

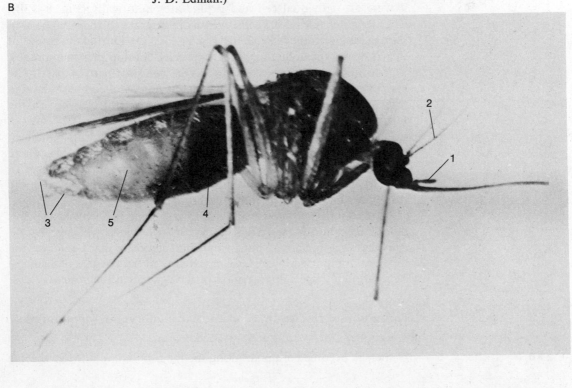

genesis" we will include those developmental events that occur between the formation of the zygote and the exit of the fully developed individual from the egg (*eclosion*). The changes that occur following hatching from the egg will be included in the discussion of post-eclosion morphogenesis. Thus morphogenesis, in this context, is taken to comprise all developmental events that occur between the formation of the zygote and the emergence of a sexually mature adult. A general description of ovarian embryogenesis follows. Hagen (1951), Johannsen and Butt (1941), Sharov (1966), and Snodgrass (1935) are recommended for further readings in the area of insect morphogenesis.

Formation of the Blastoderm and Germ Cells

The first distinct layer of cells to form in the morphogenesis of any metazoan animal is the *blastoderm,* which is composed of a single layer of cells, the *blastomeres.* The means by which this layer of cells arises varies in different kinds of animals and is correlated with the quantity of yolk material initially present in the egg. In animals with little yolk material the zygote divides into two equal parts; each of these daughter cells divides into two equal parts; and so on. This is the process of *cleavage* and eventually leads to a ball of cells, the *morula,* which subsequently develops an internal cavity, the *blastocoel,* surrounded by the blastoderm. Since in this method of blastoderm development, the entire zygote divides, the term *holoblastic* (holo = whole; blast = bud or sprout) cleavage is applied. However, in the vast majority of insects, Collembola (springtails) and certain parasitic Hymenoptera being the most outstanding exceptions, the eggs have large quantities of yolk and hence cleavage of the entire egg is inappropriate. In this group the fusion nucleus and associated cytoplasm behaves as though it were an individual cell and proliferates mitotically (Fig. 8–13B). The daughter (cleavage) cells eventually migrate to the periphery of the egg and form the *blastoderm* (Fig. 8–13C, D). During the course of this process, the individual cleavage cells develop cell membranes. This form of cleavage is referred to as *meroblastic* (mero = part). The eggs of some arthropods and a few insects (e.g., Collembola) undergo total cleavage initially and subsequently peripheral cleavage in the formation of the blastoderm. This is called *combination cleavage.*

During meroblastic cleavage, some of the cleavage cells remain in the yolk or return to it after reaching the periphery of the egg. These are the *vitellophages* (yolk cells; Fig. 8–13C, D) and are considered to be responsible for the initial digestion of the yolk, making it more readily assimilable by the other embryonic cells.

At the same time the blastoderm forms, some of the cleavage cells differentiate into *germ cells* (Fig. 8–13C–E), which will give rise to gametes in the late larval, pupal, and adult stages. In many embryos the differentiation of germ cells is correlated with the passage of cleavage cells through a specialized region of the egg called the *oosome.*

Formation of the Germ Band and Extraembryonic Membranes

Following the completion of the blastoderm, the cells on one side of the egg become columnar along the longitudinal midline of the egg

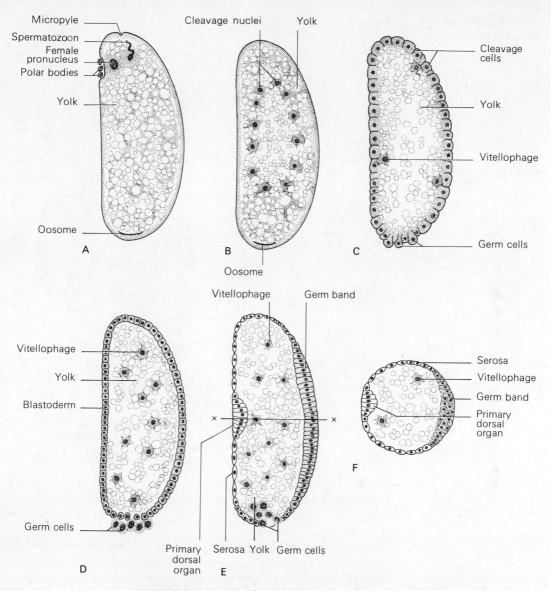

Fig. 8–13

Fertilization—germ-band formation (chorion omitted; A–E, sagittal sections). A. Just prior to fertilization (i.e., fusion of spermatozoon and female pronucleus). B. Cleavage. C. Blastoderm forming and germ cells differentiating. D. Blastoderm formation complete. E. Germ band and primary dorsal organ formation. F. Cross section at same stage as E. The location of the section is indicated in E by the line x—x. (Redrawn from Johannsen and Butt, 1941.)

(Fig. 8–13E, F). In a lateral direction from this midline, the cells become successively less columnar, finally merging with the remaining cells of the blastoderm, which tend to become squamous. This thickened area of columnar cells of the blastoderm is the *germ band* and subsequently elongates and develops into the embryo, while the remaining cells take part in the formation of the extraembryonic membranes. In most insects folds from the area outside the germ band grow over the germ band (i.e., overgrowth; Fig. 8–14A),

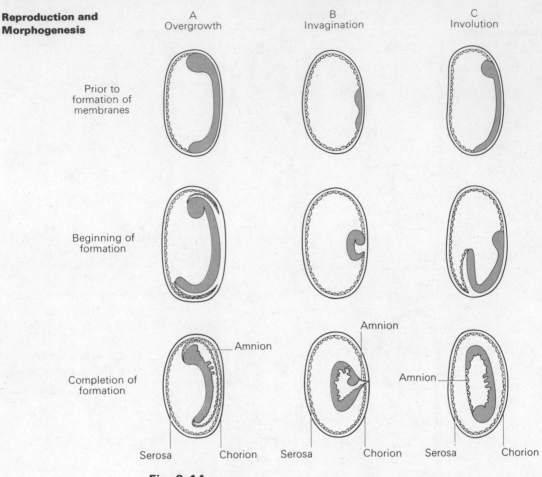

A
Overgrowth

B
Invagination

C
Involution

Prior to
formation of
membranes

Beginning of
formation

Completion of
formation

Amnion

Amnion

Amnion

Serosa   Chorion   Serosa   Chorion   Serosa   Chorion

Fig. 8–14
Formation of extraembryonic membranes (diagrammatic).

eventually meeting along the longitudinal midline. The inner and outer
layers of one fold then merge with the respective layers of the other,
forming an inner *amnion* around the developing embryo and an outer
serosa around the yolk, amnion, and embryo. In some insects the
extraembryonic membranes may form by invagination or involution
of the embryo instead of overgrowth (Fig. 8–14B, C). The first method
is found in the Apterygota; the second occurs among the orders
Odonata, some Orthoptera, and Homoptera (Snodgrass, 1935). As
the extraembryonic membranes are forming, the germ cells become
located at what will be the posterior end of the embryo.

As the germ band forms in some insects, particularly in Apterygota,
but also in some Pterygota, a cluster of cells form the primary dorsal
organ (Fig. 8–13E, F). This structure disappears at dorsal closure
and may be glandular in function.

Differentiation of the Germ Layers

In animals with little yolk material and in which a blastocoel
develops, an invagination eventually occurs which subsequently

differentiates into well-defined mesodermal and endodermal layers. These three germ layers (*ectoderm*, *mesoderm*, and *endoderm*) then differentiate into the various tissues and organs of the fully developed organism. The formation of the mesoderm and endoderm is referred to as *gastrulation*. Unfortunately this process in insects is not straight-forward and clear. Unequivocal interpretation of insect morphogenesis in terms of this *germ-layer theory* has proved to be extremely difficult; insect embryologists hold widely divergent opinions. Fox and Fox (1964) point out that there is much to be said for not attempting to distinguish between mesoderm and endoderm in the insect embryo but for simply recognizing an outer ectoderm and inner layers. For convenience of discussion, we shall label those cells which give rise to the midgut as endodermal in origin, keeping the aforementioned qualifications in mind.

Gastrulation in most insects occurs as the amnion and serosa are forming. It begins, in most species, as a longitudinal, furrowlike invagination (Fig. 8–15A–C) which runs most of the length of the venter of the germ band. Eventually the invagination flattens out and the outer edges come together and fuse, forming a longitudinal band (*inner layer* or *mesentoderm*) of cells surrounded by an outer layer, which at this point is appropriately referred to as *ectoderm*. Another type of inner layer formation (Fig. 8–15D, E) consists of a ventral longitudinal band of cells in the germ band sinking into the yolk and being overgrown by the remaining cells of the germ band. In a third and less common method of inner layer formation, a longitudinal band of cells proliferates from the germ band (Fig. 8–15F). Whatever the method of formation, the end result is essentially the same (an outer or ectodermal layer and an inner or mesentodermal layer). Eventually the inner layer differentiates into two lateral longitudinal bands (*mesoderm*) and a median strand with cell masses located at its

Fig. 8–15
Germ-layer differentiation (yolk and chorion omitted). A–C. Invagination. D. and E. Overgrowth. F. Delamination. (Redrawn with modifications from Snodgrass, 1935.

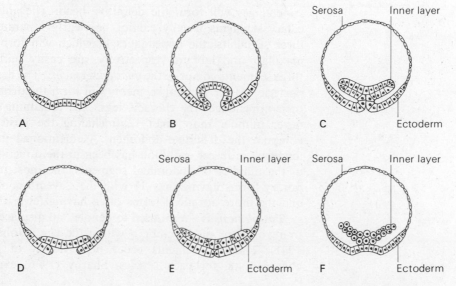

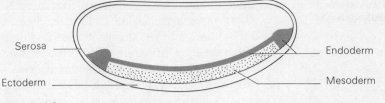

Fig. 8–16
Sagittal section of an embryo (diagrammatic). (Redrawn from Snodgrass, 1935.)

anterior and posterior end (Fig. 8–16). For our purposes the median strand with its anterior and posterior cell masses will comprise the *endoderm*.

Segmentation, Appendage Formation, and Blastokinesis

Segmentation of the embryo begins very soon after the germ band has formed and originates in the mesoderm. This segmentation later becomes quite evident in nearly all the organs of mesodermal origin (dorsal vessel, muscles, and so on) as well as those of ectodermal origin (nervous system, tracheal system, and so on). The endoderm and the portions of the ectoderm that give rise to the fore- and hindgut are unaffected by the segmentation process.

As segmentation proceeds, transverse furrows become externally evident in the ectoderm (Fig. 8–17). Soon after segmentation begins, bilateral evaginations of the ectoderm, which contain mesodermal tissue, begin to appear. These will form the various body appendages. Initially the germ band (Fig. 8–17A) is comprised only of the *protocephalon*, which is bilaterally expanded, and a narrow lobe extending posteriorly and from which the remaining body segments will develop. When the segmentation of the embryo is essentially complete (Fig. 8–17B–D) and all appendage rudiments have formed, the portions of the embryo that will form the three tagmata of the insect body can be discerned. The protocephalon, the antennal and intercalary segments, and the following three segments with their paired appendages will form the definitive head. The antennal and intercalary segments usually cannot be clearly separated, but the next three segments, the appendages of which will form the mandibles, maxillae, and labium, respectively, are easily outlined. These last three segments comprise the *gnathal segments*. The next three segments posterior to the gnathal segments will form the definitive thorax and their appendages the thoracic legs. The remaining segments, which never number more than 12, including the posterior telson, will compose the definitive abdomen. As mentioned in Chapter 2, the tendency in insect evolution has been in the direction of a reduction in the number of abdominal segments, the most primitive contemporary forms having only 11 with the eleventh containing the anus, and the more advanced forms often having fewer than this number.

Blastokinesis is a term used to denote "all displacements, rotations, or revolutions of the embryo within the egg" (Johannsen and Butt, 1941). These movements occur in a variety of predictable ways characteristic for each species. Sharov (1966) explains two likely functions for such movements: (1) they are an adaptation to the large quantity of yolk characteristic of the vast majority of insect eggs and enable the embryo to make the most efficient use of the yolk material; and (2) they probably serve a protective role since when the embryo is surrounded by yolk, its chances for desiccation would be decreased somewhat.

Organogenesis

Subsequent to the completion of the formation of the three germ layers, each undergoes further differentiation, eventually forming the various tissues and organs of the fully developed embryo.

Fate of the Mesoderm. As segmentation occurs, the intersegmental portions of the mesoderm either become very thin or separate altogether. Eventually each segmental mesodermal layer develops two lateral lumens (Fig. 8–18A, B), the *coelomic sacs,* and finally the mesoderm differentiates segmentally into the following parts

Fig. 8–17

Segmentation and appendage formation. A. Prior to segmentation. B and C. Successive stages. D. Segmentation complete.

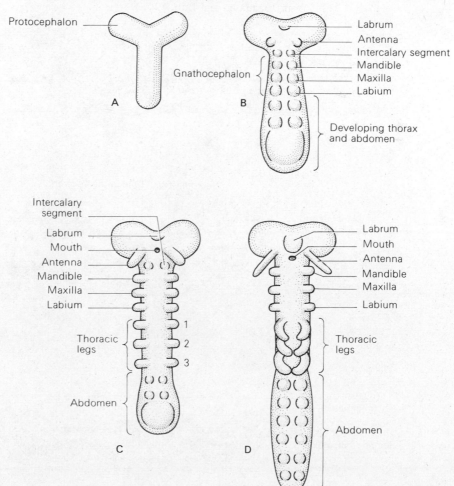

(Fig. 8–18C): (1) the *splanchnic* or *visceral* layer, (2) the *somatic layer,* (3) the fat body, (4) *cardioblasts,* and (5) the *genital ridge.* Blood cells or hemocytes arise from the middle strand, which we have labeled endoderm, but the question still remains as to whether, in fact, this strand is endoderm. As development progresses, the splanchnic layer comes to form the visceral muscles and the somatic layer, the skeletal muscles. At dorsal closure, to be discussed briefly later, the cardioblasts form the heart portion of the dorsal vessel. The aorta arises from the median walls of the antennal coelomic sacs (Johannsen and Butt, 1941). The genital ridges are later suppressed in all but the eighth and ninth abdominal segments. In these segments, the germ cells, which, you will recall, differentiated while the blastoderm was forming, migrate into the genital ridges and together with these mesodermal cells form the *anlage* (precursors) of the gonads.

Fate of the Ecto- and Endoderm. The alimentary canal begins its development as two invaginations of the ectoderm, one in the

Fig. 8–18
Cross section of an insect embryo at successive stages of differentiation (chorion and most of yolk omitted; serosa omitted in B and C). (Redrawn from Johannsen and Butt, 1941.)

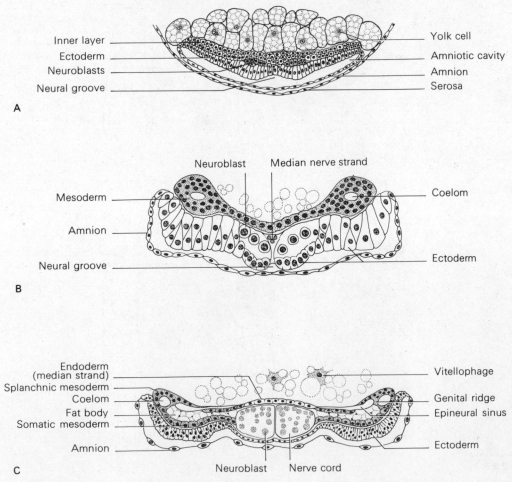

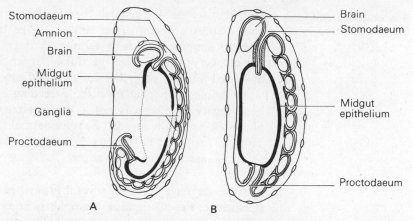

Fig. 8–19

Sagittal sections of an insect embryo at successive stages in the development of the alimentary canal. (Redrawn from Johannsen and Butt, 1941.)

cephalic region (*stomodael*) and one in the last abdominal segment (*proctodael*; Fig. 8–19A). These two invaginations subsequently form the fore- (*stomodaeum*) and hindgut (*proctodaeum*), respectively (Fig. 8–19B). The midgut (*mesenteron*) arises from the proliferation of the cell masses at each end of the median strand of endoderm (Fig. 8–16), which Johannsen and Butt (1941) designate as the "midgut epithelial rudiments." Some of the cells freed from the middle strand may also contribute to midgut formation in some species (Johannsen and Butt, 1941). The endodermal cells eventually envelope the remaining yolk material (Fig. 8–19B). The alimentary canal is completed when the membranes between the stomodaeum, mesenteron, and proctodaeum perforate. The Malpighian tubules develop as evaginations of the proctodaeum immediately posterior to the mesenteron.

The brain, subesophageal ganglion, segmental ganglia, and associated paired nerve cords are of ectodermal origin. The earliest evidence of their development is the presence of a longitudinal neural groove in the ectoderm and the differentiation of *neuroblasts* (Fig. 8–18). The neuroblasts subsequently give rise to the ganglia, nerve cord, and associated nerves. The ganglia and nerves of the stomatogastric nervous system develop from neuroblasts, which arise from the stomodaeum.

All the ventilatory structures of insects are of ectodermal origin. Spiracles, tracheae, and tracheoles arise as invaginations of the ectoderm; the various "gill" structures develop as evaginations.

Oenocytes arise segmentally in the abdomen from the ectoderm.

Dorsal Closure and the Definitive Body Cavity. As the embryo develops it spreads over the yolk material and the extraembryonic area becomes smaller and smaller. The extraembryonic membranes, the amnion and serosa, usually disappear before the embryo has completed its development. Generally, they fuse at some point ventrally and a longitudinal cleft forms. The resultant folds, which consist of amnion fused to serosa, are drawn back over the embryo and the serosa condenses, forming the *secondary dorsal organ.* Eventually the dorsal organ sinks into the remaining yolk material and the amnion provides a temporary dorsal closure. Subsequently the ectoderm grows over the dorsal organ, which is detached and absorbed into the yolk. The ectoderm effects the final dorsal closure.

As should be evident from the preceding discussion, the definitive body cavity of insects is not a true coelom but develops from the blastocoel, which is invaded by mesodermal tissue (Snodgrass, 1935). Thus, as explained in Chapter 3, the body cavity is correctly referred to as a hemocoel or blood cavity.

Polyembryony. Among several groups of parasitic Hymenoptera (Chalcidae, Proctotrupidae, Vespoidea, Braconidae, Ichneumonidae) a developing egg may divide mitotically, producing several embryos. This asexual multiplication of individuals is called *polyembryony.*

Control of Pre-eclosion Morphogenesis

Morphogenesis may be looked upon as a process in which primordial cells become increasingly differentiated in a stepwise fashion, finally becoming the specialized parts which comprise the total, functional organism. Events that occur during this process are ultimately under the control of the nucleus in the newly forming egg. However, in the completed egg, the activities of the cleavage nuclei are controlled by an interaction of the nuclei themselves and factors within the cytoplasm, and the tissues these nuclei eventually form are determined largely by their position within the egg.

Experiments that have elucidated the control of embryogenesis have been based on the removal, destruction, disruption, or separation of various parts of the egg at different times during its development. Techniques utilized in these experiments include the use of microsurgery, ultraviolet light, x-ray, cautery, and ligation. Recent reviews pertaining to control of embryogenesis are those of Agrell (1964) and Counce (1961).

Cleavage and migration of the nuclei and their accompanying islands of cytoplasm begin at the *cleavage center*, which is located in the region where the future head will form. The earliest effect of the cytoplasm on cleavage cells occurs at the *activation center,* which is initially located near the posterior pole of the egg. The interaction of the cleavage cells and the activation center results in a change that initiates further growth and development. The effects of this change then proceed anteriorly. Blocking the effects of the cleavage center by removal of exposure to ultraviolet light results in a failure of germ-band formation. Further developmental direction comes from the *differentiation center*, located near the middle of the presumptive germ band, at which point the prothorax will eventually form. From this center all subsequent changes are induced, both in an anterior and posterior direction, and differentiation not only begins here but is always at an advanced stage relative to the rest of the developing insect. It has been shown by ligation experiments that the operation of the differentiation center does not involve the diffusion of any material out into the rest of the egg, but is the point of initiation of a contraction of the yolk from the chorion, an action that creates a space into which the cells forming the blastoderm can move. (It must be mentioned that the activation-center and differentiation-center descriptions are based on the study of comparatively few species of insects, and that although they are thought to be similar in most species, one should be cautious in applying these ideas to all species.)

Earlier in this section mention was made of the specialized region near the posterior pole of the egg. This is quite similar to the "centers" described above in that it determines which cleavage cells (those which migrate through it) will form the germ cells.

Insect eggs are quite variable as to when the cleavage cells become "determined" (i.e., when they become irreversibly destined to form a specific tissue). One extreme, found, for example, in *Musca* and *Drosophila*, is the "mosaic" egg, in which determination is completed before the egg is deposited and the elimination of any part of it will result in the formation of an embryo missing whatever structure the excised or destroyed cells would have formed. In eggs at the other extreme, determination is not complete until well after deposition, the cells having retained the "potency" to form any tissue. Such eggs are referred to as "regulation," this term alluding to the fact that the capacity for a change in developmental fate (regulation) persists through a much more advanced state than in the mosaic type.

Platycnemis (Odonata, Anisoptera) is an example of an insect with a regulation type of egg, and if a recently deposited egg is ligated in the middle, a "dwarf" embryo forms, demonstrating its well-developed capacity for regulation (Wigglesworth, 1965). Many gradations between the extremes of mosaic and regulation have been described and seem to correlate with phylogenetic level, the mosaic condition occurring in such groups as Diptera and the capacity for regulation being found to an increasing extent among the "lower" orders, such as Odonata. A correlation also exists between the regulative capacity and the amount of cytoplasm; the greater the cytoplasmic volume, the less the regulative capacity (Agrell, 1964).

The control of later development in the embryo is not as well understood as in the earlier stages, but there is evidence that the ectoderm is almost completely autodifferentiating, while the mesoderm depends on the ectoderm for the "induction" of further differentiation.

Imaginal as well as adult characters are somewhat determined during embryogenesis, but apparently later than and independent of the earlier determinative change, which affects larval characters. In a manner analogous to the malformations produced in the larva by the removal or destruction of determined cells, similar removal or destruction following the determinative change that affects adult characters cause malformations in the adult.

Oviposition

In oviparous insects the eggs are propelled down the oviducts by peristaltic waves and may either be deposited in groups or singly, depending on the species. In terms of structures that serve in oviposition, there may be a well-developed, appendicular ovipositor, or the abdomen may be modified such that it can telescope into a relatively long tube and hence function effectively as an ovipositor. The latter structure is commonly referred to as an *ovitubus* and is found among the Thysanoptera, Diptera, and others. Ovipositors are reduced or lacking in the following orders (Davey, 1965): Odonata, Plecoptera, Mallophaga, Anoplura, Coleoptera, and the panorpid orders.

Eggs are deposited in a variety of ways, some merely being dropped passively (e.g., in walking sticks) and others being "glued" singly or in masses to some substratum. Several orthopterous insects (e.g., locusts, mantids, and cockroaches) deposit their eggs in packets, or *oothecae*. The locusts secrete a frothy material that encases an egg mass, which is deposited in the ground. Mantids deposit their eggs on twigs in a foamy secretion which eventually hardens to produce the ootheca. Cockroaches have oothecae with a cuticlelike surface; some species carry these around with them, while others glue them to a substratum. In most species the secretions involved with forming, covering, and gluing the egg masses to the substrate come from the accessory glands. Oothecae of various sorts are presumably designed such that they permit gaseous exchange without undue water loss. Insects such as parasitic Hymenopterans use their ovipositors to inject their eggs deep into plant tissues or a host insect.

Oviposition is evidently under neural and endocrine control and may in some instances be brought about by insemination. See Chapter 7 for a brief discussion of the behavioral aspects of oviposition.

Eclosion

Eclosion is the process of hatching or exiting from the egg. Although the details of this process vary from group to group, eclosion generally first involves the swallowing of amniotic fluid with the attendant diffusion of air into the egg. In addition to the amniotic fluid, some of this air may be swallowed by the insect. The problem faced at eclosion is the rupture of the chorion and other embryonic layers and escape from the confines of the egg. This rupture may occur irregularly about the surface of the egg or along preformed lines of weakness. In some instances the embryonic membranes may be weakened by the action of digestive enzymes. The actual force involved in the perforation of the chorion and embryonic membranes may be applied via various structures, including spines or eversible bladders or may simply involve the forced expansion of one region of the body by contraction of another, a process aided by the previous swallowing of amniotic fluid and, in some instances, air. Some insects (e.g., Lepidoptera) chew their way out of the eggshell. In one instance [*Glossina* (the tsetse fly)] in which the egg is retained and hatches within the "uterus" of the parent, the larva splits the chorion, but a parental structure, the *choriothete,* removes it from the larva.

Post-eclosion Morphogenesis

Under the heading "post-eclosion morphogenesis" we shall include those events which occur between eclosion and emergence of the adult insect.

Oviparity and Viviparity

Most insects are *oviparous*; that is, when they are released from the parent, they are surrounded by the eggshell or chorion. Depending on the species involved, the chorion ranges from very thin and delicate to very hard and thick, and the developmental stage of the insect within it varies from being in early cleavage to being ready to hatch as an active, free-living individual. On the other hand, many insects are *viviparous* (i.e., they give "birth" to individuals that do not have a chorionic covering). The terms *larviparous, nymphiparous,* and *pupiparous* are commonly used to refer to viviparous larvae, nymphs, and pupae, respectively. Although a group of Diptera are referred to as Pupipara, they are, in fact, larviparous, and this term is really inappropriate. Members of the genus *Glossina* probably represent an extreme in viviparity in insects. One larva develops at a time in the highly specialized "uterus" and receives nutriment by means of specialized *uterine* (accessory) *glands*. When released from the parent insect, the tsetse larva is ready to pupate and does so within a few hours. While in the uterus they ventilate by protruding their posterior spiracles through the parent's genital opening. The term *ovoviviparity* is sometimes used and refers to instances where an egg with a chorion develops, but the egg hatches within the parent before it is deposited (Hagen, 1951). In some insects the embryo develops in the hemocoel of its parent (e.g., members of the order Strepsiptera).

Growth

Associated with the evolution of an external, inexpansible exoskeleton has been the concurrent necessity for a mechanism that allows for increase in size. The solution to this problem is the process of *molting*, the periodic digestion of most of the old cuticle, secretion of new cuticle, and shedding of undigested old cuticle. The last-mentioned step, shedding of undigested old cuticle (*exuvium*), is commonly referred to as *ecdysis*. Pertinent details regarding molting and ecdysis are given in Chapter 2.

As a typical insect progresses from the newly hatched immature form it goes through a series of molts and ecdyses, generally increasing in size with each one. Between each molt, the insect itself is called an *instar* and the interval of time passed in that instar is sometimes referred to as a *stadium*. When an insect has secreted a new cuticle but has not as yet escaped from the confines of the undigested portion of the old cuticle, it is described as being in the *pharate* condition (Hinton, 1946). In many species of insects, especially those with a small number of instars (e.g., mosquitoes), it is possible to determine exactly the instar of a given individual by characteristic morphological traits. In others, there may be little or no change between instars other than growth, which may vary considerably with the availability of food and other environmental factors. The final instar, during which sexual maturity and functional wings (if they are going to develop at all) are realized, is the adult, or *imago*. A few insects (e.g., thysanurans) continue to molt after they reach the adult stage; however, it is unlikely that there is any growth associated with these molts. An adult of any species of insect could probably be induced to molt artificially by the introduction of the molting hormone (ecdysone); we shall consider this later. In members of the order Ephemeroptera (mayflies) there are two winged instars, the first being the *subimago* and the second the adult. Other than mayflies, adult pterygote insects have never been known to molt in the adult stage under natural conditions.

The number of instars varies between different species of insects, the majority having between 2 and 20. In some insects the number of instars is quite constant; however, in others it may be variable, in response to environmental factors (e.g., availability of food and temperature). In some species the number of instars may be different for males than for females. In very broad, general terms, the more specialized insects tend to have fewer instars.

Growth in insects occurs as the result of an increase in the number of cells by mitotic division and in many instances by an increase in cell size. The increase in weight between a newly hatched immature and a fully grown immature is usually quite pronounced. For example, according to Richards and Davies (1957) a fully grown larva of *Cossus cossus* (Lepidoptera; Cossidae, the carpenter moths) weighs 72,000 times its original weight, and it takes three years to accomplish this growth. In many other insects the magnitude of growth is somewhat smaller, varying from about 1,000 times on up. Growth in weight from instar to instar may approximate an S-shaped curve, although its continuity may be disrupted by the swallowing of water, by aquatic insects, or the ingestion of food, as in blood-feeding insects.

The growth in length and surface area appears to be discontinuous, owing to the relative inflexibility of the cuticle. Based on the observation of a degree of regularity in the extent of linear growth in various structures which has been observed in several species, certain "growth laws" have been induced. The most important of these is *Dyar's law,* which is based on the assumption that the growth as reflected in the various linear dimensions follows a geometric progression and says

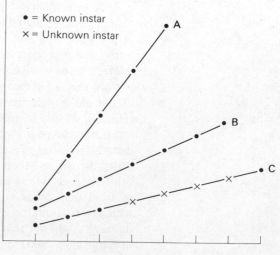

Fig. 8–20

Graph depicting Dyar's law as it might apply to three hypothetical species of insects. A, B and C are curves generated by plotting the log of a linear measurement against the number of a given instar. Given the information for the known instars in curve C, one can determine the total number of instars and the magnitude of the linear dimension pertinent to each one.

that the ratio of increase is the same for many of these dimensions at each molt and is constant for each species. Thus if we were to plot the logarithm of the measurement of some linear dimension at each instar of a given species against the number of the instar, we would expect to obtain a straight line (Fig. 8–20). Where this rule applies, it is useful in determining the number of instars of an insect, since if the dimensions of the final instar and the ratio of increase between any two successive instars are known, the dimensions of all the other instars can be deduced by interpolation and their number found by counting the number of points on the generated growth curve. However, in instances where there are a variable number of molts in response to certain environmental or other factors, Dyar's law will not allow the determination of the number of instars.

An extension of Dyar's law is *Przibram's rule*, based on the assumption that weight doubles during each instar and says that all linear dimensions at each molt are increased by the ratio of 1.26 or the cube root of 2 (Wigglesworth, 1965). Although this rule applied in the species studied when it was devised and in certain other species, it fails to apply in many other instances, for example where weight more than doubles during each molt or where growth is attained by increase in cell size instead of cell number, as, for example, the larvae of certain true flies. In addition, Przibram's rule assumes *harmonic* or *isogonic* growth (i.e., that all parts of the insect and the body as a whole increase by the same ratio during each molt). However, in the majority of insects growth during each molt is *heterogonic* or *allometric* (i.e., some parts of the insect body develop at different rates than other parts). Heterogonic growth can be described by the following equation (Wigglesworth, 1965): $x = ky^a$, where x is the dimension of the whole, y the dimension of a part, a the growth constant, and k the constant of proportionality.

Metamorphosis

Most insects at eclosion are morphologically different from the adult. The degree of difference varies from relatively slight to pronounced with many intermediates. The developmental process by which a first-instar immature stage is transformed into the adult is called *metamorphosis*, which means literally "change in form." This process may take place gradually, with the immature being in general appearance comparatively similar to the adult, or it may be quite abrupt, the immature instars being drastically different from the adult with the transformation from the immature to the adult form occurring in a single stage.

In general terms the class Insecta can be divided into groups based on the type of metamorphosis or whether it is present or not. Members of the subclass Apterygota do not undergo any change in form, the immature instars differing from the adults only in the development of the gonads and external genitalia (Fig. 8–21A). These insects are sometimes grouped as Ametabola and are referred to as having ametabolous development.

Members of the subclass Pterygota can be divided into two groups relative to the degree of metamorphosis that occurs: *hemimetabola*

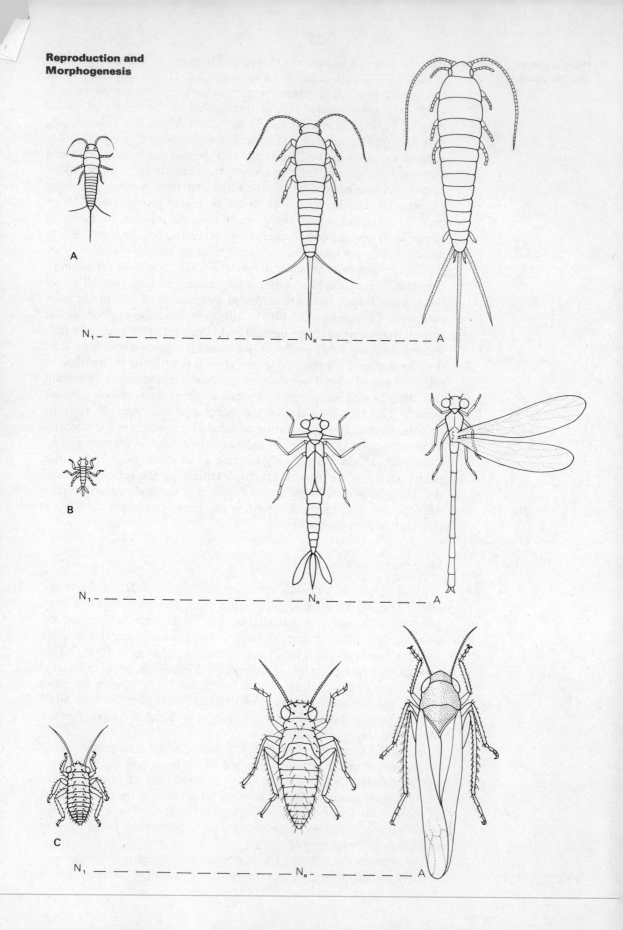

N_1 —————————— N_n ————— A

N_1 —————————— N_n ————— A

N_1 —————————— N_n ————— A

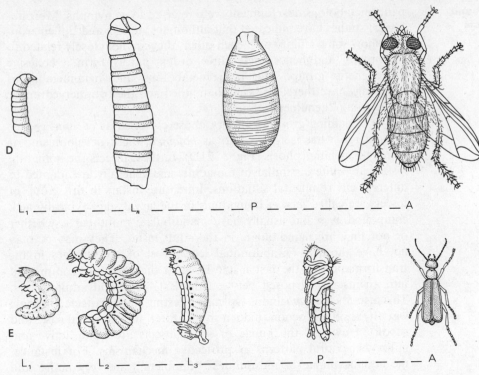

Fig. 8–21

Types of insect metamorphosis. A. Ametabolous, silverfish. B. Paurometabolous, damselfly. C. Hemimetabolous, leafhopper. D. Holometabolous, house fly. E. Hypermetamorphosis, beetle, *Epicauta cinerea* (Coleoptera; Meloidae). A, adult; L_1, first instar larva; L_2, second instar larva; L_3, third instar larva; Ln, nth instar larva; N_1, first instar nymph; Nn, nth instar nymph; P, pupa. (C redrawn from Essig, 1958; D and E, redrawn from Packard, 1898.)

and *holometabola*. Among the hemimetabolous insects the immatures resemble the adults in many respects, including the presence of compound eyes, but they lack wings, gonads, and external genitalia. During the course of their development (Fig. 8–21B, C) the developing wings are externally apparent as "wing pads." Thus the orders that fall into this group are those which are classified as *exopterygotes*. The hemimetabolous form of development is often referred to as simple, direct, or incomplete metamorphosis. The immature instars in this group of insects are commonly referred to as *nymphs*.

In the past insects with hemimetabolous development were subdivided further. Those insects (Odonata, Ephemeroptera, and Plecoptera) that pass the immature instars in an aquatic environment and the adult instar in a terrestrial/aerial environment and which at least superficially appear to be quite different from the adult stage (e.g., immature stoneflies, with highly specialized ventilatory gills, versus adults, which lack these structures and have well-developed wings) were classified as being *hemimetabolous*, or undergoing an incomplete metamorphosis (Fig. 8–21C). Those insects which pass both the immature and adult instars in essentially the same environment were classified as being *paurometabolous*, or undergoing gradual metamorphosis (Fig. 8–21C). The immature instars of hemimetabolous insects were called *naiads*; those with paurometabolous development were referred to as nymphs. Morphological studies have indicated that although Odonata and Ephemeroptera show some affinity with each other, they are not closely related to Plecoptera, and hence these three orders hardly form a cohesive phylogenetic group. The paurometabolous and hemimetabolous groupings have therefore been abandoned and the two merged under the heading "hemimetabolous."

The remaining group of orders, those classified as *endopterygotes,* all undergo what is referred to as *holometabolous* development, or complete metamorphosis (Fig. 8–21D). In these insects the immature instars are quite dissimilar to the adults and generally are adapted to different environmental situations. Immature instars in this group of insects are called *larvae.* Individuals in the larval stage typically lack compound eyes and usually have mandibulate mouthparts, whether or not they are mandibulate in the adult instar. They may or may not have thoracic or abdominal legs. Most of the changes in the transformation of the last instar larva to the adult are compressed into an instar interposed between the last larval and adult instars. This instar is the *pupa* and is typically a resting stage protected in some way (by a silken cocoon, hidden in leaf litter, in a puparial case, and so on). However, the pupae of some insects are quite active and utilize behavioral patterns as protective mechanisms. For instance, the aquatic pupae of mosquitoes ("tumblers") are very active and capable of diving by means of specialized paddles at the end of their abdomen in response to potentially threatening stimuli. In most insects the larval instars resemble one another except for a minor morphological details, which are useful in distinguishing one instar from another. However, some holometabolous insects pass through one or more larval instars which are distinctly different from the others (Fig. 8–21E). This phenomenon is called *hypermetamorphosis* and has been described in certain species in the orders Hemiptera, Neuroptera, Coleoptera, Diptera, and Hymenoptera, and in all species in the order Strepsiptera.

The grouping of insects based on metamorphosis as described above, although useful in general pragmatic terms, fails somewhat when reexamined in the light of other information. For example, members of some orders that are included in Hemimetabola are secondarily wingless in the adult stage and their development more closely resembles that of the apterygotes in that it is essentially ametabolous. More important are certain of the *hemimetabolous* insects, which, in fact, have pupal stages. For example, thrips (order Thysanoptera) and certain members of the order Homoptera (members of the family Aleyrodidae, whiteflies, and certain scale insects, e.g., *Pseudococcus* spp.) have a distinct resting "pupal" stage between the last "larval" and the adult instars (Hinton, 1948). Even members of the "primitive" order Odonata undergo considerable morphological change during the transformation from the aquatic nymph to the terrestrial/aerial adult. These changes include loss of rectal gills, restructuring of the labium, modifications in the head and abdomen, complete reconstruction of the alimentary canal, and nearly complete replacement of the abdominal musculature (Clements, 1963, citing the work of Snodgrass).

It is difficult not to look upon the development of these insects as being *holometabolous*, indicating that holometabolism has probably evolved on separate, independent occasions. However, Hinton (1963) recognizes a distinct difference between the holometabolous exopterygotes and the endopterygotes. He points out that since the wings develop internally in endopterygotes, there is insufficient room in the larval thorax for the development of both the wings and the wing muscles. Therefore, a molt is required to evaginate the wings (the larval–pupal molt) and provide space for the development of wing muscles. A second molt (pupal–adult) is then required to release the adult from the confines of the pupal cuticle. A pupal stage, in this sense, is unnecessary in exopterygotes since the wings develop externally. Hinton accounts for the "pupae" in certain exopterygote insects as follows:

> In some exopterygotes such as the Aleyrodidae and male Thysanoptera and Coccoidea, the general structure of the feeding larval instars has departed very widely from that of the adult. The structural differences between the feeding larval instars and the adult are normally bridged in the last larval instars. As the differences between the two stages became greater and involved a greater degree of re-organization of the internal tissues, it would seem that the last or last two larval instars became more and more quiescent and eventually ceased to feed. It may be noted here that the structural re-organization required to bridge the gap between the feeding larval stages and the adult of some exopterygotes is greater than in the primitive endopterygotes, e.g. some Megaloptera. No difficulty necessarily arises if these quiescent or semi-quiescent stages of exopterygotes are called pupae provided that it is recognized that their origin is quite independent from that of the endopterygote pupa and their initial functional significance is different.

Wigglesworth (1965) and others confine usage of the term "metamorphosis" to those changes which occur when an insect becomes an adult, regardless of whether the adult stage is reached by a single molt of the last immature instar or in two molts with a pupal stage in between. Wigglesworth (1954) points out that metamorphosis is commonly looked upon as a renewal of embryonic development. However, he regards the larvae and adults of holometabolous insects as essentially two organisms which are latent within the genome of the embryo and which are expressed in sequence. He views this as a sort of temporal polymorphism. The evolution of a pupal stage in which comparatively drastic changes can occur has evidently enabled the divergent evolution of larvae and adults, which are usually adapted to radically different environmental modes of existence. As will be explained later, the insect developmentalists and endocrinologists have found a mechanism that probably underlies all development and through which the various degrees of metamorphosis evolved.

The immature stages of insects take on a wide variety of form. In most instances the immatures (nymphs) of hemimetabolous species closely resemble the adults, but in the holometabolous species, the larvae are commonly drastically different from the adults. Although there is considerable variation in the appearance of

Fig. 8–22
Larval types.
A. Campodeiform, alderfly, *Sialis* sp. (Neuroptera; Sialidae).
B. Carabiform, ground beetle, *Harpalus* sp. (Coleoptera; Carabidae).
C. Elateriform, click beetle (Coleoptera; Elateridae).
D. Eruciform, clear-winged moth (Lepidoptera; Aegeriidae).
E. Platyform, aquatic beetle, *Eubrianax edwardsi* (Coleoptera; Dascillidae).
F. Scarabaeiform, branch and twig borer, *Polycaon confertus* (Coleoptera; Bostrichidae).
G. Vermiform, flesh fly, *Sarchophaga* sp. (Diptera; Sarcophagidae). (A, B, D, and G redrawn from Packard, 1898; C, E, and F redrawn from Essig, 1958.)

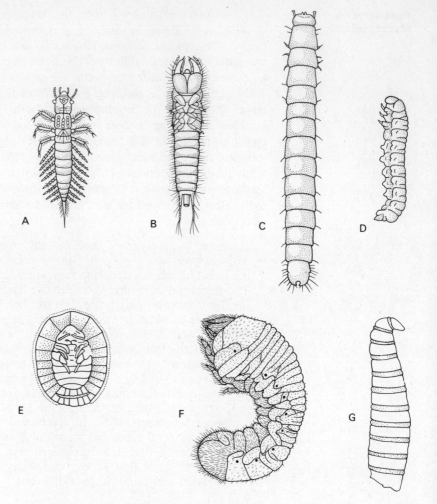

larvae of the different holometabolous groups, there are sufficient similarities to allow recognition of distinct "larval types." Some commonly used terms for types of larvae (Fig. 8–22) are

1. *Campodeiform* larvae show a resemblance to diplurans in the genus *Campodea* in having flattened bodies, long legs, and usually long antennae and cerci (e.g., several beetles, Neuroptera, and Trichoptera).
2. *Carabiform* larvae resemble the larvae of carabid beetles, which are similar to the campodeiform type but have shorter legs and cerci (e.g., several beetles).
3. *Elateriform* larvae resemble click beetle larvae (Coleoptera; Elateridae) and have cylindrical bodies with a distinct head, short legs, and a smooth, hard cuticle (e.g., certain beetles).
4. *Eruciform* larvae are the typical "caterpillars" of members of the order Lepidoptera or the caterpillarlike larvae of certain Hymenoptera and Mecoptera and have a cylindrical body with a well-developed head, short antennae, and short thoracic and abdominal legs (prolegs).
5. *Platyform* larvae have flattened bodies with or without short

thoracic legs (e.g., certain Lepidoptera, Diptera, and Coleoptera).

6. *Scarabaeiform* larvae are the "grubs" and have a cylindrical body typically curled into a C shape, a well-developed head, and thoracic legs (e.g., several beetles).

7. *Vermiform* larvae have wormlike bodies with no legs and may or may not have a well-developed head (e.g., several Diptera, Coleoptera, Hymenoptera, Siphonaptera, and Lepidoptera).

It must be emphasized that the above classification of larval types is a pragmatic one and reflects only gross similarities and not phylogenetic affinities. Obviously one should expect to find numerous examples of larvae which would resemble two or more of these groups. Berlese (Richards and Davies, 1957) classified insect larvae based on the assumption that they are essentially free-living embryos which are released to the external environment in different stages of development in different insect species. He recognized three such stages: *protopod, polypod,* and *oligopod.* Protopod larvae are very uncommon and represent a very early state of development in which even comparatively little segmentation has occurred. These larvae are found, for example, among certain parasitic Hymenoptera which larviposit in the hemocoel of other insects, placing the "embryo" in the only kind of environment possible for survival—a protected one in which it is surrounded by nutriment. Polypod larvae resemble caterpillars, and hence the eruciform type described above would fall readily into this group. Oligopod larvae lack the abdominal prolegs and may or may not have well-developed cerci. The campodeiform, carabiform, elateriform, platyform, and scarabaeiform types listed above fall nicely into this category. The vermiform (apodous) type of larva is considered to have developed secondarily from the polypod or oligopod type, depending on the kind of insect, and hence is not recognized as a separate group.

If one subscribes to Berlese's view of holometabolous larvae, he must recognize a fundamental difference between these larvae and the nymphs of hemimetabolous insects, which in this context represent postoligopod stages of development. However, one can as easily look upon holometabolous larvae as specialized nymphs with a concurrent concentration of the changes to the adult form in a single stage, the pupa. From this point of view there are no fundamental differences between larvae and nymphs, and the terms could appropriately be used as synonyms. As will be explained subsequently, the results of comparatively recent investigations of the physiological mechanisms that control metamorphosis have supported this latter conception of the relationship between hemimetabolism and holometabolism.

As with larvae, pupae have been grouped according to similarities. Hinton (1946) classified pupae based on whether or not they have articulated mandibles used in escaping from a cocoon or pupal cell. Those pupae which do have functional mandibles and use them for such escapes he described as being *decticous.* Examples of decticous pupae include members of the orders Neuroptera, Mecoptera, Trichoptera, and certain lepidopterous families. Pupae without

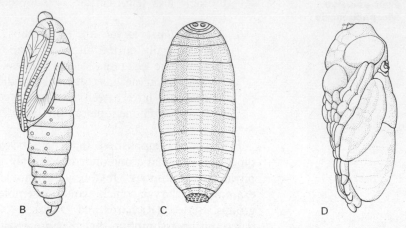

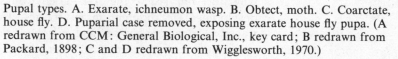

A   B   C   D

Fig. 8–23

Pupal types. A. Exarate, ichneumon wasp. B. Obtect, moth. C. Coarctate, house fly. D. Puparial case removed, exposing exarate house fly pupa. (A redrawn from CCM: General Biological, Inc., key card; B redrawn from Packard, 1898; C and D redrawn from Wigglesworth, 1970.)

functional mandibles used in escape from cocoon or pupal cell are termed *adecticous* and are found in all or some members of the orders Strepsiptera, Coleoptera, Hymenoptera, Diptera, and Siphonaptera. Another approach to grouping pupal types (Fig. 8–23) is based on whether the appendages are free or adherent to the body. *Exarate* pupae (Fig. 8–23A) have the appendages free and are usually not covered by a cocoon. In *obtect* pupae (Fig. 8–23B) the appendages adhere closely to the body and they are commonly covered by a cocoon. All decticous pupae are exarate, whereas the adecticous types may be either exarate or, a third type, *coarctate* (Fig. 8–23C). In this type the pupa is encased in the hardened cuticle of the next-to-last (penultimate) larval instar, the *puparium*. However, the pupa itself is of the adecticous exarate form.

A number of histological changes usually occur during the transition from immature to adult. In hemimetabolous insects these changes are comparatively gradual, being spread throughout the nymphal instars, although there are generally more changes during the last instar than during earlier ones. In holometabolous insects, these changes occur mostly in the pupal stage. The extent of change during the pupal period is variable depending on species, and in some a large amount of tissue change occurs. These changes are accomplished by means of tissue breakdown (histolysis) and tissue reorientation, growth, and differentiation (histogenesis). Histolysed tissues may simply dissolve in the hemolymph or may be ingested by phagocytic hemocytes. Many adult tissues are formed from masses of cells that have persisted in an undifferentiated state throughout the larval instars even though they increase in number by mitotic divisions. These masses of embryonic cells are generally referred to as *imaginal buds,* or *discs* (Figs. 8–24 and 8–25), and are involved in the formation of structures such as mouthparts, antennae, wings, and legs. Rings of embryonic cells are generally found at the posterior extremity of the

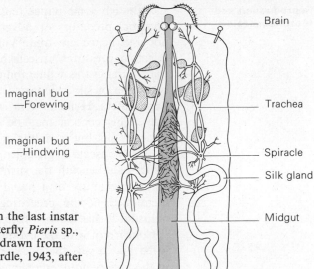

Brain

Imaginal bud
—Forewing

Imaginal bud
—Hindwing

Trachea

Spiracle

Silk gland

Midgut

Fig. 8–24

Imaginal buds in the last instar
larva of the butterfly *Pieris* sp.,
dorsal view. (Redrawn from
Folsom and Wardle, 1943, after
Gonin.)

foregut (*anterior imaginal ring*) and the anterior extremity of the
hindgut (*posterior imaginal ring*). These undergo considerable growth
and differentiation during the pupal stage and contribute significantly
to the formation of the adult alimentary canal. In some instances,
certain larval tissues persist into the adult stage, and these vary with
the different groups of insects.

Fig. 8–25

Imaginal leg bud in a fourth instar larva of the mosquito *Aedes triseriatus*.
1, Imaginal leg bud; 2, cuticle. (Scale line = 50 microns.)

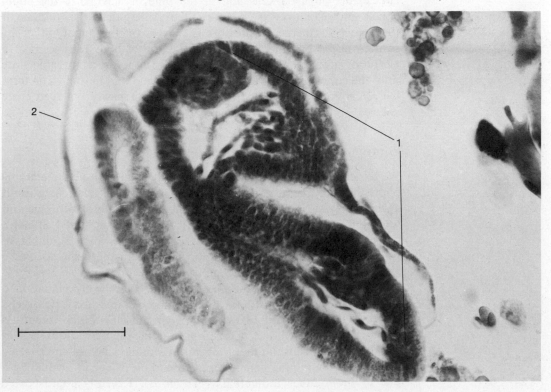

Although some pupae (e.g., mosquitoes) are active and can evade potential predators or adverse environmental conditions, most are not and hence are quite vulnerable. In response to such negative selection pressures, insects have evolved a number of mechanisms that decrease this vulnerability. Probably the most common of these is the use of silk in constructing a cocoon of some sort (e.g., many Lepidoptera, Hymenoptera, Neuroptera, Trichoptera, and Siphonaptera). This cocoon may be composed solely of silk or it may be basically silk but have bits of environmental debris incorporated into it. Many insects (e.g., coleopterous and lepidopterous larvae) construct cells beneath the surface of the soil and pupate within them. Another example of a pupal protection mechanism is the puparium mentioned in the preceding paragraph. Obviously the pupae of parasitic forms derive protection from being within another insect.

Concurrent with the evolution of the pupal protection mechanisms listed above was the necessary development of means for escaping at the time of adult emergence. The decticous pupae of Hinton (1946) chew their way out of their cocoon or cell and may be aided in this process by posteriorly directed spines. Escape methods for adecticous pupae include a variety of spines or other hard and sharp protuberances and an eversible bladder, the *ptilinum*, in the head which is used to force open the "cap" of a puparial case. The latter mechanism is typical of the cyclorrhaphous Diptera.

At least a portion of the pupal period is spent in the *pharate* condition. This is where the adult is fully formed but is still within the confines of the pupal cuticle. The term *pharate* also applies to other stages. When an adult is nearly ready to emerge from the pupal cuticle, the pupa darkens and at the time of emergence the pupal cuticle splits in various places, particularly along the dorsal midline of the body. At emergence pressure is exerted on the pupal cuticle by swallowing air or water and the contraction of body muscles. For a period of time the newly emerged adult is in a very vulnerable state until the cuticle hardens and the wings expand and harden. The term *teneral* is commonly used to refer to such a newly emerged, pale colored, soft-bodied individual regardless of stage.

Control of Growth and Metamorphosis

The search for the mechanisms controlling growth and metamorphosis has occupied the time of investigators for decades. As a result a reasonably clear picture of at least the major patterns of tissue and hormone involvement has emerged. Wigglesworth (1954, 1964b), Gilbert (1964), Schneiderman and Gilbert (1964), and a number of others consider this topic at length. Here we merely want to give a short outline to complement the preceding sections.

A wide variety of techniques has been utilized in the aforementioned studies. Early workers utilized ligation and decapitation extensively, and these techniques still find wide application today. Later techniques employed have included microsurgical removal of various tissues (see Fig. 4–7), transplantation of various tissues, hemolymph transfusion, and joining the hemolymph circulation of one individual with one or more other individuals (*parabiosis*). Recently sophisticated

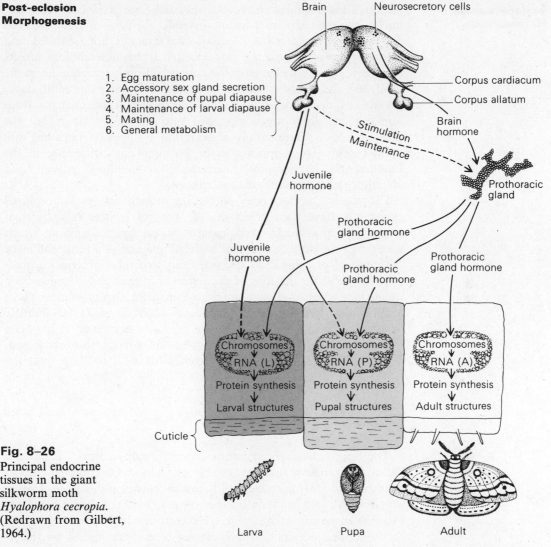

Brain   Neurosecretory cells

1. Egg maturation
2. Accessory sex gland secretion
3. Maintenance of pupal diapause
4. Maintenance of larval diapause
5. Mating
6. General metabolism

Corpus cardiacum
Corpus allatum

Stimulation
Maintenance

Brain hormone

Prothoracic gland

Juvenile hormone

Juvenile hormone

Prothoracic gland hormone

Prothoracic gland hormone

Prothoracic gland hormone

Chromosomes
RNA (L)
Protein synthesis
Larval structures

Chromosomes
RNA (P)
Protein synthesis
Pupal structures

Chromosomes
RNA (A)
Protein synthesis
Adult structures

Cuticle

Larva   Pupa   Adult

Fig. 8–26
Principal endocrine tissues in the giant silkworm moth *Hyalophora cecropia*. (Redrawn from Gilbert, 1964.)

243

biochemical techniques have made possible the extraction and, in the case of ecdysone, purification, of the various hormones, and cytological techniques have given rise to at least a vague idea of the nature and modes of action of these hormones.

The tissues involved in the control of growth and metamorphosis (Fig. 8–26) have been shown to be (1) the *median neurosecretory cells* in the *pars intercerebralis* region of the brain, (2) the *corpora cardiaca*, (3) the *prothoracic glands* located in the prothorax in close association with tracheae, and (4) the *corpora allata*. The neurosecretory cells in the brain secrete the *brain hormone*, which accumulates in the corpora cardiaca and is subsequently released into the hemolymph. The prothoracic glands secrete the *prothoracic gland hormone* (*molting hormone* or *ecdysone*), which initiates the growth and molting activities of cells. The brain hormone stimulates the secretory activity of the prothoracic gland. The corpora allata secrete the *juvenile hormone*, which stimulates larval development while preventing or retarding development of adult characteristics. During the immature instars both molting hormone and juvenile hormone are produced. However, during the last immature instar, juvenile hormone is not produced and hence the expression of adult characters is not inhibited and metamorphosis to the adult stage occurs. In hemimetabolous insects there is a gradual decrease in the concentration of juvenile hormone produced and hence a gradual progression toward the adult stage. On the other hand, in holometabolous forms the changes from immature to adult are compressed into the pupal stage, but the principle of discontinuation of secretion of juvenile hormone still applies. Thus there appears to be no significant difference in the fundamental physiological mechanisms controlling growth and metamorphosis in either of these groups.

Ultimately the hormones regulating growth and metamorphosis apparently have their effect on the control centers of individual target cells (the nuclei) and apparently act by determining which genes are brought into action at a given time and as a result influence the kinds of proteins (both enzymes and structural proteins) synthesized. Support for this idea of the action of at least ecdysone has been found in studies of *chromosomal puffs* in giant chromosomes. These are localized swellings in chromosomes which are taken to indicate heightened gene activity at the point of their occurrence. It has been shown that injection of pure ecdysone into *Chironomus* larvae produces puffing patterns in chromosomes identical to those patterns which occur during pupation.

Polymorphism

The term *polymorphism* means literally "many forms." From this definition it follows that any number of phenomena could be cited as examples of polymorphism: differences between the sexes (*sexual dimorphism*), any differences between individuals caused by the external environment, differences between developmental stages (i.e., the "temporal polymorphism" referred to earlier), and so on. However, the more restricted definition of Richards (1961) seems more useful, although there is still a certain amount of equivocation regarding the meaning of polymorphism. Richards' definition is as follows: Polymorphism exists when "one or both sexes of a species occur in two or more forms which are sufficiently sharply distinct to be recognizable without a morphometric analysis; the occurrence is regular or recurrent; the rarer of the two forms makes up a reasonable proportion of the population (say, at least 5 per cent) or, as in some social species, the rarest type is at any rate essential to the survival of the species." Actually, since form and function are inseparable, functional as well as morphological differences should be included in one's definition and concept of polymorphism.

Following Richards's definition, there are basically two types of polymorphism, both of which are controlled by genes (Wigglesworth, 1965). In one form the relative abundance of the different forms depends on the selection pressures acting on a given population. Hence different populations may show significant differences in the relative abundance of the different forms. An example of natural selection influencing the relative abundance of different forms by differential action on genotypes is found in the variations in the relative numbers of different forms of mimetic butterflies in different regions depending on the abundance of the species which they mimic (Wigglesworth, 1965). In the other type of polymorphism, expression of form, although ultimately under genic control is determined by the effect various environmental factors have in stimulating or inhibiting the influence of certain genes, all individuals in a population having essentially the same genotype. A number of environmental influences have been shown to play a role in polymorphic expression in different species. For example, nutrition, both in terms of quantity and dietary balance, plays a significant role in polymorphism in honey bees. Female larvae (from fertilized eggs) all have the potential to develop into either workers or queens, depending on the period of time they are fed royal jelly, a highly nutritious substance produced by the salivary glands of workers. If larvae are fed royal jelly for only two or three days, they develop into worker adults, but if they are fed royal jelly throughout the larval stadia, they develop into queens. By experimentally varying the length of time larvae are fed royal jelly, it has been possible to produce intermediates between workers and queens (Wigglesworth, 1965).

Another environmental factor that is known to influence morphologic expression is the invasion by certain parasites. For instance, mermithid nematodes cause worker or soldier ants to show female traits. Parasitization of certain insects by strepsipterans may also cause such changes in form. Another example of an environmental factor that influences the expression of particular forms and functions is found among the termites, many of which have an elaborate caste system. This caste system is at least partly mediated by substances, transferred from one individual to another, which regulate the numbers of a given caste in a colony. For example, in many termite species supplementary reproductive castes do not appear as long as the original king and queen are present. Evidently the king and/or queen produce a substance that circulates throughout the colony and inhibits the formation of supplementary reproductives. Death of the royal pair would then result in the loss of this inhibitory substance, and hence the supplementary reproductives would appear, assuring the continued survival of the colony. Substances that are passed between individuals and mediate some function like the example just given are called *social hormones*, or *pheromones*.

As outlined by Richards (1961), species in several orders exhibit one or more types of polymorphism. He lists the following orders as those in which no examples of polymorphism are known to him: Thysanura, Diplura, Protura, Ephemeroptera, Grylloblattoidea (included with the Orthoptera in this text), Embioptera, Mallophaga, Thysanoptera, Mecoptera, and Trichoptera. Polymorphism is expressed in a variety of traits in different insects. Examples of these traits include coloration and patterns of color, as for example in many butterflies; the presence, absence, or attenuation of wings, as found in aphids and other insects; chromosomal differences; the relative size of different structures; and various integumental structures, including horns, spines, and other protuberances.

The Royal Entomological Society's Symposium on Insect Polymorphism (Kennedy, 1961) should be consulted for several interesting
and stimulating papers regarding this topic.

Regeneration

Most, if not all, insects are probably capable of at least a certain
amount of regeneration, particularly with regard to wound healing.
This process often involves hemolymph clotting and hemocytes, but
the epidermal cells play the major role (Wigglesworth, 1965). Wigglesworth (1965), studying *Rhodnius* (Hemiptera; Reduviidae), observed
that dead or injured epidermal cells apparently produce a substance
that has an attractive effect on the surrounding epidermal cells, which
migrate to a wound and lay down new cuticle. Concurrently, mitoses
in the regions from which the epidermal cells migrated restore the
original density of cells in those regions. Some insects [e.g., *Carausius*
(Orthoptera; Phasmidae; Wigglesworth, 1965)] can be decapitated
and the head replaced with the result that the epidermis and gut, but
not neural tissue, grow together again. However, this is not the case
in other insects that have been studied (e.g., *Cimex* and *Rhodnius*)
(Wigglesworth, 1965), in which only the continuity of the integument
(cells and cuticle) is reestablished.

If appendages of developing larvae are removed, they often are
regenerated during later instars, indicating that at least some of the
cells surrounding an appendage are sufficiently undifferentiated,
retaining the ability to reform that appendage. Some insects (e.g.,
walking sticks and certain other Orthoptera) exhibit the ability to
spontaneously amputate a leg, generally between the trochanter and
femur. This phenomenon is called *autotomy* and has the obvious
adaptive value of allowing escape from the grasp of a predator. Many
insects (e.g., walkingsticks and mantids) can completely regenerate a
lost appendage, and this regeneration usually requires a molt.
Evidently the capacity for regeneration of the external form of an
appendage is wholly within the epidermal cells, since removal of
associated ganglia has failed to block regeneration of form, at least
in the species studied (Wigglesworth, 1965).

In some instances abnormal regeneration may occur. Abnormalities
include the duplication or triplication of an appendage at its tip and
heteromorphous regeneration, where the regenerated appendage is like
another appendage on the body but unlike the one it replaces (Wigglesworth, 1965). An example of the latter would be the growth of a leg
where an antenna had originally been.

Aging

Brief consideration of those processes which ultimately lead to the
termination of life of an insect is an appropriate way to complete our
discussion of post-eclosion morphogenesis. The review by Clark and
Rockstein (1964) covers aging at length and has served as the source
for the information presented here.

Aging includes all those changes in structure and function which occur from the beginning of life to its termination. These changes are predictable and reproducible. *Senescence* refers to all those changes in structure and function which decrease an individual's capacity for survival and ultimately cause the death of the individual.

The course of senescence is influenced in general terms both by hereditary and environmental factors. Probably in the majority of situations these two forces act together to produce the observed senescence. According to Clark and Rockstein (1964), hereditary factors that give rise to senescence are as follows: "(1) cessation of growth and, therefore, failure to replace regressing cells, (2) failure of a juvenile or growth substance, (3) the increasing production or accumulation of an aging factor or hormone, (4) depletion of essential substances, or (5) accumulation of substances which may be chemically or mechanically harmful to the aging organism." Under environmental influences they include "(1) cumulative radiation effects, (2) pathological effects, and (3) physically traumatic influences of a changing environment."

Where information is available, species have a constant distribution of individual life spans under defined conditions (genetically and environmentally). Between species, life spans are quite variable and the overall range is probably fairly accurately reflected by 1 day for mayflies to more than 25 years for certain termite species. Differences in mean life span have been found between different genetic strains, between populations with different diets, and between populations exposed to different temperature regimes. Some examples of mean adult life span in different insects are given in Table 8–1.

Diapause

Although diapause could very well have been mentioned at several points in this chapter, discussion of this topic has been saved for the end since it applies to all life stages of insects and serves quite well as a lead-in to the chapter on insects and their environment.

Insect development is subject to two basic kinds of suppression. In one type, sometimes called *quiescence*, an insect responds to adverse environmental conditions by entering a developmental arrest and immediately resumes development when conditions are no longer adverse. In the other type, *diapause*, an insect enters into a state of developmental arrest in response to certain environmental conditions, which may or may not be adverse in themselves but are usually correlated with the onset of adverse conditions, and does not necessarily resume development with the return of favorable conditions.

The texts by Wigglesworth (1965) and Andrewartha and Birch (1954) both contain excellent discussions of diapause, and the monograph by Lees (1955) provides a useful and extensive review of this topic.

The Occurrence of Diapause

Diapause has been described in all the life stages in different insect species. In the adult stage it is characterized by the failure of the

Table 8–1
Mean Adult Life Span (in Days) of Insects[a]

Insect	Male	Female	Reference[b]
Diptera			
Drosophila melanogaster			
Wild (line 107)	38.1	40.1	Gonzalez (1923)
Vestigial mutant	15.0	21.0	
D. subobscura			
9 inbred lines			
(average)	40.0	36.4	Maynard Smith (1959)
4 outbred populations			
(average)	56.8	60.0	
Musca domestica	17.5	29.0	Rockstein and Lieberman
			(1959)
M. vicina	20.8	23.3	Feldman-Muhsam and
Calliphora			Muhsam (1945)
erythrocephala	35.2	24.2	
Aedes aegypti		15	Kershaw et al. (1953)
Lepidoptera			
Acrobasis caryae	6.5	7.3	Pearl and Miner (1936)
Bombyx mori (unmated)	11.9	11.9	Alpatov and Gordeenko (1932)
B. mori (mated)	15.2	14.2	
Fumea crassiorella			
(unmated)		5.5	Matthes (1951)
F. crassiorella (mated)		2.3	
Samia cecropia	10.4	10.1	MacArthur and Baillie (1932)
S. californica	8.7	8.8	
Tropea luna	5.9	6.0	
Philosamia cynthia	5.9	7.1	
Telea polyphemus	8.1	10.0	
Callosamia promethea	4.6	7.0	
Pyrausta nubilalis	13.0	17.4	
P. penitalis	6.8	7.7	
Carpocapsa pomonella	9.4	10.2	
Hymenoptera			
Apis mellifera			
Summer bees		35	Ribbands (1952)
Winter bees		350	Maurizio (1959)
Habrobracon juglandis			
Wild type	24	29	Georgiana (1949)
Small wings, white			
eyes, mutant	20	24	
H. serinopae	62	92	Clark and Rubin (1961)

Table 8–1
Mean Adult Life Span (in Days) of Insects[a] (Continued)

Insect	Male	Female	Reference[b]
Orthoptera			
Blatta orientalis	40.2	43.5	Rau (1924)
Periplaneta americana	200	225	Griffiths and Tauber (1942)
Schistocerca gregaria	75	75	Bodenheimer (1938)
Coleoptera			
Tribolium confusum	178	195	Park (1945)
T. madens	199	242	
Procrustes	374	338	Labitte (1916)
Carabus	323	386	
Necrophorus	232	291	
Dytiscus	854	740	
Hydrophilus	164	374	
Melolontha vulgaris	19	27	
Cetonia aurata	57	88	
Lucanus cervus	19	32	
Dorcus	327	375	
Ateuchus	338	467	
Sisyphus	198	266	
Copris	497	623	
Geotrupes	700	642	
Oryctes	37	55	
Blaps mortisaga	848	914	
B. gigas	700	728	
B. magica	700	728	
B. edmondi	700	728	
Akis	854	951	
Pimelia	669	714	
Timarcha	135	182	

[a] From Clark and Rockstein (1964).
[b] Full citations can be found in Clark and Rockstein (1964).

gonads to enlarge, while in the egg, larval, and pupal stages it consists of a developmental arrest. In many insects, diapause is obligatory (i.e., insects enter this state during every generation in spite of variations in environmental conditions). Other species show a facultative diapause and can go on for generation after generation without developmental arrest and then enter diapause in response to appropriate environmental conditions. Species or strains with obligatory diapause have only one generation per year (i.e., they are *univoltine*). On the other hand, those with facultative diapause may complete two or more generations per year (i.e., they are *bivoltine, trivoltine, quadrivoltine,* and so on).

The stage in which an insect enters diapause is a species characteristic. However, even closely related species can enter diapause in quite different stages. Usually diapause occurs only during one life stage, but there are exceptions where diapause is entered twice. In a given species, there may be strain differences with regard to diapause. One strain may show an obligatory diapause, another facultative diapause and still another not enter diapause at all. For example, univoltine, bivoltine, and quadrivoltine races of the silkworm *Bombyx mori* are known in addition to races that undergo no diapause whatsoever. It is not uncommon for geographical races to show differences in the number of generations per year, depending on locality. Both environmental characteristics acting directly or through natural selection of genotypes are responsible for these differences in voltinism. An example of such racial differences was found in the European corn-borer, *Ostrinia nubilalis* (Lees, 1955). Several years after the introduction of this species into the eastern United States it was found that the race in the Great Lakes States was univoltine with an obligatory diapause, whereas the New England race was bivoltine.

Induction, Duration, and Termination of Diapause

As mentioned in the definition of diapause, the environmental conditions responsible for its induction are not necessarily adverse, but may at least signal the imminent onset of adverse conditions. The most important environmental factor that induces diapause is the length of daylight, and compared to other environmental factors, this one is the most accurate and invariable in terms of its usefulness as a means of attaining synchrony with seasonal changes. *Aedes triseriatus* is an insect (mosquito) that completes several generations during the summer, but in the fall the eggs of the last generation enter a diapause state in response to changing day length despite the fact that temperature at the time of entering this state may be higher than in the spring when diapause is terminated and development commences (Andrewartha and Birch, 1954). In a few instances the length of the dark period is the determining factor in the initiation of diapause.

Temperature is another important factor in the induction of diapause in certain species. Generally, low temperatures favor the initiation of diapause, although this is not always the case [e.g., high temperatures favor diapause in *Bombyx mori* (Lees, 1955)]. Temperature usually acts in association with photoperiod in producing diapause. Other environmental factors that may be involved with diapause induction include unfavorable nutritional conditions, low moisture content of food, desiccation, crowding, maternal diet, maternal age at oviposition, and maternal exposure to low temperatures during oogenesis (Wigglesworth, 1965). Diapause in parasitic insects may be induced by diapause in the host, environmental factors acting on the host, or environmental factors acting directly on the parasite. In some parasitic species development is closely attuned to the hormonal changes in the host, and in some host species the presence of a parasite disrupts a diapause that would otherwise have occurred.

Andrewartha and Birch (1954) introduce the idea that development can be divided into two aspects: *morphogenesis* (the morphological changes during development) and *physiogenesis* (the physiological changes during development). They then look upon diapause as a stage in physiogenesis during which morphogenesis is disrupted and during which time certain physiological changes must occur before morphogenesis can be resumed. They call these physiological changes "diapause development." Low temperature seems to be the environmental factor that most affects diapause development. For example, several species (e.g., *Bombyx mori*) must be exposed to low temperatures for certain intervals of time before diapause can be terminated. When diapause occurs during a hot dry season, exposure to high temperatures for a period of time may be necessary for diapause development. Wounds and various kinds of shocks, such as pricking with a needle, electrical stimulation, and so on, may cause the termination of diapause in certain species (Andrewartha and Birch, 1954).

Control of Diapause

Three hypotheses have been advanced to explain the control of diapause induction (Andrewartha and Birch, 1954). One is that growth is arrested by the accumulation of inhibitory substances that require a period of "rest," during which time metabolism is reduced while excretion continues, resulting in the eventual disappearance of inhibitory substance with the subsequent resumption of growth. A second hypothesis relates diapause to desiccation of tissues. The remaining hypothesis relates the onset of diapause to the absence of growth hormones. The latter hypothesis is generally considered to best explain the diapause induction during all the post-eclosion stages.

Adaptive Significance of Diapause

Diapause is looked upon as a physiological timing mechanism that provides for (1) the induction of a comparatively resistant state during periods of adverse environmental conditions, particularly low temperatures; (2) the resumption of development when adverse conditions have disappeared and food is available (e.g., synchrony with host plants); and (3) synchrony of adult emergence in species with a short adult life.

Adkisson (1966) and Mansingh (1971) both consider the adaptive significance of diapause in some detail and also provide good, up-to-date bibliographies on insect dormancy. In addition, Mansingh (1971) proposes a new classification of the various types of dormancy in insects.

Selected References

REPRODUCTION
Davey (1965); DeWilde (1964a, 1964b); Engelmann (1968, 1970); King (1970); Schaller (1971).

Reproduction and Morphogenesis

MORPHOGENESIS

Agrell (1964); Bodenstein (1971); Clark and Rockstein (1964); Counce (1961); Gilbert (1964); Hagan (1951); Highnam (1964); Johannsen and Butt (1941); Kennedy (1961); Lees (1955); Richards (1961); Schneiderman and Gilbert (1964); Wigglesworth (1954, 1964b).

9

Insects and Their Environment

Consideration of the insect and its environment represents the culmination of our treatment of the levels of structural and functional organization of insects. The study of organisms in relation to their environment comprises the modern science of ecology, which has developed in recent years from the earlier, essentially descriptive endeavor, natural history. The Odums' (1959) comment that ". . . it is more in keeping with the modern emphasis to define ecology as the structure and function of nature" is consistent with the inclusion of this chapter under the general heading "Structure and Function."

Ecology can be looked upon as being comprised of two subdivisions, depending on the approach one uses: *autecology* and *synecology*. The autecological approach is to study the interrelationships between a single organism or species and its environment, the organism or species being the focal point. An example would be a study of the influences of temperature on a population of a particular insect species. Synecology is concerned with the interactions of communities of organisms. An example of a synecological study would be an investigation of the interrelationships of the plants and animals in a woodland pond. A number of excellent general ecology texts are available (see Selected References) as up-to-date treatises on this subject.

The basic functional unit of ecology can be considered to be the *ecosystem,* which is composed of a community of interacting organisms (plants and animals) and the nonliving components of their environment. Ecosystems may vary a great deal in size and complexity, ranging from a woodland pond to a conifer forest. The ecosystem is one of the basic units used in the study of the circulation of matter and energy between living organisms and their nonliving environment. Odum and Odum (1959) describe four basic constituents of an ecosystem: (1) abiotic substances, (2) producers, (3) consumers, and (4) decomposers. *Abiotic substances* are the inorganic and organic compounds found in the environment. The *producers* are mainly the green plants, which synthesize food from carbon dioxide and water by utilizing the energy derived mainly from solar radiation. The *consumers* are those organisms, the vast majority of which are animals, which ingest particulate organic matter and other organisms (plants and animals). The bacteria, fungi, and similar forms which break down complex organic compounds to simple substances usable by the producers comprise the last category, the *decomposers*. Insects, of course, fall into the third category, consumers.

The Life-System Concept

Clark et al. (1967) introduce a particularly useful concept for our purposes, the concept of the life system. They define a life system as "that part of an ecosystem which determines the existence, abundance, and evolution of a particular population." In other words, the life system of an insect is that part of the environment ("effective environment") which actually determines the fate of a given population plus the population itself. It is obvious that not all the parts of an ecosystem necessarily directly influence a given animal population, although the situation becomes rather hazy when one considers indirect influences. The essential components of a life system are presented in Figure 9–1. In the following sections we shall discuss the two major entities involved in the functioning of a life system: a population and its effective environment.

Environmental Components

In this section we shall discuss the various components of the environment which have been found to be significant in their influence

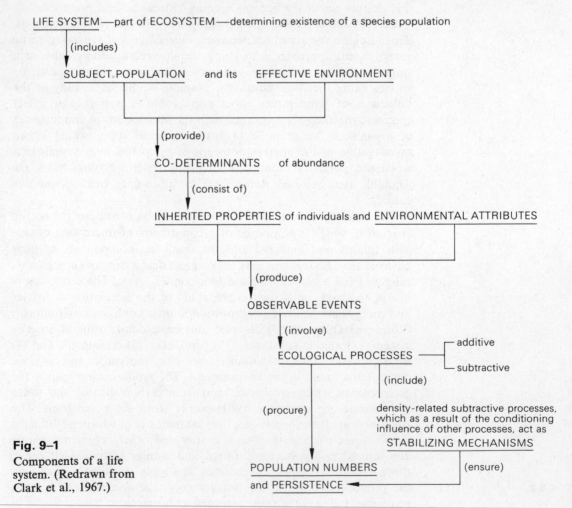

Fig. 9–1
Components of a life
system. (Redrawn from
Clark et al., 1967.)

on insect or other animal or plant populations. Any combination, or all of these components, may at some time act to varying degrees upon a given population. When one of them approaches or exceeds the species-specific limits of tolerance, whether these limits be expressed in terms of survival, development, fecundity, and so on, it becomes a limiting factor. That is, it becomes the environmental component that is directly responsible for limiting the extent of survival, degree or rate of development, fecundity, and so on. In a natural situation it is often exceedingly difficult to pinpoint a given environmental component as a limiting factor since obviously several components are likely to be interacting to produce the "effective environment" of any population. For the same reason, it is difficult to relate studies of environmental effects under the carefully controlled conditions of a laboratory to what actually occurs in a natural situation.

Another problem further complicates the picture. Until fairly recently environmental components, in particular temperature, moisture, and light, have been measured based on the tacit assumption that they are uniform throughout an ecosystem. For example, relationships between changes in the size of a population and temperature changes have been sought using a measurement of temperature changes in the general location of the population. However, it is now clear that the temperature on even one part of a plant may be quite different from that on another part. Thus, at least in many instances, to obtain a truly accurate picture of any relationship that might exist between population size and temperature changes one must measure the temperature in the specific part of the ecosystem (e.g., bottom side of a leaf) in which the members of a population are found. Such measurements have become possible with the advent of modern, sophisticated instrumentation. Realization of this problem has led to the development of the concept of "microenvironment," which implies the recognition of distinct horizontal and vertical (spatial) and temporal differences in environmental components within an ecosystem. Figure 9–2 portrays such differences in terms of relative humidity and water vapor pressure at different levels above the ground (vertical) and at different times during the 24-hour cycle (temporal).

Andrewartha and Birch (1954) divide environmental components into four basic groups as follows: (1) weather, (2) food, (3) other organisms, and (4) a place in which to live. Rolston and McCoy (1966) modify this scheme slightly and prefer to call the last category listed "a particular place in which to live" or habitat. We shall use Andrewartha and Birch's classification as modified by Rolston and McCoy as a basis for discussion of the various environmental components. Other ecologists have preferred to divide the components of the environment into two major categories: physical (abiotic) and biological (biotic).

Weather

Weather may be considered to be "the result of the combined action of all the physical factors of the environment at any given time" (Graham and Knight, 1965). It varies continually throughout

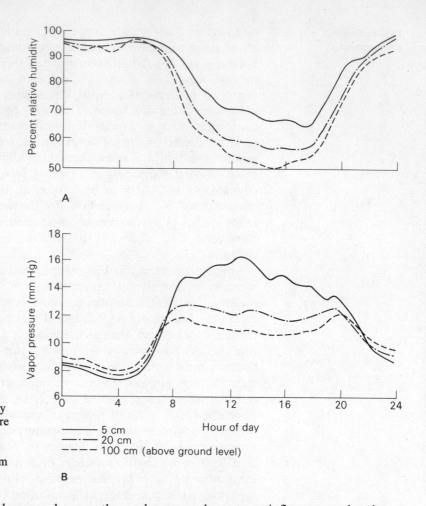

Fig. 9–2
Changes in relative humidity (A) and water vapor pressure (B) at different levels above the ground and at different times of day. (Redrawn from Chauvin, 1967.)

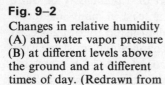

days, weeks, months, and years and exerts an influence on the abundance, longevity, rate of development, and so on, of insects from one year or season to the next. Climate, on the other hand, is the average course or condition of the weather in a locality over a period of several years. Weather undergoes rapid, often violent changes while climate tends to remain pretty much the same or change very slowly over a period of many years. The main elements of weather are temperature, moisture, and light. Each of these will be considered in turn and will be followed by discussion of various other physical environmental factors known or thought to exert a degree of influence on insects.

Temperature. Every insect species has a fairly well defined range of temperature within which it is able to survive. Exposure to temperatures above the high or below the low extremes of this range results in death. The extremes of this tolerable range of temperatures, of course, vary from species to species and also with the physiological state of an individual. Thus there are times or stages during the life cycle when an individual may be able to survive exposure to high or low temperatures that at other times would kill it. For example, many insects are able to survive much lower temperatures in the fall and winter than in the spring or summer. Tropical species are generally less tolerant of cold than those in temperate zones. Terrestrial insects usually have a somewhat wider range of tolerable temperatures than do aquatic insects, and not surprisingly the usual range between high and low temperatures in terrestrial habitats is usually substantially greater than that in aquatic habitats.

The range of survival relative to temperature for most insects probably lies somewhere between 0 and 50°C, although it is likely that no one species can thrive throughout this entire range. There are, however, exceptional species that are able to survive at temperatures well beyond these extremes. For instance, some dipteran larvae apparently thrive at temperatures of 55°C or higher, while certain species of beetles go through their entire life cycle in ice grottos at temperatures slightly below and slightly above 0°C (Andrewartha and Birch, 1954). The firebrat, *Thermobia domestica*, can live indefinitely at temperatures of 42°C and higher (Andrewartha and Birch, 1954). It typically inhabits such places as ovens, hot-water pipes, and similar "hot" environments. If insects of a given species are exposed to a gradient of temperatures, they will move along it until they reach the "preferred temperature," at which point they will tend to congregate. In zones of temperatures several degrees below the upper and lower limits of survival, insects are dormant, being in states of *estivation* and *hibernation*, respectively. In the range between these zones of dormancy, insects are active.

The actual cause or causes of death at the limits of the temperature range are not entirely clear. At the lower limit the submicroscopic structures of cells may be disrupted by the formation of ice crystals or the metabolic balance may be thrown off. At the upper limit protein denaturation and the disruption of ordered molecules (as, for example, in the wax layer of the epicuticle) are likely to be involved. The phenomenon of insect cold hardiness has been reviewed by Salt (1961). Many insects that hibernate or enter diapause in temperate regions are able to survive low temperatures for considerable periods of time. Most of these hibernate or diapause in a stage that is more cold hardy than the preceding one. Some are capable of long exposures to low temperatures and display a certain amount of cold acclimation but succumb to freezing of the body fluids. For example, *Aedes aegypti* larvae reared at 30°C are killed by exposure to −0.5°C for 17 hours but survive such an exposure if preconditioned for 24 hours at 18 or 20°C (Bursell, 1964).

Other insects are able to avoid freezing because they possess the ability to become supercooled (i.e., cooled below the point of freezing but without freezing actually taking place). Individual species vary as to the temperature to which they can supercool. Glycerol has been found in the tissues of many hibernating and diapausing insects. Andrewartha and Birch (1954) explain "a close relation between the concentration of glycerol and the temperature to which individuals of *Bracon cephi* can supercool has been established." Glycerol begins to appear in the tissues in correlation with the advent of hibernation or diapause and usually disappears rapidly when the insects come out of these states (Salt, 1961). A number of insects (e.g., many caterpillars) are able to survive the formation of ice within their bodies. All members of this group of "freezing-tolerant" insects have been found to possess high concentrations of glycerol in their hemolymph.

The lethal effects of high temperatures are difficult to study because of the influence of moisture changes on the results, since these two environmental parameters interact with one another. We shall consider this interaction later. It has, however, been established that lethal temperatures may vary depending upon the temperature to which an insect, or more properly a sample of a population of insects, has been exposed previously. This has been demonstrated in *Calliphora*, *Phormia*, and others (Bursell, 1964). This high-temperature acclimation is the counterpart of the phenomenon of cold acclimation, which occurs in some insects at the opposite end of the temperature range of a species. Acclimation to both high and low temperatures may or may not occur in the same species. Like cold acclimation, high-temperature acclimation is likely to be of value to insects under natural circumstances, since the seasonal and daily high temperatures are generally preceded by a gradual transition from somewhat lower temperatures. This acclimation at both ends of an insect's tolerable range of temperatures may be looked upon as promoting the insect's survival against the effects of extreme daily and seasonal temperature fluctuations.

Temperature also affects the duration of life. For example, a clear relationship has been shown to exist between the duration of life in male and female *Drosophila subobscura* and temperature (Bursell, 1964).

Temperature may also affect the survival of an insect by influencing the rate of utilization of food reserves when food is present in limited quantities (Bursell, 1964). This problem is particularly acute in bloodsucking insects such as the tsetse flies, *Glossina* spp., which depend on stored reserves of food between blood meals. Increases in temperature would effectively shorten the survival period between blood meals. Obviously if a fly uses up its reserves from one blood meal before it is able to obtain another, it will perish.

Temperature exerts a strong influence on the reproduction and rate of development of insects. As with lethal limits of temperature, insects also have definite tolerable ranges of temperature in terms of reproduction and development beyond which neither will occur. For example, the temperature range in which the eggs of the beetle *Ptinus* will develop is between 5 and 28°C (Bursell, 1964). *Pediculus* fails to lay eggs below 25°C (Wigglesworth, 1965). For a given species the range in which development will occur is probably somewhat broader than that in which reproduction will be successful. Within the tolerable ranges, egg development, oviposition rate, and the rate of larval and pupal development will usually increase with increasing temperature. A good example is the effect of temperature on the pupal development of the mealworm beetle, *Tenebrio*. The duration of pupal life is decreased by 180 hours (from 320 to 140) as the temperature is increased by 12°C (from 21 to 33°C) (Wigglesworth, 1965).

The specific ranges of tolerance and influence of temperature on rates vary among species. In addition, the span of the ranges will vary from species to species. Within each range of tolerance will be an optimum zone in which the rates of reproduction and development are maximal. For example, the oviposition rate of *Toxoptera graminum* increases with increasing temperature to a maximum of approximately 25°C and then falls off. Most laboratory studies of the effects of temperature on reproduction and development have been carried out at constant temperature. The results of such studies do not necessarily reflect what would occur under natural, uncontrolled conditions. The periodic fluctuations in temperature that are likely to occur under natural conditions tend to stimulate higher rates of development than at a constant temperature. For example, "grasshopper eggs kept at a variable temperature showed an average acceleration of 38.6% and nymphs an acceleration of 12%, over development at comparable constant temperature" (Odum and Odum, 1959).

Rolston and McCoy (1966) present a very general, but quite useful graph (Fig. 9–3) illustrating the ideas of tolerable temperature ranges as they pertain to survival, mobility, development, and reproduction.

The distribution, horizontal and vertical, of an insect species is often greatly affected by temperature. In temperate zones the northern extreme of a given insect's distribution is commonly determined by low-temperature extremes. When northern limits are determined in this manner, there is usually a zone somewhat below the extreme limits in which the overwintering stage is killed but which is repopulated during the warm season. Proceeding southward from such an area, a greater and greater percentage of the overwintering individuals survive. For example, the corn earworm, *Heliothis zea* (order Lepidoptera, family Noctuidae), in eastern North America must become completely reestablished during the warm season each year in Canada and to a progressively lesser extent in the United States proceeding southward (Andrewartha and Birch, 1954). Of course the degree of cold hardiness displayed will also play an important role in determin-

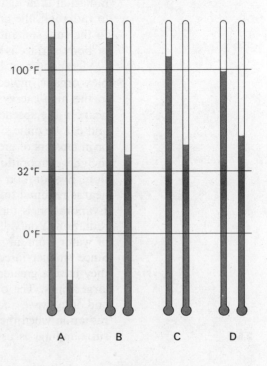

Fig. 9–3
Tolerable temperature ranges for various aspects of a hypothetical insect's life. Within each pair of thermometers, the left-hand one registers the maximum tolerable temperature, and the right-hand one, the minimum tolerable temperature. A. Survival. B. Mobility. C. Development. D. Reproduction. (Redrawn from Rolston and McCoy, 1966.)

ing the northernmost limits of the distribution of a particular species.
The number of generations per year often varies between the more
northern and more southern parts of an insect's range. For example,
the corn earworm completes a generation approximately every 36 days
as long as weather permits (Rolston and McCoy, 1966). Thus in its
northern range it may only complete two or three generations per
year, whereas in the southern portion it may complete several more,
breeding the year round. One would expect a gradation in the actual
number of generations between the northern and southern extremes
of the range. Moisture also exerts an influence on distribution.
Temperature also affects, to a greater or lesser extent, the rate of
dispersal of an insect species.

To this point we have been discussing the various ways temperature
influences insects. We now want to consider the actual temperature
of an insect and the factors that may influence it. We have established
the idea that the temperature, or any of several other parameters, may
vary substantially even within a small area (e.g., different parts of a
plant). Likewise, the temperature of an insect in one part of a habitat
may not be anywhere near the same as one in another and may not
be the same as the ambient temperature. On the other hand, the
temperature of an insect under controlled laboratory conditions is
usually reflected by the ambient temperature.

The temperature of an insect arrives at a given point depending
on the sources of heat gain and loss operating under a given set of
circumstances. The major sources of heat gain are solar radiation and
metabolic heat. Solar radiation may cause the temperature of an
exposed insect to be significantly different from the ambient temperature (Fig. 9–4). The effect of solar radiation on the temperature of an
insect is influenced by such factors as size, larger insects being more
affected than smaller ones; color, darker colors absorbing more
radiation than lighter ones; shape, the more surface directly exposed
to radiation, the greater the absorption; and orientation with respect
to the sun, some insects orienting such that a large or small amount
of body surface is exposed.

Metabolic heat is that which results from the breakdown of complex organic molecules. Approximately half of this energy is stored
in the high-energy bonds of ATP; the remaining half is released as
heat. In the absence of solar radiation, this is the sole source of heat
and can be quite significant in heat balance, particularly during flight
or in clusters of gregarious forms. Sources of heat loss from an insect
include evaporation, convection, conduction, and long-wave radiation. Evaporation of water from an insect has a cooling effect since
heat is required to propel a molecule of water from the body surface.
Evaporation is the major cause of heat loss in the absence of solar
radiation. You will recall from Chapter 3 that the rate of evaporation
of water from an insect is dependent partly on the size of the insect.
Since smaller insects have a larger ratio of surface area to volume,
they have a greater tendency to lose water through evaporation than
larger ones. The other causes of heat loss—convection, conduction,
and long-wave radiation—operate mainly in the presence of solar
radiation when the temperature differential between an insect and its
surroundings is usually greater than in the absence of sunlight. Air

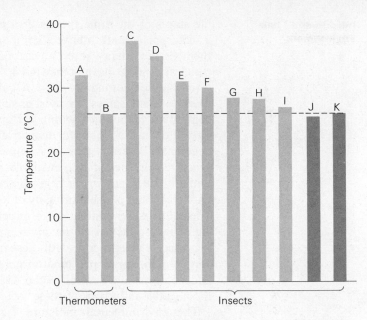

Fig. 9–4

Effect of radiation on the body
temperature of selected insects.
A. Thermometer painted black.
B. Air temperature. C. *Bombus*
(bumble bee). D. *Xylocopa*
(carpenter bee). E. *Apis* (honey
bee). F. *Anisoplia* (beetle) G.
Asilus (robber fly). H. Damselfly.
I. Butterfly. J. Butterfly. K.
Damselfly. Light grey, sunlight;
dark grey, shade. (Based on
data from Masek Fialla, 1941.)

movement may accentuate the effects of all these causes of heat loss,
including evaporation. Dense coverings of hairs and scales, on the
other hand, may serve to retard heat loss.

Many insects exert a degree of control of heat gain and loss. For
example, many insects are gregarious and tend to form clusters under
conditions of low temperature. Overwintering honey bees form such
clusters within their hives. The combined effects of the metabolic heat
of the component individuals may actually accelerate the rate of
development, as in the case of certain caterpillars (Bursell, 1964). The
nests of social insects provide good examples of temperature control.
For instance, the nests of some species of termites, because the
structure of the cells permit them to act as an air blanket, have an
effect analogous to storm windows. During a hot day, the outer
surface of a nest may be too hot to be touched by a human hand,
whereas deep within the nest the temperature may be somewhere
around 30°C. On a cold winter day, the internal temperature of the
same nest may be 9 or 10°C warmer than the air surrounding it
(Skaife, 1961). Lüscher (1961a) describes the evolution of such "air
conditioning" in termite nests. The nests of some wasps (e.g., hornets)
are so constructed that they have an outer "paper" envelope covering
the layers that contain the rearing cells, resulting in an air-blanket
effect. Other social wasps (e.g., *Polistes*) do not construct such an
envelope. If a nest gets too hot, a number of its inhabitants will fan
their wings vigorously at the nest entrance and may even bring in
drops of water, which has the effect of increasing the cooling from
fanning (Richards, 1953). The behavioral mechanisms involved in
orientation with respect to the sun are important in temperature
regulation, as are those which enable an insect to respond to a gradient
of temperature and locate an "optimal" zone.

Moisture. The water content of insects varies from less than
50% to more than 90% total body weight (Wigglesworth, 1965).
Variation occurs both between different species and between different

life stages of the same species. For example, soft-bodied insects such as caterpillars tend to have a comparatively large amount of water in their tissues; many insects with hard bodies (i.e., relatively thick cuticles) tend to have somewhat lesser amounts. Active stages of a given species commonly have a considerably higher water content than dormant stages. In most instances it is critical that the water content be maintained within certain limits which are generally not as well defined as those for temperature and are influenced to a large degree by several other environmental factors (e.g., temperature, pressure, air movement, available surface water). However, if the limits of tolerance under a given set of circumstances are exceeded, an insect either perishes or many of its activities are seriously impaired. The actual water content of an insect is the result of the influences of the moisture factors of the environment and all the means by which the insect is able to cope with these influences. In this section we shall consider the principal environmental moisture factors, how they affect insects, and several of the ways insects are adapted to combat the negative influences of these factors.

The environmental moisture factors of major significance are precipitation, humidity, condensation, and available surface water. Rainfall is the most common and widespread form of fluid precipitation. Snow is the most common form of solid precipitation; hail and sleet are less common forms. The annual and seasonal amounts of precipitation are primarily determined by the movements of large masses of air and by topographical characteristics. Thus there are very "wet" and very "dry" regions and innumerable gradations between. Humidity refers to the amount of water vapor in the air and depends on temperature and atmospheric pressure.

Condensation (dew, fog, and white frost) occurs when the atmosphere becomes saturated with water vapor. Saturation is the result of the relative humidity approaching 100% or the temperature dropping below the dew point (the point at which the relative humidity becomes 100%).

Available surface water is related to all the other moisture factors and to the nature of the substrate (soil, leaf surface, bark, and so on). In reference to the latter, it is well known that soils vary tremendously in their water-holding capacity and the rapidity with which water runs off or soaks in.

All these moisture factors influence the water balance of terrestrial insects. What, then, are the environmental factors that influence the water balance of aquatic insects? The "wetness" or "dryness" of an aquatic environment is a function of the osmotic pressure. Thus insects living in fresh water must cope with a comparatively "wet" environment; those in brackish and salt water (e.g., many of the salt-marsh forms, such as the mosquito *Aedes sollicitans*) are living in a very "dry" environment.

Compounding the water-balance problem for insects is the fact that under other than carefully controlled laboratory conditions the moisture factors are continually changing. Obviously an insect living in a given locale must be able to survive the limits of these changes. As with temperature and the other weather parameters, we must again think in terms of microenvironments. For example, the amount of rainfall measured at the edge of a forest hardly reflects the actual amount of water reaching an insect living in a tree hole or on the underside of a leaf. All the moisture factors vary both temporally and spatially. Humidity is useful in the illustration of this point. It may vary with location, time of day or year, topography, vegetation, and so on. For example, relative humidity often tends to be comparatively high during the night and lower during the day and is commonly different at different heights above the ground (Fig. 9–2).

It was mentioned earlier that the moisture content of the environment may determine whether or not an insect survives, that the tolerable limits are not as clearly defined as those for temperature, and that these limits may vary between species and between different life stages of the same species. Within the tolerable limits, there is of course an optimal zone in which a given species thrives. Failure to survive may occur under either conditions of excessively low environmental moisture content for the active stages of most insects and under conditions of excessively high moisture content for many. Death under very dry conditions is generally due to excessive water loss. Under conditions of excessive moisture the causes of death are more variable and may be direct or indirect. For example, drowning may be the result of excessive moisture, as is the case with overwintering pupae of the moth *Heliothis zea* during wet years in the southeastern United States (Andrewartha and Birch, 1954).

Survival is indirectly affected by very wet conditions by the spread of viral, fungal, and bacterial diseases and by any negative effects the excessive moisture might have on the food of a given insect species. For example: "The fall webworm, *Hyphantria cunea*, the armyworm, *Pseudaletia unipuncta*, and the gypsy moth, *Porthetria dispar*, succumb most readily to viruses when the weather is warm and the humidity is high. Whether high humidity affects the host or the development of the pathogen is not clear. However, many viruses do not develop rapidly unless temperatures of 21 to 29.4°C and relative humidities of 50 to 60% are experienced" (Subcommittee on Insect Pests, 1969). If excessive moisture doesn't kill an insect, it may seriously affect the length of its life. For example, newly emerged adult migratory locusts. *Locusta migratoria,* live longer the lower the humidity (Andrewartha and Birch, 1954). Since environmental moisture content may determine survival, it is commonly a major factor along with temperature in determining the geographic distribution of an insect species. For example, according to Rolston and McCoy (1966).

The High Plains grasshopper reproduces and develops in an area of about 50,000 square miles, and from this center the adults may spread over an area perhaps twice as large. The region of endemic infestation lies in the short-grass belt of the High Plains where the average winter temperature generally falls between 28° and 38°F and the average annual precipitation ranges from about 15 to 18 inches. That the High Plains grasshopper flourishes in only a part of the short-grass belt indicates that climatic conditions rather than host availability, limit the range of the insect.

Extremes of the moisture content of the environment directly influence many of the activities of insects, including feeding, re-

production, and development. Spruce budworm larvae stop feeding
when the air becomes saturated with water (Graham and Knight,
1965), and the tsetse fly, *Glossina tachinoides*, does not feed on its
vertebrate hosts when the relative humidity is above 88%
(Andrewartha and Birch, 1954). Newly emerged adult migratory
locusts do not produce eggs below about 40% relative humidity
(Andrewartha and Birch, 1954). Generally low humidities adversely
affect the rates of oviposition, which increase as humidity increases.
The rate of development may be decreased by extremes of moisture
content. For example, the spruce budworm larvae mentioned above
would likely take a relatively long time to develop under wet conditions since these conditions cause cessation of feeding. Under very
moist conditions silkworm larvae fail to pupate (Andrewartha and
Birch, 1954). According to Bursell (1964) the incubation time for eggs
of the spider beetle *Ptinus*, under a constant temperature of 20°C, is 15
days at 30% relative humidity and 10 days at 90%. He further points
out that generally higher humidities are more favorable for embryonic
stages than low humidities. Andrewartha and Birch (1954) present two
sets of hypothetical curves (Fig. 9–5) that summarize well the points
made in this and the preceding paragraphs.

The effects of the environmental moisture content are often strongly
modified by other weather factors. Temperature and moisture, in
particular, interact to a large extent in their effects. Thus temperature
exerts a relatively great effect on insects under the extremes of

Fig. 9–5
Influence of humidity
on various aspects of a
hypothetical insect's
life. A. Insect not
harmed by high
humidity. B. Insect
adversely affected by
high humidity. Zone 1,
lethal dryness; Zone 2,
favorable moistness;
Zone 3, lethal wetness.
(Redrawn from
Andrewartha and
Birch, 1954.)

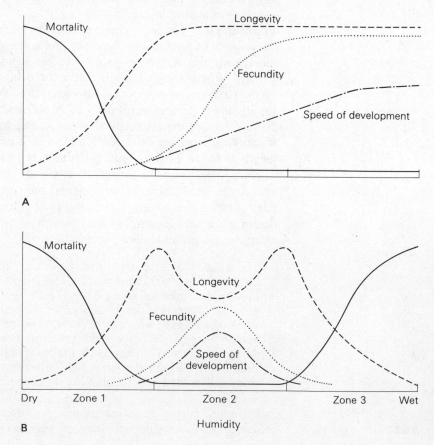

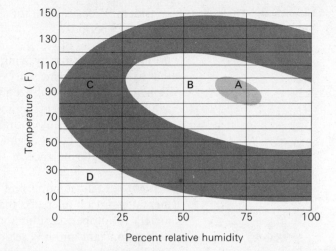

Fig. 9–6

Interrelationship of temperature and
humidity as they affect the rate of
development of a hypothetical insect.
A. Region of most rapid development.
B. Region of favorable development.
C. Region of retarded development.
D. Region of no development. (Redrawn
from Graham and Knight, 1965.)

moisture conditions and vice versa. For example, the boll weevil is
more tolerant of higher temperatures at comparatively low humidities than at high humidities (Graham and Knight, 1965). Another
good example of the interactions of temperature and moisture is
their combined effect on the rate of development of insects. Graham
and Knight (1965) present a graph that illustrates this example well
(Fig. 9–6). As mentioned earlier, air movement, atmospheric pressure,
availability of surface water, and other factors also influence to a
greater or lesser extent the effects of moisture conditions on insects.

Insects are adapted to varying degrees to cope with changes in
environmental moisture conditions in a number of ways. Quiescent
stages (eggs and hibernating, aestivating, and/or diapausing stages)
are often very well suited for exposure to prolonged drought. The eggs
of many species of mosquitoes (e.g., *Aedes, Psorphora,* and others)
can withstand prolonged drying in air (Clements, 1963). Many
insects become dormant (summer—aestivation; winter—hibernation)
in response to different environmental conditions and in this state are
able to withstand long periods of drought. For example, the potato
beetle, *Leptinotarsa decemlineata,* becomes quiescent in response to
dryness and can survive in this resting, desiccated state for months.
Such a response ensures that when eggs are laid, the larvae will be
exposed to environmental conditions more conducive to survival
(Andrewartha and Birch, 1954). The water content of diapausing
and anabiotic stages is reduced to a fraction of that in active stages.
Insects in these states may survive for months or even years. Hinton
(1951, 1960) describes a chironomid larva, *Polypedilum vanderplanki,*
that is able to survive almost complete dehydration for several years.

Active stages of insects are also able to cope with moisture changes.
It is well established that insects will respond to a moisture gradient
in such a way that they move to an "optimal" zone (see Chapter 7).
This obviously promotes survival. Other factors, of course, may
influence exactly how an insect behaves in a moisture gradient. If the
flour beetle, *Tribolium,* has been starved, it will demonstrate a preference for moist air, but if it has been given access to food and water,
it prefers dry air. Time is an important element in the survival of
active insects under adverse moisture conditions. Many can survive

considerable desiccation for short periods of time or slight desiccation for longer periods.

Most insects offset the influence of dryness and other factors promoting water loss by drinking water or taking it in with food. Some that live under extremely dry conditions, such as the mealworm beetle, *Tenebrio*, are able to utilize metabolic water (i.e., water that becomes available as a result of the metabolic breakdown of food materials). Other insects (e.g., thysanurans) are able to absorb moisture directly from the atmosphere (Wigglesworth, 1965). See Chapter 3 for a discussion of water balance in aquatic insects.

Light. Light is important to insects directly more as an environmental point of reference than as a survival factor in that its parameters (photoperiod, illuminance, wavelength, and so on) are more or less constant and it is seldom, if ever, directly lethal under natural conditions. In the nonequatorial region of the earth there is a regular change in photoperiod due, of course, to the fact that the earth's axis of rotation is tilted $23\frac{1}{2}°$ from vertical to the imaginary plane that passes through the sun and the earth's orbit. This regular change in photoperiod serves as an annual clock for insects and is used by many of them to maintain synchrony with the seasons and their host plants. Photoperiod is one of the major stimuli that induces diapause. The daily cycle of dark and light with crepuscular periods in between also serves as a clock by which the feeding, mating, and so on, activities of most insects are regulated. As with temperature and moisture, the reactions of insects to photoperiod and other light parameters varies both among different species and among different life stages of the same species.

The wavelength of light is utilized by many plant-feeding insects as a means of host finding. The position of the sun and degree of polarization of light in different parts of the sky are important to many insects in orientation and navigation. More information on the response of insects to the various light parameters is to be found in Chapters 5 and 7. Although light is not itself lethal, its effect on both aquatic and terrestrial plants may indirectly influence the activities of insects. For example, the amount of light reaching submerged aquatic vegetation will affect the oxygen-generating photosynthesis of this vegetation and in turn affect the oxygen concentration in the water. As we shall see, oxygen absorbed in water exerts a considerable influence on aquatic insects.

Other Factors. Other environmental factors that under some circumstances influence insects include currents in air and water, gases dissolved in water, air composition, and electricity.

Currents in air and water are determined to a large extent by physiographic conditions. Thus air movement is considerably modified by trees and various other types of vegetation and by any other physical situations that may block or redirect it. Currents in water are influenced by such factors as the volume of water, slope of stream bed, temperature differences between the surface and various depths, and so on. Wind is a very effective agent in the distribution of insects—aphids, leafhoppers, and others being blown for hundreds or even thousands of miles. Wind determines the direction taken by migratory locusts (Johnson, 1963).

Air movement may be directly responsible for the death of insects in two ways. First, severe wind and heavy rain may together cause considerable mortality. Second, movement of air above a surface where evaporation is occurring (e.g., insect cuticle) maintains a very sharp gradient of water vapor concentration and hence tends to increase the rate of evaporation. Other factors being constant, the rate of evaporation is proportional to air movement. On the other hand, air movement may be beneficial if humidity is exceedingly high. Water currents are very significant in that they often determine which species of insects will live in a given area. For example, the various genera of mayflies (order Ephermeroptera) may be classified as to whether they are still or rapid water forms (Needham, Traver, and Hsu, 1935). The legs and bodies of these insects are appropriately adapted (e.g., legs capable of clinging, and streamlined bodies for those which must cope with fast-moving water). Various other insects also show interesting adaptations to maintain their position in moving water. For example, black fly larvae fasten themselves to stones or other stationary material in the water. Several caddisfly species attach their cases to various submerged objects. Many aquatic insects (e.g., mosquito larvae) are unable to survive in moving water. Another important aspect of currents in the aquatic environment is their function in the circulation of dissolved gases, salts, and nutrients.

The insect fauna in a given location is often determined to a large extent by the amount of dissolved oxygen available. For instance, caddisfly and mayfly larvae may be found under conditions of relatively high oxygen concentration, Chironomid and Simuliid larvae (order Diptera) at somewhat lower concentrations, and certain mosquito and other fly larvae at very low concentrations (Usinger, 1956a). The amount of dissolved oxygen is determined by water movement, wind, water splashing, the photosynthetic activities of aquatic plants, and so on. Other dissolved gases that may be of importance are carbon dioxide and nitrogen. The former is very soluble in water and a by-product of cellular respiration.

According to Odum and Odum (1959), the gaseous composition in air is remarkably constant and is probably not a significant limiting factor for terrestrial insects and other animals. However, Chauvin (1967) points out that the gas content in cavities in flowers may be quite different from the surrounding air and that a large number of insects live in such cavities. He also explains that the composition of air in the middle of a thick canopy of vegetation is not necessarily the same as in open air, nor is it necessarily the same throughout the day since temperature and the photosynthetic activities of plants vary during the course of a day. Chauvin (1967) cites an example where the carbon dioxide above fields of wheat and clover was measured at different times of day. The results were that the carbon dioxide decreased toward the middle of a day and reached a maximum between midnight and 6 A.M.

Electrical factors have seldom been taken into account, although it is known that electrical forces may have a directive effect on insects

(Folsom and Wardle, 1934). Under natural conditions, the ionization of air and atmospheric potential may vary and may affect certain activities of insects. For example, the ionization of air has been found to modify the flight activity of the blow fly, *Calliphora* (Chauvin, 1967). An increase in the number of ions causes a temporary increase in activity. It has been found that the flight of *Drosophila* is abruptly reduced for a short period of time by sudden exposures to a potential gradient of 10 to 62 volts/cm (Chauvin, 1967). Under natural conditions the atmospheric potential falls rapidly between the ground and a few meters above the ground (Chauvin, 1967).

Food

The feeding habits of insects were considered in Chapter 7. In this section we shall discuss how insects may be influenced by variations in both the quantity and quality of their food resources. As a group, insects are essentially omnivorous. As individual species, however, they exhibit much variation, some being extremely selective and perhaps feeding on only one or a few kinds of food (including host plants and animals and prey) and others taking advantage of many kinds. Whatever the particular habits of a given species may be, the abundance and quality of food may play an important role in its survival, longevity, distribution, reproduction, speed of development, and so on.

Quantity of Food. According to Andrewartha and Birch (1954) the animal populations that consume all or most of their food resources are few compared with those which do not. Thus any "absolute" shortage of food is probably not an important limiting factor for most populations. However, such shortages may occur in patches throughout the distribution of a given population. Under some circumstances there may not be an absolute shortage of food, but for a variety of reasons portions of potential food for a given species may not be available for consumption. In these circumstances there is an "effective" shortage of food.

There are, no doubt, many causes of "absolute" and "effective" shortages of food. Among these causes are large numbers of individuals per unit quantity of food, more than one species consuming the same food material(s), species that influence the food of other species without consuming the food, and other environmental factors. The first instance might occur, for example, among insects during the actively growing larval stages. When there is a relatively small number of individuals per unit quantity of food, there is enough for all to grow to adulthood. However, as numbers increase, a point is reached where there is insufficient food for all individuals present to grow to maturity and only those which develop most rapidly reach adulthood; the slower developers do not. In reference to the second possible cause of food shortage listed, it is obvious that two or more species consuming the same food will reduce the total amount available for any one of them.

Species that do not actually consume the food supply of another may exert a positive or a negative influence on that supply. Many of the sanitary practices carried out by man are appropriate examples. For instance, incineration of garbage and treatment of waste materials are in effect destruction of the food supply of many insect species, especially flies. On the other hand, man, in failing to carry out these procedures, is increasing the same food supply and thus enhances the spread of insects that thrive in such materials. Microbes which are pathogens of an insect that is the prey of another may reduce the population size of the prey species sufficiently to influence the predator species. For example, "an epizootic of the fungus *Entomophthora* so reduced the numbers of the caterpillar *Plutella* that the predators (*Angitis* spp. and others) suffered severely from a shortage of food where it had been abundant before the outbreak of disease" (Andrewartha and Birch, 1954). A similar situation may exist among a plant species, the insects that feed on it, and microorganisms that are pathogenic to members of the plant species. In fact, any environmental factor, biotic or abiotic, which has the effect of reducing populations of particular plants or animals is likely to indirectly cause the reduction of the populations of animals which utilize these plants or animals for food and in turn the reduction of the parasites and predators of these animals. Monophagous insects or other animals (those which eat only one kind of food) are more likely to be affected by such reductions in food supply than those polyphagous species which feed on a wide variety of food materials.

Examples of causes of "effective" shortages of food include accidental separation from the food source, certain behavioral traits, and effects of feeding activities on animal hosts. It is not difficult to imagine an ectoparasite starving to death when it is separated from its host. Behavioral mechanisms keep tsetse flies in close proximity to tall bushes. These flies will feed on vertebrates that are in the vicinity of these bushes but will not feed on the same or other vertebrates in open grassland, although they may be quite abundant (Clark et al., 1967). Where a single or a few blood-feeding insects might quite successfully feed to satiation on a host, larger numbers may elicit an avoidance or destructive response on the part of the host, resulting in very few, if any of them, being able to feed to satiation. Thus, although a given host may be quite able to furnish a blood meal for several hundred tsetse flies, mosquitoes, and so on, it may tolerate the feeding activities of only a small fraction of that number.

When there is an absolute shortage of food, the food plant or animal may be completely destroyed within a given region, but it is unlikely that total destruction of a food plant or animal could occur in this way. A shortage of food has the obvious effect of reducing populations of the insects that utilize it. If a food shortage is not too acute, all feeding individuals may survive but be much smaller than they would have been under a situation of greater abundance of food. However, as numbers of individuals increase, a point is eventually reached where fewer and fewer individuals are able to survive and finally none survive to adulthood. Another frequently encountered result of food shortage is cannibalistic behavior on the part of the starving species. One well-known species in which cannibalism is known to occur is the confused flour beetle, *Tribolium confusum* (Clark et al., 1967). Cannibalism may also occur when there is overcrowding, accidental injury to an individual, or during the vulnerable period immediately following ecdysis and before the cuticle has undergone the sclerotization process.

A number of characteristics may be of value to insects in circumventing the problems caused by a shortage of food. Probably one of the most important is well-developed powers of dispersal. Species that respond to food shortages by dispersing actively and widely are much more likely to discover a fresh food source than those species which do not. Thus, in these terms, food shortages would be of much greater significance to insects with weak powers of dispersal than those with strong powers. Many adult insects lay their eggs in situations where the food supply is extremely abundant relative to the numbers of individuals feeding on it. This is characteristic of such insects as fruit flies and house flies. Thus the behavior of the adults in egg deposition essentially guarantees abundant food for their offspring. Polyphagy, whether on the part of a parasite, predator, or phytophagous insect is another adaptation that ensures against the likelihood of being exposed to dangerous food shortages.

Quantity of Food. The nutritional requirements of insects in general were discussed in Chapter 3. Those activities which are particularly well known in terms of the influence of the nutritive quality of food are egg production and the development of immature stages to adulthood. Survival, longevity, and size may also be determined to a large extent by the quality of available food. Many insect species are able to store sufficient nutrients during the larval stages to accomplish the major activities carried out by adults (copulation and egg production and deposition). The adult lives of these species are comparatively short since they commonly do not feed at all. The prefix ephemero- in the ordinal name Ephemeroptera (mayflies) is of Greek origin and means "temporary" or "for a day," referring to the fact that the adult members of this order live for only a short period of time. They live only long enough to copulate and deposit their eggs, the nutrients for egg development and for sufficient energy to allow a fairly brief period of flight and copulation having been obtained during the larval stages. In other species the larval stages may store proteins sufficient for egg production, but the adults must ingest water and carbohydrates to carry out their activities.

Many mosquitoes and other true flies fall into a third category, in which adults usually must ingest a complete diet (especially protein) in order to sustain life and produce eggs. For the mosquitoes and other blood-feeding forms, blood meals afford the necessary protein. The queens of the social Hymenoptera and termites also fall into this third category. These insects lay eggs for almost the entirety of their adult lives and require food constantly in order to do so. The rate of development of larval stages may be influenced to a considerable extent by the quality of food.

An excellent example of the effect of quality of food on the rate of development is given by Folsom and Wardle (1934). Both bananas and horse manure are suitable foods for house fly maggots. However, at a temperature of approximately 21°C, the maggots complete their development on horse manure one to nearly two weeks sooner than on bananas. The quality of ingested food in some instances influences the outcome of development. The best-known example of this is the difference between worker and queen honey bees. The differences between these two castes depend entirely on the diet that each receives during its period of larval development, notwithstanding the fact that each caste is reared in a cell unique to that caste (Richards, 1953). Both receive the same food for the first two days of larval life—royal jelly, a substance secreted by glands that open into the mouths of workers. After the third day, larvae destined to become workers receive a diet containing increasing amounts of honey, whereas the queen larvae continue to be fed royal jelly throughout their immature lives.

Other Organisms

The other-organisms category of environmental components includes individuals of the same or different species as part of each other's environment, in other words, intraspecific interactions and interspecific interactions.

Intraspecific Interactions. The ways in which members of the same species interact are largely related to population density. There seem to be advantages associated both with relatively low and relatively high densities. Examples of circumstances in which high population density may be advantageous include mate finding and survival of potential predation (Andrewartha and Birch, 1954). In a situation where the probability of finding a mate is decreased due to low population density, the result would be fewer fertilizations and hence a lower average fertility (see under "Populations" in this chapter).

If an insect can serve as prey for a predator that usually preys on a different species, the probability of its being eaten may well be increased by its being with a few as opposed to being with many others of the same species. In addition, the production of large numbers of insects may be a species way of offsetting the negative influence of certain environmental factors. For example, when the reproductive stages of termites swarm in preparation for pairing and establishing new nests, they are subject to a tremendously high rate of predation, but, owing at least partly to the fact that they are present in such vast numbers, a few survive to maintain the species. Whether a given population density is advantageous or disadvantageous is of course relative and depends upon the degree of other environmental influences acting. This is particularly apparent with regard to such available resources as food and shelter. Clearly the maximum level of population of a given species an area could support depends on the abundance of these resources in that area. Thus given a situation of limited resources, intraspecific competition becomes increasingly pronounced as the maximum population level is approached and may result in the death, directly or indirectly, or emigration of individuals. Other environmental influences may, however, serve to keep a population below a level where competition for resources would become a limiting factor.

Interspecific Interactions. Interspecific interactions may be conveniently divided into competition, symbiosis, predation, and indirect effects.

Interspecific competition comes about when the needs of two or more different species for a resource (e.g., food or shelter) coincide. In some instances one species whose life needs are essentially identical to another, an *ecological homologue,* may literally cause the disappearance of the other species in a given area. This is the phenomenon of *competitive displacement,* the practical applications of which are considered in Chapter 12. Interspecific competition, at least in insects, apparently involves more subtle interactions than individual confrontations to the death.

Symbiosis is used here to mean a close association between two different species, whether this association be mutually advantageous *(mutualism)*, disadvantageous for one of the members *(parasitism)*, or advantageous for one without harm to the other (*commensalism*).

There are many examples of mutualistic relationships among insects—between different kinds of insects [e.g., ants that actively care for and protect aphids, which in turn excrete honeydew, which is ingested by the ants (Way, 1963)] or between insects and other organisms. The relationships described in Chapter 7 between the yucca moth and yucca plant and between the fungus-growing ants and the fungus they culture are good examples of mutualism between insects and other organisms, as are the relationships between microbes and insects relative to digestion and nutrition mentioned in Chapter 3. Another excellent example of a mutualistic relationship is that which exists between the bull's-horn acacia, *Acacia cornigera,* and the ant *Pseudomyrmex ferruginea.* Janzen (1967) describes this relationship as follows:

> The bull's-horn acacia . . . is a representative swollen-thorn acacia with well-developed foliar nectaries, enlarged stipular thorns, and small nutritive organs (Beltian bodies) borne at the tip of each leaf segment. The colony of *P. ferruginea* living in the enlarged stipules obtains sugars from the foliar nectaries, and oils and proteins by eating the Beltian bodies.

> . . . The workers patrol and clean the surfaces of the acacia, and bite and sting animals of all sizes that contact the plant. The workers maul any other species of plant that contact the acacia and in many cases, any that grow under the acacia. The colony attains a very large size and up to 25 per cent of the workers may be active on the surface of the acacia both day and night. The larger the colony becomes, the smaller is the damage sustained by the plant from defoliators. The colony enhances its own probability of survival by protecting the acacia on which it is completely dependent for food and domatia.

Among the organisms that parasitize insects are insects themselves (Fig. 9–7, see also Chapter 7), pathogenic microorganisms, mites, and nematodes.

The interest in microbes that are pathogens of insects has largely been focused on their potential use in biological control. In the past several years the specialty of insect pathology has come into its own. Readers interested in a detailed treatise of this area are invited to

A

Fig. 9–7

Scanning electron micrographs. A. Bee
louse, *Braula coeca* (Diptera, Braulidae),
an example of an insect that parasitizes
another insect. B. Dorsal view of the
tarsus, showing the comblike claws used
to cling to the characteristic branched
hairs, C, of the honey bee. The larva of
this insect lives beneath the honey
cappings and the adult lives on the body
of an adult bee and takes food directly
from the mouth of its host. An adult bee
louse is about 1.5 millimeters long.
(Courtesy of A. Dietz, W. J. Humphreys,
and J. W. Lindner.)

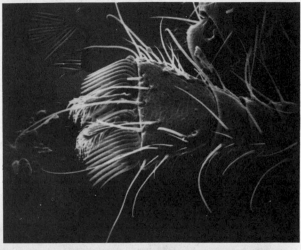

B

C

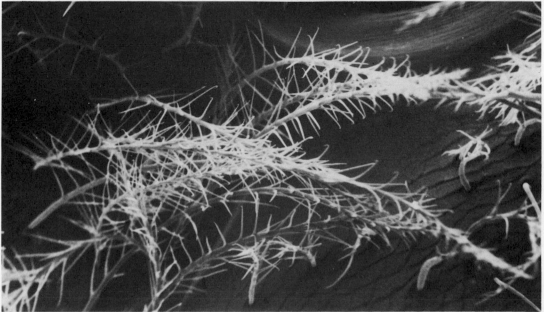

consult Steinhaus (1949, 1963). In addition, the work by the Subcommittee on Insect Pests (1969) includes an excellent chapter on the microbial control of insects with several pertinent references. In certain instances the microbes may be both insect and vertebrate parasites, in which case the insects may serve as vectors.

Well over 1,000 microbes, most of them pathogens, have been described in insects and new ones are being added regularly. According to the Subcommittee on Insect Pests (1969), the current list includes ". . . 90 species and varieties of bacteria, 260 species of viruses and rickettsiae, 460 species of fungi, 255 species of protozoa . . .." These microbes gain entrance into insects orally, via wounds in the integument, or are capable of actively penetrating the integument themselves (particularly fungi). In some instances pathogens that have gained entrance may be passed *transovarially* (via the egg) to offspring.

Among the species of bacteria known to infect insects, the best known is *Bacillus thuringiensis,* a spore-forming bacterium that has been used with some success in biological control. Another member of the same genus, *Bacillus larvae,* is the causative agent of American foulbrood, a serious disease of honey bee larvae. In addition to members in the genus *Bacillus,* certain species in the genera *Clostridium* and *Streptococcus* have also been found to be insect pathogens.

Viruses (Fig. 9–8) are obligate cellular parasites composed in part of DNA or RNA. Those associated with insects are generally not as well known or understood as are the bacteria. They are classified on the basis of the part of the host cell in which they develop, the presence or absence and morphology of an inclusion body, and the kind of nucleic acid (DNA or RNA) of which they are made. Some

Fig. 9–8
Electron micrograph showing the virus of Venezuelan equine encephalitis in thoracic tissue of the mosquito *Aedes aegypti.* The virus is approximately 450 angstroms in diameter and is composed of RNA surrounded by a protein coat. The arrow indicates a single virus. (Courtesy of R. F. Ashley.)

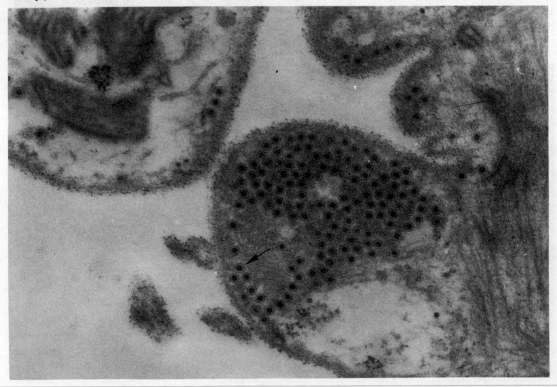

insect viruses produce granular inclusion bodies; most produce crystalline polyhedral bodies in the nucleus or cytoplasm of an infected cell. Virus particles may be alone in an inclusion body or in packets. Several rather tentative genera of viruses that are pathogenic for insects have been identified and described. The principal insect hosts thus far have been found in the orders Lepidoptera, Hymenoptera, Diptera, Coleoptera, and Neuroptera. By far the most have been associated with the Lepidoptera. It is interesting that a virus may act in conjunction with a bacterium to produce a disease syndrome not produced by the virus or bacterium separately. Such is the case in two silkworm diseases, gattine and flacherie, *Streptococcus bombycis* and *Bacillus bombycis*, respectively, being the bacteria involved (Rolston and McCoy, 1966).

Some rickettsiae have been observed to be insect pathogens. In size they lie somewhere between viruses and bacteria and, unlike viruses, they are capable of independent metabolism *in vitro*. Those rickettsiae which are pathogenic for insects are very slow in killing their hosts, have been grown in mammalian-tissue culture, and have killed white mice upon injection. Owing to this slow kill in insects and potential danger for mammals, it is not likely that they will ever be used in insect control (Subcommittee on Insect Pests, 1969).

There are more species of fungi which are known to be pathogenic to insects than in any other group. These fungi are often referred to as being *entomogenous*. These entomogenous fungi have been described in all four classes of Thallophytes, but most belong to the Phycomycetes and Ascomycetes. Species in the genus *Beauveria* (class Deuteromycetes, or Fungi Imperfecti) have been used in attempts at biological control. However, the success of fungus infections in insects (i.e., whether they become established and kill an insect or not) depends to a great extent on weather conditions, and hence these microbes would likely be undependable as agents of biological control.

A large number of protozoa are known to kill insects, but, like rickettsiae and fungi, they act very slowly. No doubt many of those slow-acting microbes weaken their hosts and make them more susceptible to the effects of other environmental components. Insect pathogens have been described in the protozoan classes Sarcodina (amebas), Mastigophora (flagellates), and Sporozoa (sporozoans). The microsporidian *Nosema bombycis* (class Sporozoa) causes the "pebrine" disease of silkworms. The elucidation of this relationship by Louis Pasteur in the nineteenth century stands as a classic in microbiological research. Other well-known microsporidian insect pathogens include *Nosema apis* and *N. pyraustai,* which cause disease conditions in the honey bee and European corn borer, respectively. A mosquito, *Culex erraticus,* parasitized by the microsporidian, *Thelahania minuta,* is shown in Figure 9–9.

Mites are very small arthropods in the class Arachnida, order Acarina. Many species are known to parasitize insects (Fig. 9–10). For example, the larvae of members of the family Erythraeidae are parasitic upon insects and other arthropods and have been found attached to wing veins and other locations on the insect body (Baker et al., 1956). *Acarapis woodi* (family Scutacaridae) causes the acarine disease of bees. This species either lives in the tracheae or on the

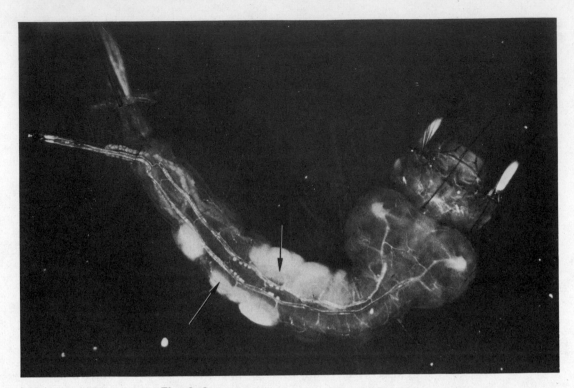

Fig. 9–9

Microsporidian, *Thelahania minuta* (indicated by arrows) infecting the fat body of the mosquito *Culex erraticus*. (Courtesy of H. C. Chapman and J. J. Petersen.)

external parts of the bee's body and weakens the host to the point where it can no longer fly and eventually succumbs (Baker et al., 1956; Gochnauer, 1963).

About 100 species of nematodes (phylum Aschelminthes, round worms; Fig. 9–11) have been described as being associated with approximately 1,500 insect species (Subcommittee on Insect Pests, 1969). In most instances, these "entomophilic" nematodes damage and kill their host. Welch (1965) reviews quite adequately the area of "entomophilic nematology." The nematodes show some promise in the biological control of insects, but much work needs to be done.

No doubt many relationships between two or more insect species or between insects and other organisms are commensal. For example, probably many of the microbes associated with insects neither harm nor particularly help their host but gain a place to live and food from it. Two good examples of commensalism, as the term is used in this text, are *phoretic relationships* and *inquilines*. In a phoretic relationship an individual of one species attaches to an individual of another and gains a mode of transportation. Insects may be the transporters, riders, or both. Some chewing lice attach themselves by their mouthparts to louse flies (family Hippoboscidae), some of which are also parasites of birds and in this way move from one host to another (James and Harwood, 1969). The torsalo fly, *Dermatobia hominis* (Fig. 9–12), deposits eggs on other flies, such as mosquitoes, black flies, and house flies. When these insects come into contact with a potential vertebrate host (including man), larvae emerge from the eggs and penetrate the skin (James and Harwood, 1969). Certain female members in the family Scelionidae (order Hymenoptera) are phoretic on adult grasshoppers and mantids and parasitic on their eggs (Borror and DeLong, 1971). In other words, they ride along until the host lays its eggs. Pseudoscorpions are known to be phoretic on certain flies and beetles and on other arthropods. In some instances they may eventually eat their transporter, but they usually prey on smaller organisms. Some pseudoscorpions in the family Chernetidae live under the elytra of large beetles and prey upon parasitic or phoretic mites which are also living on the beetles (Weygoldt, 1969).

Inquilines are animals that live in the habitations of other animals. Several insects are house guests of social insects, particularly ant colonies, without doing any apparent damage, but sharing some of the benefits to be obtained (e.g., food and shelter). Certain galls "con-

Fig. 9–10
Scanning electron micrograph of a mite (Arachnida; Acari) on the head of a termite (highly magnified). The mite is indicated by the arrow and is shown at a still higher magnification in the insert above. (Courtesy of Walter J. Humphreys.)

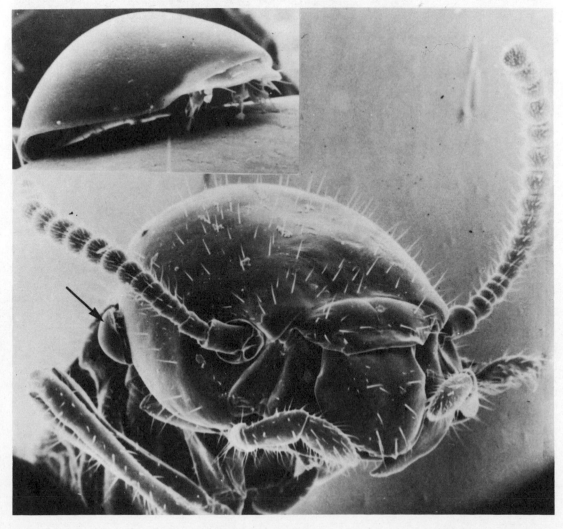

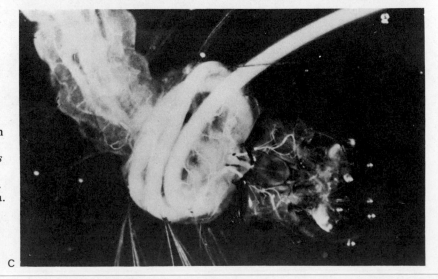

Fig. 9–11
Nematode parasite, *Reesimermis nielseni*, in the thorax of the mosquito *Culex pipiens quinquifasciatus*. A. Several infected larvae. B. Single infected larva. C. Postparasitic stage of the nematode escaping. (Courtesy of H. C. Chapman and J. J. Petersen.)

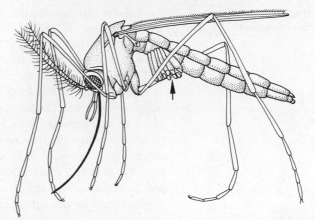

Fig. 9–12

Mosquito, *Psorophora* sp., carrying the eggs
(indicated by arrow) of the torsalo fly,
Dermatobia hominis. (Redrawn from USDA
photo.)

structed" by one species are shared by larvae of another species
without any apparent damage to the original owner.

Predators of insects include insects themselves; other Arthropoda,
such as mites, spiders, scorpions, pseudoscorpions, and so on; and
vertebrates, including birds, reptiles, amphibians, fish, and mammals.
Insect predators were discussed in Chapter 7, and comments regarding
vertebrate predators are included in Chapter 12. Sweetman (1936a)
discusses predatory invertebrates in some detail. In addition to
animals, a few species, such as the Venus's-flytrap, pitcher plants, and
the aquatic *Utricularia vulgaris*, trap, digest, and accumulate insects
(Frost, 1959).

In addition to the essentially direct effects that other organisms,
including insects, have on insects, there are a number of indirect
effects. No doubt the greatest effects are those produced by man
purposely against pest populations and inadvertently against the vast
majority of "innocent" ones (see Chapter 12). The effects of modern
agriculture, urban sprawl, and superhighways on various habitats are
obvious and the far-reaching and unfortunate effects of highly residual
pesticides have only in the last decade or so begun to be recognized.
Air and water pollution and their effects on the environment are
beginning to get the kind of attention they deserve. Insects themselves
are sometimes polluters. Clark et al. (1967) cite the work of Dunn
(1960): "Predatory anthocorid bugs enter open galls and feed on a
few of the aphids therein. The aphids which are not attacked die
from the fumigant effect of chemicals produced by the stink glands of
the bugs."

Habitat

Habitat may be defined as the place where a given insect (organism)
lives. For one species it may be an entire field or forest; for another
the shoreline of a stream; and for still another, the undersides of
leaves of a particular kind of plant. However it may be, a given
species tends to be found living in pretty much the same kind of
surroundings throughout its distribution. Obviously there may be a
certain amount of variation within the limits of tolerance of the
species. The degree of suitability will vary with the extent to which

the habitat satisfies the needs within these limits of tolerance.

The concept of habitat cannot be adequately discussed without a consideration of the concept of ecological *niche*. Within a given habitat, each species has a characteristic position spatially, temporally, and functionally (i.e., each species occupies a particular physical location at a given time and does a particular thing). The parameters space, time, and function as they apply to a given species define the ecological niche of that species. Thus for two different species to have the same niche, they must do the same thing at the same time and in the same place. Two animals that do occupy the same niche (ecological homologues) are, as mentioned earlier, competitors, generally one of which eventually overshadows the other (i.e., there is competitive displacement of one species by another).

Populations

Insect species can usually be recognized as being subdivided into more or less discrete groups, or populations, throughout their natural distribution. Obviously, when studying a given population, an investigator could not possibly, in the vast majority of instances, study an entire population, much less an entire species. Thus he must rely on "sampling"; that is, he must obtain and study a comparatively small number of individuals from a population, and hope that the characteristics they exhibit represent those of the whole population. Whether or not this hope is realized depends upon the adequacy of the sampling method. Sampling methods vary with the kind of insect, its habitat, and the kind of information to be derived from the sample. Many highly sophisticated sampling techniques have been developed, and elaborate statistical analyses have been made easier by the use of computers. It is well recognized that studies of populations must be made over several generations and that a given population may have a wide variety of individuals of different ages and in different life stages.

Relationships Between Environmental Components and Populations

Populations and their environments have little meaning when discussed separately. In this section we shall consider how these components of a life system relate to one another. The concepts covered in this section are largely from Clark et al. (1967).

Short-Term Numerical Changes. The long-term trends of numerical change in an insect population vary with several factors and may be quite different for different species. Thus it is difficult, if not impossible, to make broad generalizations regarding insect populations in the long term. However, if we consider numerical change over a very short interval of time, we can make certain generalizations. During this short interval of the population curve (numbers of individuals plotted against time) can be defined by the following equation:

$$N_t = N_0 e^{(b-d)} t - E_t + I_t$$

where t = the short interval of time
N_t = the number of individuals at the end of time t
N_0 = the initial number of insects (all life stages)
b = number of births during time t
d = number of deaths during time t
e = a constant (base of Naperian logs)
E_t = number of insects that emigrate from the area of study during time t
I_t = number of insects that immigrate into the area of study during time t

The emigration and immigration rates (E_t and I_t) must be considered since, as pointed out earlier, we are limited in that it is usually not feasible or possible to study an entire population or species and therefore in a given study area individuals are likely to come and go. Holding these two terms constant it becomes obvious that change in numbers is dictated by the relationship between the birth and death rates. If the birth rate exceeds the death rate, the number of insects will increase during t; if it is less than the death rate, the number of individuals will decrease during t. If the two rates are equal, N_t will equal N_0.

The birth rate, death rate, and emigration–immigration rates are influenced by a variety of factors. Obviously weather, in particular temperature, will influence all these rates. In addition, it can be argued that any factor which exerts an influence on one of these rates influences the others as well. For example, an environmental change, say in temperature, might cause a higher mortality among insects in one stage than in another or in insects of a certain age. If this age-specific mortality happened to occur in ovipositing females, one could say that both the birth and death rates were being affected.

It is easier to look at birth rate, death rate, and the emigration–immigration rates separately while assuming that the others are zero during the short time interval t. Looking at it in this way the major factors that influence the birth rate are the average fecundity of the females, the average fertility of the females, and the sex ratio. The average fecundity represents the average number of offspring that would be produced by each female under ideal environmental conditions; average fertility represents the average number of offspring actually produced, the difference between fecundity and fertility being related to such environmental factors as quality and quantity of available food, weather components, and population density. The sex ratio is the fraction of the total population that is female. In most insects the sex ratio is 0.5, but in species where parthenogenesis occurs there may be drastic deviations from this 50–50 balance of sexes. As with fecundity and fertility, the sex ratio may be influenced by several environmental factors.

The death rate is affected by such factors as adverse weather conditions, temperature extremes in particular; predators, parasites, and pathogens; accidents; low vitality; food shortage; and lack of adequate shelter (Clark et al., 1967).

The section "Migration and Dispersal" in Chapter 7 should be consulted for a brief discussion of immigration and emigration. The occurrence rates of these activities are also influenced by many environmental factors, such as weather and quality and quantity of food, or crowding.

More Regarding the Life-System Concept. Several different theories have been proposed to explain the numerical behavior of insect populations. Clark et al. (1967) have provided a very useful integrating concept, which brings together the ideas of several investigators (i.e., the "life-system" concept introduced earlier in this chapter). Our objective at this point is to elaborate somewhat on this concept. Figure 9-1 summarizes the various components of a life system.

Clark et al. recognize two kinds of "ecologocial events" which occur as a result of the interaction of the "co-determinants of abundance" (the "subject population" and its "effective environment"): *primary* and *secondary*. Primary events are those which are reflected in the demographic equation explained above (i.e., births, deaths, emigrations, and immigrations). Secondary events are those which exert an influence on the "magnitude, extent, frequency, or duration of primary events." Our earlier discussion of environmental components has been essentially a review of the many different contexts (temperature, food, and so on) in which secondary events may occur.

Clark and his colleagues further explain that primary and secondary ecological events result from the action of basically two kinds of ecological processes on populations, additive and subtractive, or more clearly a given ecological process may have either an additive or subtractive effect on a given population at a given time. Additive processes are those which act in a positive way on a population. Examples of additive processes include immigration and weather conditions that reduce populations of natural enemies. The planting of a monoculture such as corn may very well be looked upon as an additive process for those species, such as the European corn borer, which thrive on this crop, or those insects which are parasites of predators of the corn-eating species. Subtractive processes act in a negative way on a population, causing the death of individuals and/or a decrease in the number of progeny produced; these processes include emigration and adverse weather conditions. The planting of corn would constitute a subtractive process for insects that could not survive in the corn agroecosystem.

Additive and subtractive ecological processes may be either density independent or density related (i.e., the effects of the action of a given ecological process on a population may or may not be related to the density of that population). Subtractive density-related processes are of special interest since they may act, in some instances, as regulators of the level of abundance in a manner analogous to negative feedback.

Long-Term Numerical Changes. Over the long term, populations and their effective environments interact with resultant fluctuations in the size of the populations. Some populations fluctuate irregularly, apparently in response to changes in environmental components such as food supply and weather. Other populations tend to fluctuate regularly about a mean level of abundance. This may be a reflection of the action of some subtractive density-related process that is serving to regulate the level of abundance. It has recently become more and more apparent that, in a given life system, certain "key" ecological processes that have age-specific effects are acting to influence the major population trends, while other processes exert somewhat lesser effects. Elaborate "multifactor" studies have been carried out for a number of insect species. The objective of these studies has been to derive "life tables" that contain information as to population densities at different times during the life cycle and age-specific information regarding the "key" ecological processes that account for mortality. Such a life table is presented in Table 9–1. The development of one of these tables requires the procurement of population samples for many generations. Another approach that has been used where preliminary studies have indicated a few key influences involved in a life system is the *key-factor method*. In this type of study, population density is measured at only one point in each generation. This approach requires much less sampling and enables investigators to concentrate on critical periods of time in the life cycle when the "key" process or processes are active without spending much time on other periods.

Increased understanding of the ecology of insect populations, of pest species in particular, will have considerable practical application:

Table 9–1
Life Table for Second Generation of the Diamondback Moth on Early Cabbage, Merivale, Ont., 1961[a]

		Mortality		
Age Interval	**Numbers per 100 Plants**	**Causative Factors**	**Per 100 Plants**	**Per Cent**
Eggs	1,580	Infertility	25	1.6
Larvae				
Period 1	1,555	Rainfall	1,199	77.1
Period 2	356	Rainfall	36	10.1
		Parasitism by *M. plutellae*	52	14.6
Period 3	268	Parasitism by *H. insularis*	69	25.7
Pupae	199	Parasitism by *D. plutellae*	92	46.2
Moths	107	Sex (49.5% ♀ ♀)	1	1.0
Females × 2	106	Photoperiod	78	73.6
"Normal" females × 2	28	Adult mortality	20	71.4
Generation totals			1,572	99.5

Trend index,[b] 0.55. Stable number rate,[c] 99.1 %.

[a] From Harcourt and Leroux (1967).

[b] Trend index = $\dfrac{\text{eggs laid in new generation}}{\text{eggs laid in old generation}}$.

[c] Stable number rate = the mortality rate necessary for the population to remain the same size from generation to generation. Since the percent mortality for the generation represented in this table was 99.5%, it is apparent that the population was declining.

(1) The prediction of the population level of each pest from generation to generation. This information, based on trend index data, helps to guide the growers in effective advance planning of spray programs and in more intelligent use of pest-control measures. (2) The feedback of information on the economic thresholds of a pest and on the crop damage caused by the active immature stage. For example, it is now known that five bud moth larvae, or fewer, per 100 leaf clusters on apple do not constitute an economically important population. Hence, sprays are omitted at these densities of the insect, because foliage or fruit damage is below the 5% economic level generally tolerated by the industry. In the spring, chemical control of the bud moth is omitted if winter lows have been −21°F or below, because such lows kill 90 to 95% of the overwintering larvae. However, these lows scarcely affect larvae of the pistol case-bearer, which overwinter near the bud moth and receive similar protection, and the omitting of spray applications against the bud moth favors the survival of parasites that control and regulate populations of the casebearer. (3) The utilization of information gained through ecological studies to govern, when possible, the use of insecticides as a possible substitute for other density-independent control factors, such as weather, in integrated pest management practices.—(Subcommittee on Insect Pests, 1969).

Selected References

GENERAL

Allee, Emerson, Park, Park and Schmidt (1949); Andrewartha and Birch (1954); Florkin and Schoffeniels (1969); Kendeigh (1961); Kormondy (1969); Lewis and Taylor (1967); Macan (1963); McGill (1965); Odum and Odum (1959); Pielou (1969); Southwood (1966); Whittaker (1970).

INSECT ECOLOGY

Brian (1965); Bursell (1964); Chauvin (1967); Clark, Geier, Hughes and Morris (1967); Cloudsley-Thompson (1962); Folsom and Wardle (1943); Frost (1959); Graham and Knight (1965); Harcourt (1969); Hynes (1970); Macan (1962); Muirhead-Thomson (1968); Rolston and McCoy (1966); Southwood (1966).

PART TWO

Unity and Diversity

10

Evolution and Systematics

Insect Evolution

Our objective in this section will be to outline in rather broad terms the patterns that are thought to have occurred during the evolution of insects and their relatives. We shall first consider the origin of the phylum of which insects are important members, Arthropoda, followed by the evolution of the class Insecta, and finally discuss briefly the phylogeny of insects themselves. Origins preceding Arthropoda are the concern of general invertebrate zoology, but the interested reader may consult Sharov (1966), Meglitsch (1967), or any of several invertebrate zoology texts. At least a rudimentary knowledge of evolutionary theory is assumed in this discussion. Readers unfamiliar with the theory of organic evolution, the concept of geologic time, and so on, should consult a good general biology or zoology text or any of several excellent texts available which deal specifically with organic evolution (see Selected References).

Origin of Arthropoda

Comparison of present-day forms has lead to the recognition of structural and developmental affinities among the phyla Annelida, Arthropoda, and Onychophora.

Annelida, the segmented worms, apparently evolved from the ancient protostomes, which also gave rise to the mollusks (Meglitsch, 1967). Annelids (Fig. 10–1) are soft-bodied, elongate, cylindrical, bilaterally symmetrical animals. Their bodies are composed of a series of repeating segments or metameres capped anteriorly by the *prostomium* or *acron* and posteriorly by the *periproct* or *telson*. The mouth is on the venter between the prostomium and first metamere. The anus opens from the periproct. Some annelids have several bilateral pairs of segmental appendages called *parapodia* which arise as evaginations of the body wall. Segmentation of the body is evident in nearly all tissues save the alimentary canal, which is essentially a longitudinal tube running from the mouth to the anus. The circulatory system is a closed system of tubes, usually one dorsal longitudinal and one ventral longitudinal vessel, connected by lateral trunks which provide vascularization for the various body tissues. The blood commonly carries a respiratory pigment. The body cavity is a coelom divided into segmental compartments. Excretion is accomplished by paired segmental *nephridia*, tubes that communicate between the coelom and

287

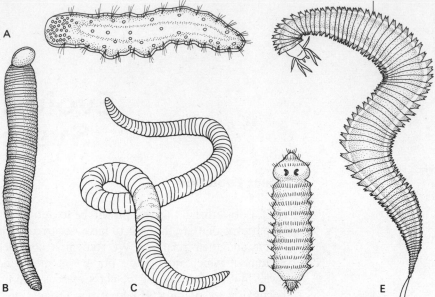

Fig. 10–1

Representatives of four classes of segmented worms, phylum Annelida.
A. Chaetogaster (Oligochaeta). B. Leech (Hirudinea). C. Earthworm
(Oligochaeta). D. Dinophilus (Archiannelida). E. Lugworm (Polychaeta).
(Redrawn from Stiles, Hegner and Boolootian, 1969.)

exterior. Ventilation is cuticular or by means of gills evaginated from
the body wall. The nervous system is composed of an anterior pros-
tomial ganglion which lies dorsal to the alimentary canal. It is connected by a pair of bilateral connectives to a ventral chain of paired
segmental ganglia linked by paired longitudinal connectives. Some
annelids have a *trochophore* larval stage, which has played an

Fig. 10–2

Scorpion, illustrating the major
chelicerate features. (Courtesy of the
U.S. Public Health Service (redrawn).)

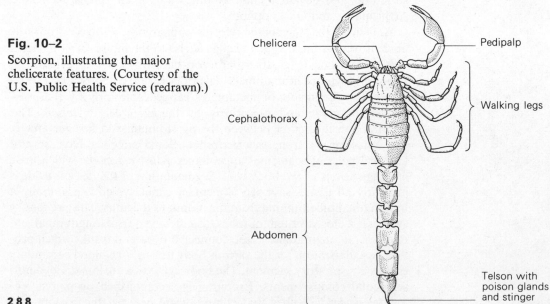

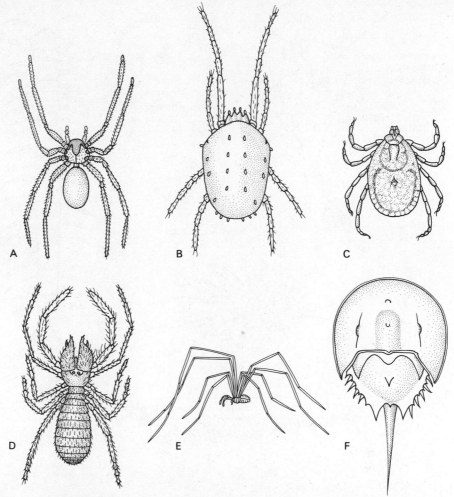

Fig. 10–3
Representatives of the subphylum Chelicerata. A. Spider, *Loxosceles* sp.
(Arachnida; Araneida). B. Mite (Arachnida; Acari). C. Tick (Arachnida;
Acari). D. Wind scorpion (Arachnida; Solpugida). E. Harvestman or daddy
longlegs (Arachnida; Phalangida). F. Horseshoe crab, *Limulus* sp.
(Xiphosura). (A–E courtesy of the U.S. Public Health Service (redrawn).)

important role in assessing their evolutionary relationships with members of other invertebrate phyla.

Arthropods, the joint-footed animals, are a highly successful diverse group of animals unrivaled by any group in the animal kingdom in diversity of structure and function and have radiated into nearly every imaginable ecological niche. A feature that has played a central role in their success is their relatively impermeable exoskeleton. Such an impermeable covering was a prerequisite for invasion of the terrestrial environment, which poses a continuous challenge to the water-holding capabilities of animals, a threat accentuated in arthropods by their generally small body size.

Arthropods (Figs. 10–2, 10–3 and 10–4) have segmented bodies made up of chitinous exoskeletal plates of varying degrees of hardness which are separated by less-hardened "membranous" areas.

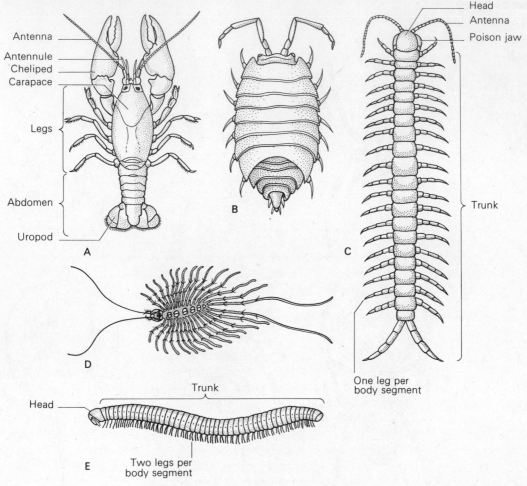

Antenna
Antennule
Cheliped
Carapace

Legs

Abdomen

Uropod

A

B

Head
Antenna
Poison jaw

Trunk

C

One leg per
body segment

D

Trunk

Head

Two legs per
body segment

E

Fig. 10–4

Representative mandibulates. A. Crayfish (Crustacea). B. Sowbug
(Crustacea). C. Centipede, *Scolopendra heros* (Chilopoda). D. Eastern house
centipede, *Scutigera cleoptrata* (Chilopoda). E. Millipede, *Narceus
americanus* (Diplopoda). (Courtesy of the U.S. Public Health Service
(redrawn).)

Paired segmental appendages are usually present on at least some
body segments. Arthropods share several fundamental traits with
annelids. Particularly outstanding are their segmented bodies and
the ventral chain of segmental ganglia. On the other hand, the
arthropods have an open circulatory system and the body cavity is a
hemocoel, coelomic sacs forming during embryogenesis, but not
forming the body cavity. Also most arthropods ventilate by means
of a tracheal system and have well-developed appendicular mouthparts.

Onychophorans are a very small contemporary group of terrestrial
animals but are quite significant in that they show morphological
affinities with both annelids and arthropods and are considered to be
related in some way to members of both these major phyla. Onychophorans (Fig. 10–5) have elongate wormlike bodies with several
bilateral pairs of lobelike legs. They are not segmented as adults, but

are clearly segmented during their embryogenesis. They have paired nephridia along the body which open near the leg bases. This fact, together with other morphological traits, is suggestive of a relationship with the annelids. On the other hand, they have a number of traits in common with the arthropods, including an open circulatory system with a hemocoel, tracheal ventilation, and appendages associated with the mouth. Onychophorans also have several characteristics not shared with either the annelids or the arthropods. They have been variously classified as annelids, arthropods, or a separate phylum. They will be grouped as a separate phylum in this text.

The basic similarities among annelids, arthropods, and onychophorans strongly suggest common ancestry at some point in metazoan evolution. Some authors have, in fact, classified all three groups as members of the superphylum Annulata (Snodgrass, 1952). Sharov (1966), in a similar vein, supports Cuvier's concept of the phylum Articulata with annelids, onychophorans (Malacopoda), and Arthropods being recognized as subphyla. The details of the evolutionary relationships among these three groups are purely hypothetical. Unfortunately, knowledge of the fossil record is of little help, since each group is distinctly defined in the Cambrian strata, from which our oldest "good" fossils come. This, of course, means that they arose from their supposed common ancestor in pre-Cambrian times. There is general agreement that the common ancestor was an Annelid or annelidlike creature. The polychaete annelids are considered to be the oldest of the contemporary annelids and hence the closest to this hypothetical ancestor.

Some investigators have suggested that the primitive arthropod evolved directly from a primitive polychaete annelid (Meglitsch, 1967). Snodgrass (1952), later supported by Sharov (1966), reasonably conceived of a segmented wormlike creature which produced two branches, one that ultimately gave rise to the polychaete annelids and the other to a form with paired, undifferentiated lobelike legs (lobopod). He envisioned the lobopods as subsequently branching

Fig. 10–5

Onychophoran, *Peripatus*. A. Dorsal view. B. Ventral view. (Redrawn from CCM General Biological, Inc., key card.)

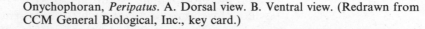

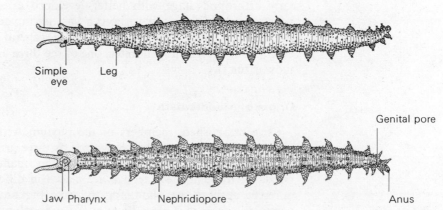

Preantenna   Oral papilla

Simple eye   Leg

A

Genital pore

Jaw Pharynx   Nephridiopore   Anus

B

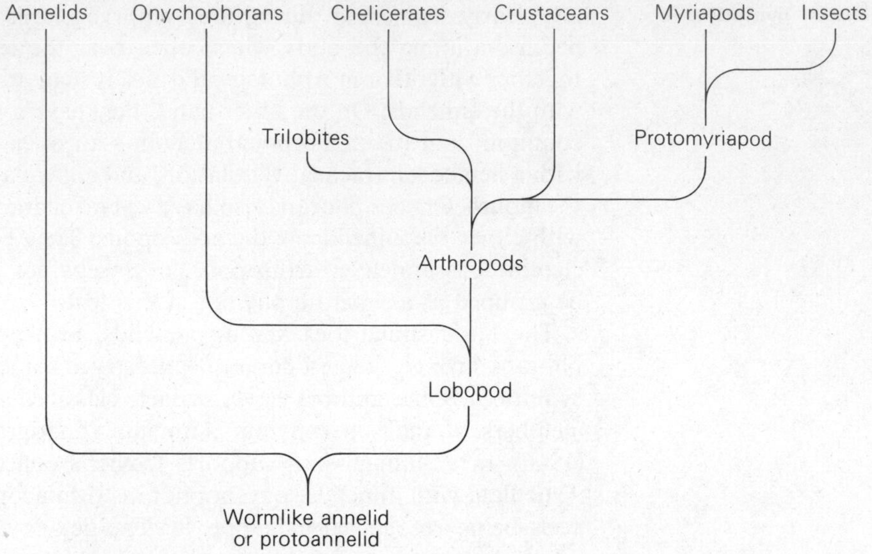

Annelids   Onychophorans   Chelicerates   Crustaceans   Myriapods   Insects

Trilobites   Protomyriapod

Arthropods

Lobopod

Wormlike annelid
or protoannelid

Fig. 10–6
Hypothetical phylogeny of arthropods, annelids, and onychophorans.
(Redrawn from data of Snodgrass, 1952, and Sharov, 1966.)

into arthropodan and onychophoran lines. These ideas are summarized in Fig. 10–6.

A diagram (Fig. 10–7) originated by Snodgrass (1935) and modified
by Ross (1965) is very useful in outlining the hypothetical major
steps in arthropodan and insectan evolution. The top three forms
(A, B, and C) concern us at this point. The first represents the segmented, legless, wormlike annelid or annelidlike stage. The undifferentiated body is composed of a series of somites or metameres
capped anteriorly by the prostomium (acron) and posteriorly by the
terminal body segment, the periproct (telson). The mouth is located
between the prostomium and the first body segment; the anus
opens in the periproct. Fig. 10-7 B represents the evolutionary
stage in which paired, bilateral, lobelike appendages were developed
on the somites as well as a pair of antennal appendages and simple
eyes on the prostomium. According to Ross (1965), members of the
phylum Onychophora are thought to represent this stage of evolutionary development. The next stage (Fig. 10–7C) represents an
early arthropod stage with better-developed eyes and jointed legs
which is considered to have been a vast improvement in locomotor
potentialities. Ross (1965) points out a general similarity of this
evolutionary stage and trilobites, one of the three arthropod subphyla
(now extinct).

Origin of Mandibulata

Once established, members of the phylum Arthropoda branched
out into what are generally recognized as three groups: the trilobites
(Trilobita), chelicerates (Chelicerata), and mandibulates (Mandibulata). Although these groups are placed in different categories by
different authors, we shall, for the sake of convenience, do as is
commonly done and place them at the subphylum rank.

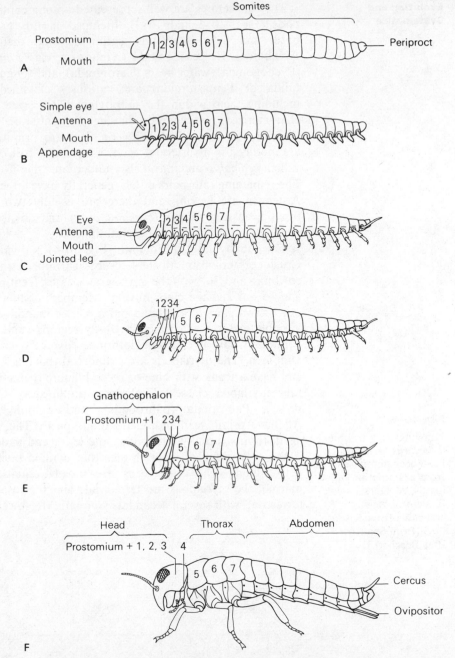

Fig. 10–7
Hypothetical stages in the evolution of the insect form. (See text for explanation.) (Redrawn from Ross, 1965.)

The trilobites (Fig. 10–8) became extinct in the Devonian period and therefore are known only from the fossil record. They were marine animals and their bodies were made up of three tagmata: *head, thorax,* and *pygidium.* The name Trilobita is in reference to these three tagmata. The head and first few thoracic segments were covered by a *carapace,* most of the body segments had a bilateral pair of jointed legs, and the head carried a single pair of antennae.

The chelicerates are well represented among contemporary arthropods (Figs. 10–2 and 10–3). Included in this group are spiders, mites and ticks, scorpions, solpugids, and their relatives (class Arachnida). Horseshoe crabs (Xiphosura, e.g., *Limulus*), sea spiders (Pycnogonida), water bears (Tardigrada), and tongueworms (Linguatulida or Pentastomida) are variously classified, some authors including them within the subphylum Chelicerata. Chelicerates are characterized by generally having six pairs of appendages; the anterior pair are called *chelicerae* and are primitively jawlike but are modified in a variety of ways. The second pair of appendages are called *pedipalps* and have also undergone manifold modifications. The remaining four pairs of legs generally have a locomotor function. Antennae are lacking and there are usually two body regions, a fused head and thorax (*cephalothorax*) and an abdomen. The appendages are borne by the cephalothorax.

The subphylum Mandibulata (Fig. 10–4) is divided into the class Crustacea (crayfish, lobsters, crabs, shrimp, barnacles, copepods, sowbugs and so on), the myriapod classes (centipedes, millipedes, and so on) and the class Insecta. Members of this subphylum have one or two pairs of antennae and a pair of mandibles that are highly variable in form and function. Body regions vary. The crustaceans commonly have two, a cephalothorax, covered by a carapace, and an abdomen. The myriapods are cylindrical, have a well-defined head, and have a trunk with nine or more bilateral pairs of legs. There are four myriapod classes: Diplopoda (millipedes), Chilopoda (centipedes), Pauropoda (pauropods), and Symphyla (symphylans). All have primitive mandibulate mouthparts. The insects possess a head, thorax (usually with wings and legs), and abdomen. The latter is the only group of invertebrates that contain members capable of active flight. The crustaceans are mostly aquatic, freshwater or marine; the myriapods are terrestrial; and the insects are primarily terrestrial, with several secondarily aquatic (freshwater) forms.

Fig. 10–8

Drawings of a trilobite, *Triarthrus becki*, which have been composed from examination of fossils. A. Dorsal view. B. Ventral view. [Redrawn from Ross, 1965 (from Schuchert after Beecher).]

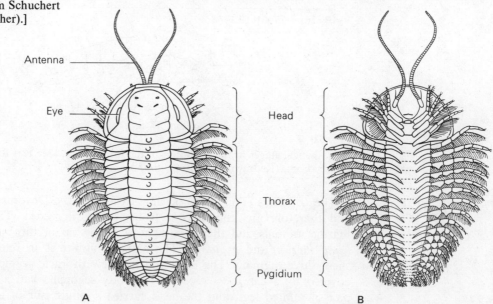

Antenna

Eye

Head

Thorax

Pygidium

A

B

The evolutionary relationships among the trilobites, chelicerates, and mandibulates are unclear, although there have been a number of opinions expressed. Some investigators argue that the arthropods are a polyphyletic group and that many of their similarities have arisen as a result of convergent evolution. However, others see them as clearly monophyletic, having evolved from a common ancestor. All three subphyla are well represented in the Cambrian strata which indicates that the splits between them occurred prior to the period of time represented by these rocks. It is generally agreed that the trilobites are of the most ancient lineage, preceding representation of the others in the fossil record. Sharov (1966) suggests that the primitive arthropodan group Dicephalosomita, related to contemporary Pycnogonida and classified together as Prosciferans, predates the trilobites. He envisions these arthropods as arising from the lobopod ancestor concurrently with the Onychophora and splitting into two main branches, one later branching into the trilobites and their relatives and the chelicerates and the other evolving into the mandibulates. This stage in arthropod evolution might in very general terms be related to Fig. 10–7C, where the first postoral pair of appendages were to become the chelicerae of the chelicerates and the first pair of walking legs of the trilobites and their relatives.

Origin of Insecta

Sharov (1966) feels that an early crustacean line branched off in the Silurian period and gave rise to the myriapods and insects, which he groups as Atelocerata. Other investigators hypothesize a direct trilobite origin for the myriapods and insects, and still others place the Onychophora in this position. Figure 10–7D might be considered to represent in a very generalized way the base of the crustacean–atelocerate branch. In the crustacean line, the paired appendages of the first postoral segment became the second antennae and hence homologous with the chelicerae of the chelicerates. These appendages were apparently lost in the myriapod–insect (atelocerate) direction.

As mentioned above, there are varying opinions as to the origin of the myriapod and insect group, but there is little doubt of a close relationship between these two groups, particularly based on the comparative morphology of the head. Figure 10–7E can be looked upon as representing this stage, bilateral appendages being retained on most body segments and the appendages of the second, third, and fourth body segments, *gnathocephalon,* becoming the typical, primitive mandibulate mouthparts. The prostomium and first postoral body segment, which has lost its primitive pair of appendages, forms the *protocephalon.*

The class Insecta is generally considered to have evolved from a myriapod or protomyriapod of some sort during the Devonian period. The final body changes (Fig. 10–7F), from a myriapod or protomyriapod to the primitive insect, consisted of the retention of legs for locomotor purposes on the three segments posterior to the gnathocephalon and the coincident loss of locomotor appendages on the remaining body segments.

Carpenter (1953) outlined four important stages in the evolution of insects. The following discussion is based on his paper. The first was the appearance of primitive wingless insects, which are represented among contemporary insects by the thysanurans. These primitive apterygotes are thought to have arisen during the Devonian period. The fossil record of apterygotes is poor, probably mainly because they were and are comparatively soft-bodied forms. Other insect groups with harder body parts and particularly wings have been more amenable to fossilization processes. Despite lack of extensive fossil evidence, insects are generally thought of as originally having been wingless.

The second major step was the development of wings (see Fig. 6–9), which is thought to have occurred during the Lower Carboniferous. Insects were the first animals to invade the air, doing so some 50 million years before the reptiles and birds. Since predatory amphibians, reptiles, and various arthropods were abundant at this time, the advent of wings apparently provided a definite selective advantage to their bearers by enabling them to escape when threatened. The early winged insects had very simple wing articulations that allowed flight, but at rest the wings were held out from the body, not being designed to be flexed posteriorly out of the way over the abdomen. These *paleopterous* ("primitive-winged") insects became the dominant group of insects during the Carboniferous and Permian but are represented today by only two orders, Odonata and Ephemeroptera.

The next evolutionary advance occurred during the Upper Carboniferous. This was the development of the wing-flexion mechanism, which enabled insects to flex the wings posteriorly over the abdomen, the *neopterous* ("new-winged") condition. This added the advantage of being able to easily run and hide from predators as well as move into niches inappropriate for forms with continuously outstretched wings. The neopterous insects readily became the dominant group of insects as they are today. According to Carpenter, they comprise 90% of the contemporary orders and 97% of the total number of species. The neopterous orders are all those except the Apterygotes, Odonata, and Ephemeroptera.

The final major evolutionary step outlined by Carpenter was the development of metamorphosis during the Upper Carboniferous. The first to develop were the forms with simple metamorphosis, the *hemimetabolous neoptera*. These were followed during the late Upper Carboniferous and the Permian by the development of forms with complete metamorphosis, the *holometabolous neoptera*. In certain groups of both the hemimetabolous neoptera (Odonata and Ephemeroptera) and the holometabolous neoptera the immature stages no doubt were and still are spent in different habitats than the adults. Metamorphosis in these instances enabled a species to take advantage of the positive aspects of both environments. For example, an insect with aquatic larval stages and terrestrial–aerial adults could enjoy the abundant and readily available nutrients of the aquatic habitats and the potentialities for dispersal offered by the terrestrial–aerial environment.

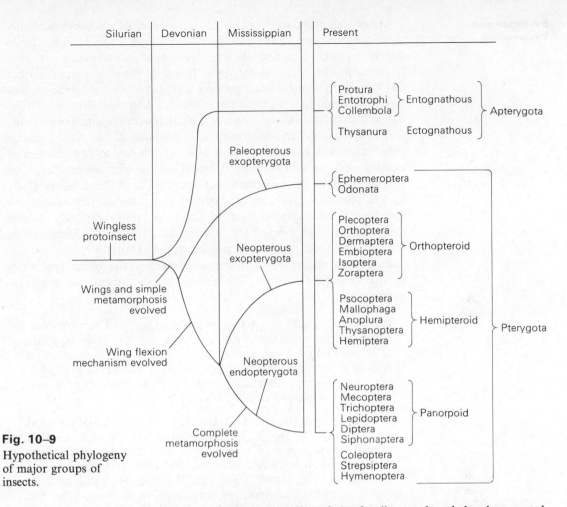

Fig. 10–9
Hypothetical phylogeny of major groups of insects.

Based primarily on studies of the fossil record and developmental and comparative morphology the various contemporary orders of Insecta have been placed in groups on the basis of affinities between them. One such system of grouping is illustrated in Fig. 10–9. The characteristics of these groups will be described in Chapter 11. No attempt has been made to illustrate the hypothetical relationships of the orders that fall into a given grouping. Opinions vary regarding these relationships and the interested reader should consult the literature on insect evolution for information in this area. Sharov (1966) and Brues, Melander, and Carpenter (1954) would serve as good starting points in seeking this information.

Systematics

Having gained some idea of the general sequence of events that have given rise to the members of the class Insecta as we know them today, we are in a position to consider the threads of unity that run through the diversity shown by this group. Finding these threads of unity—morphological, physiological, or other features—and utilizing them to produce meaningful and useful groupings of insects is the objective of insect *systematics*.

Many definitions of systematics have been offered, but basically there are two. Some authors (e.g., Mayr, Linsley, and Usinger, 1953) define systematics simply as the science of classification, making it synonymous with another widely used term, *taxonomy*. Other authors use it in a broader, general way, expanding it to include the study of evolution. For instance, Blackwelder and Boyden (1952) define systematics as "the entire field dealing with the kinds of animals, their distinction, classification, and evolution." In this context, then, taxonomy is defined specifically as "the day-to-day practice of dealing with the kinds of organisms. This includes the handling and identification of the specimens, the publication of the data, the study of the literature, and the analysis of the variation shown by the specimens" (Blackwelder, 1967). Simpson (1961) similarly defines taxonomy as follows: "The theoretical study of classification, including its bases, principles, procedures, and rules."

Regarding the place of systematics among the other specialties in the biological sciences, the following quotations are most appropriate.

It is probably safe to say that in point of years of service systematic biology is the senior branch of the whole field of the study of living things. It can hardly be otherwise, for discrimination among or between facts or things is one of the very first steps in the organization of our knowledge of these things or facts and our use of them.—(Ferris, 1928)

Systematics is at the same time the most elementary and the most inclusive part of zoology, most elementary because animals cannot be discussed or treated in a scientific way until some systematization has been achieved, and most inclusive because systematics in its various guises and branches eventually gathers together, utilizes, summarizes, and implements everything that is known about animals, whether morphological, physiological, psychological, or ecological.—(Simpson, 1961)

The major tasks of the systematist can be divided into five areas: (1) identification, (2) description, (3) classification, (4) concern with and proper application of the rules of nomenclature, and (5) study of speciation. Identification and description, classification, and speciation are sometimes referred to as *alpha, beta,* and *gamma systematics,* respectively. In general terms, the stages identification and description, classification, and studies of speciation occur in sequential order for a given group of organisms. For a few groups of animals systematics has reached the speciation level, while most groups are still in the identification, description, and classification stages of development. In this chapter we shall discuss identification, description, and classification and leave the complex, interdisciplinary studies of speciation for more specialized treatises (see the references cited in the evolution section of this chapter). In other words, in the context of our definitions of systematics and taxonomy, we will from this point on be considering taxonomy.

Obviously collection and preparation of specimens for study must precede identification and the other steps. Discussion of collection and preparation techniques would be inappropriate in this text, and the reader should refer to Borror and DeLong (1971), Borror and White (1970). Cummins et al. (1965), Knudsen (1966), Peterson (1959), Ross (1962a), or any of several other publications which deal entirely or in part with these techniques. In addition to collection of one's own specimens for study, it is a common practice to borrow specimens from other investigator's collections for the purpose of study or to trade specimens with other investigators. Another means of gaining access to collections of specimens is to visit a museum and study them there or borrow them for a period of time. By these approaches one can obtain or at least study a greater number of specimens from a much wider area of the world than he could possibly collect himself.

It is often necessary or highly desirable to culture certain species. Systematists can derive a wide variety of very useful biological data (e.g., biochemical, serological, physiological, developmental, genetic) from such cultures which would be entirely unavailable from dead, pinned, or preserved specimens. In order to arrive at the best possible description and classifications, the systematist tries or should try to gather as much pertinent information as possible. Information regarding culture methods for insects can be found in the following references: Peterson (1959), Needham et al. (1937), and Siverly (1962).

Identification

Purpose of Identification.   The purpose of identification is basically to determine what kind of organism a given specimen is. The meaning of "kind" depends largely upon one's objectives. A student in a general zoology course may be perfectly happy to identify a given organism as a fly, whereas a professional working in a specialized area of zoology or entomology may need to know exactly what species he is dealing with. The professional reasons for wanting to know the species to which an organism (insect) belongs fall into two categories: the nonsystematists reasons and the systematists reasons. The nonsystematist researcher studying some fundamental aspect of an animal or the applied zoologist (entomologist) would need to identify a specimen in order to find pertinent literature regarding basic information or appropriate control measures. It is not only critical for the researcher to know what species he is studying as a key to the literature, but also to have a name with which to associate any information he derives through his own investigations. In other words, the species name of an organism serves as the "file" category under which any and all information pertinent to this species is stored. Information representing the morphology, physiology, genetics, and so on, of an organism would be essentially useless or certainly less useful if it was not published as pertaining to a given species.

The systematist, in addition to using the species name as a key to the literature and a category for data storage, needs to identify an organism to species or at least to the point where he realizes that the specimen (or specimens) he is attempting to identify is an undescribed, "new" species. Mayr, Linsley, and Usinger (1953) refer to this aspect of identification as *taxonomic discrimination*. Further, the identification of a specimen must necessarily precede the fitting of this specimen into a scheme of classification.

1A two body regions (head and trunk) 2
1B three body regions (head, thorax, and abdomen) . . . . .an insect, CLASS INSECTA, 3

2A one pair of legs per body segment; poison jaws present . . . . .a centipede, CLASS CHILOPODA
2B two pairs of legs per most body segments; poison jaws lacking . . . . .a millipede, CLASS DIPLOPODA

3A wings absent . . . . SUBCLASS APTERYGOTA
3B wings present or secondarily absent, i.e., winged ancestors . . . . SUBCLASS PTERYGOTA

Fig. 10–10
Sample dichotomous key that differentiates the subclasses of insects from centipedes and millipedes.

Methods of Identification. Although superficially there appear to be several different methods available for identifying an organism, all ultimately are based on comparison. One of the most commonly used methods of identification is the use of a key. A key is essentially a printed information-retrieval system into which one puts information regarding a specimen-in-hand and from which one gets an identification of that specimen to whatever level the key is designed to reach. Most keys are *dichotomous* (double branching; Fig. 10–10) and at any point one is presented with two alternatives (a *couplet*), each of which leads to another pair of alternatives, and so on, until one reaches a terminal pair, one member of which presents the identification name. At each point the user of a key picks the alternative that best describes some aspect of the specimen he is trying to identify. Keys are available for use by people with different needs. The simplest are designed for the layman and, although somewhat useful, generally offer only superficial identification. More complex keys are designed for the professional nonsystematist, who needs to identify an organism to a more specific level than the layman. These keys are more technically oriented and require a certain amount of training in general insect morphology before they can be used effectively. Finally, there are keys designed for the specialist, which are very technical and require knowledge of morphologic traits of a particular group of organisms. These keys are usually the ones that carry identification to the species level.

Another method for identification is the use of pictures in the form of color plates, black-and-white photos, or line drawings. These pictures may be of entire organisms or parts of organisms and are often used in conjunction with keys. "Pictorial keys" offer alternatives in the form of pictures with or without accompanying verbal alternatives. Pictures are particularly useful when the organism under consideration is highly patterned or characteristically colored.

One of the best means for identifying a specimen is to compare it with another specimen or a series of specimens which have previously been identified. This approach can be used to obtain an identification at any level. Type specimens are those on which an original species description is based. These specimens may be extremely useful in making species identifications. Original verbal descriptions of species based on type specimens are the permanent records of the attributes of a given species and are particularly useful when the type specimens are available and constitute the only original record when type specimens are destroyed, lost, or unavailable.

In the final analysis, combinations of all or many of these methods are most often used in the identification of a given organism, whether the person making the identification is a layman or professional specialist. It must be emphasized that, at least in dealing with insects, although the layman or professional nonspecialist can, in many, if not most instances, identify a specimen to family, or in some cases below the family level, the best and safest way to obtain a species identification is to consult a specialist. Because of his experience and his familiarity with pertinent literature, the specialist is in the best position to make use of the highly specialized keys, original descriptions, and type specimens. It is safe to say that in view of the vast number of insect species already described, a given individual is likely to be a specialist in the taxonomy of only a very few families, if not a single family.

Problems Encountered in Identification. Unfortunately the identification methods discussed above are not without problems. Keys may have any of several drawbacks, including failure to include the organisms being identified, failure to include the stage or sex of the organism being identified, and failure to take into account any of a large number of possible variations in morphologic or other characters used. Pictures can be misleading unless very clear and accurate, and they may cause faulty identifications if the species of the organism being identified very closely resembles the one depicted. Verbal descriptions can be very ambiguous, particularly when such traits as color and texture are described, and quite often they require a rather extensive understanding of terminology to be of any use. Original species descriptions are often in old, obscure journals which may be difficult to obtain and may be in a foreign language. A collection of accurately identified organisms in the group being identified is not always generally available, and even a good collection may not contain specimens of the same species or even genus or family as the specimen being identified. Type specimens that have provided the basis for the original species description may be very difficult or impossible to obtain and may have been lost or destroyed. The major difficulty with the method of consultation with a specialist is that a specialist on the taxonomy of a given group of organisms may no longer be living or may never have existed. Even if a specialist does exist, there is still the problem of getting the specimen(s) to be identified to him, since he may reside in any part of the world.

Description

Although nonsystematists and systematists alike carry out the procedures involved in identification (although generally at different levels), they diverge at the point of successful identification. The nonsystematist may proceed to a line of nonsystematic investigation with his newly identified organism(s) and/or he may delve into the literature in his area of interest as it applies to this organism. On the other hand, the systematist, having completed an identification, will follow one of two courses. If the species he has identified has already been described, he will probably put it aside for possible future study with other specimens of the same species. However, if he finds during the course of his identifications that no description of the species he is studying has ever been made (i.e., it is different from all published descriptions of the species in a given group) he will proceed to describe it as a new species and place it in whatever grouping he sees as appropriate. His description will then serve to identify this species for other investigators. The production of such descriptions is the job of descriptive systematics.

Subjects of Description. However they are obtained, the subjects of a description are individual organisms. These individuals may be one or several in number. If more than one in number, they may have been obtained by a carefully planned and executed procedure such that they constitute a statistical *sample* of a population (i.e., they theoretically show the same distribution of traits as do the members of the entire population from which they were drawn). The other, and more common, alternative is that a group of specimens of supposedly the same species are obtained in a more-or-less erratic fashion, perhaps with several individuals taken in the same locality or with the different individuals having been taken from different localities at different times and under different circumstances. A group of specimens obtained in this fashion is termed a *series*. Although many species descriptions are based on series and samples, several are based on a single specimen or on a series of a few individuals.

For each species description, there is a type individual or group of type individuals. A *type* is an individual or group of individuals upon which a description is based. It is the reality that bears the name of a species. There are several kinds of types, depending upon the relationship a given individual (or individuals) has with the original single specimen set aside as the *holotype* or with a group of individuals designated as *syntypes*. The location at which a type is collected is called the *type locality*.

Features Described. We shall define a feature as any possible trait an individual could possess. This could be anything from the shape of a particular sclerite (morphological feature), to a particular kind of amino acid in the hemolymph (biochemical feature), to a particular mechanism for excretion (physiological feature), to a specific way of responding to a change in photoperiod (behavioral feature). Obviously since we have as yet no way of arriving at perfect knowledge of anything, all features of a given organism are not used in a description of that organism. It follows, then, that any species description plus all information gained since the discovery of that species is only a partial picture and that a species description in this sense is never really completed. In reality the vast majority of species descriptions are based on selected morphological features, and although the modern trend is definitely in the direction of including other sorts of data, particularly physiological, biochemical, serological, and behavioral, morphological features will undoubtedly continue to be at the center of most descriptions at least of insects.

Potentially there is a wide variety of features to choose from in a description, and certainly the more of these features, morphological and otherwise, that can be utilized for descriptive purposes the better. Mayr, Linsley, and Usinger (1953) list several categories from which useful descriptive features can be derived:

1. Morphological characters
   a. General external morphology
   b. Special structures (e.g., genitalia)
   c. Internal morphology
   d. Embryology
   e. Karyology (and other cytological differences)
2. Physiological characters
   a. Metabolic factors
   b. Serological, protein, and other biochemical differences
   c. Body secretions
   d. Genic sterility factors
3. Ecological characters
   a. Habitats and hosts
   b. Food
   c. Seasonal variations
   d. Parasites
   e. Host reactions
4. Ethological characters
   a. Courtship and other ethological isolating mechanisms
   b. Other behavior patterns
5. Geographical characters
   a. General biogeographical distribution patterns
   b. Sympatric–allopatric relationship of populations

It is extremely important to know how much or how little a given feature used in a description varies within a species. For features to be of any value in identification and classification they must be reasonably constant or vary predictably. Failure to recognize variation can lead to extreme confusion. There are any number of potential kinds of variation that can occur within a population of a given species. Linsley and Usinger (1961) outline these major types as follows:

1. Extrinsic or noninherited variation
   a. Variation due to age
   b. Seasonal variation
   c. Castes in social insects
   d. Variation due to habitat
   e. Variation due to crowding
   f. Climatically induced variation
   g. Heterogonic variation
   h. Traumatic variation
2. Intrinsic or inherited variation
   a. Primary sex differences
   b. Secondary sex differences
   c. Alternating generations
   d. Gynandromorphs
   e. Intersexes
   f. Mutations resulting in continuous variation
   g. Mutations resulting in discontinuous variation
   h. Genetic polymorphism

Future of Descriptive Systematics.   There can be no doubt whatsoever that the job of descriptive systematics is anywhere nearly finished,
particularly with the class Insecta, the members of which make up
the majority of the animal kingdom in terms of numbers of different
species. Blackwelder (1967) describes the situation graphically:

> There have been estimates of the total number of species of
> insects in the world as high as ten million. It seems likely, there-
> fore, that much less than half the species of the world are now
> known, in the sense of having been segregated and named.
> Most of those described so far are known from a few collections
> and only as adults. Only a small fraction of the families or genera
> of the world have been monographed, and only another small
> fraction have been treated in any other sort of comprehensive
> taxonomic work. Regional studies are common in Europe and
> North America, but the bulk of insects of the world are still
> poorly known or unknown.

Steyskal (1965) carried out an interesting study which illustrates
very well that much descriptive work, in fact, remains to be done
for most of the animal kingdom. By using recent lists of species, the
Zoological Record, and other sources, Steyskal generated a series
of trend curves of species descriptions by plotting the number of
species known against time beginning in the year of the publication
of Carl Linné's *Systema Naturae* (1758). He makes two quite acceptable assumptions: "the existence of a finite and practically fixed
number of species on the earth" and "that we have been gradually
approaching a state in which all the species would be known, at least
to the degree at which a specific name could be applied." Theoretically
the trend curve for a hypothetical group in which all or nearly all
species have been described would be sigmoid in shape. A gradual
rate of increase would occur initially when the study of a group was
getting started; a period of comparatively rapid increase in rate of
description follows when the techniques for study become well
developed, collections are made from vast areas, many investigators
actively study the group (e.g., due to realization of economic importance), and so on, and finally as the group becomes better and better
known the rate of description would begin to decline and level off as
only "the last few, rare, out-of-the-way, obscure, or cryptic species."
would remain. However, a leveling off of the curve could also indicate
temporary wane of interest in a given group and the rate could
increase again.

A good example of a group of animals that are known well enough
to provide a "good" example of a sigmoid curve are the birds of
North America (Fig. 10–11A). Steyskal presents curves for four
groups of insects (Fig. 10–11B–E), one of which, Rhopalocera
(Lepidoptera), demonstrates the leveling off indicative of species
description approaching completion. In addition, he provides a
curve (Fig. 10–11F) which constitutes a composite of a roughly
estimated 0.4 percent of the Animalia totaling 8,045 species, which he
points out is "an indication of the long road still ahead."

In addition to the description of new species, "old" ones are continually being reexamined in the light of new information and new

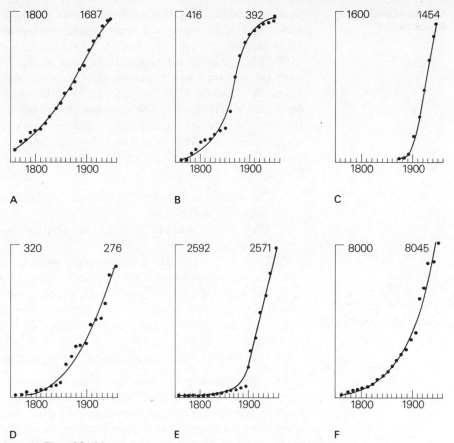

Fig. 10–11

Trend curves of rate of species description. A. North American birds. B. North American Rhopalocera. C. Siphonaptera. D. North American Vespidae. E. Culicidae. F. Composite of 8,045 species of Animalia. The numbers in the upper right-hand and upper left-hand corners of each graph indicate the number of species described and the number of species scaled. Each graph, except C, begins approximately at 1758, the year in which the tenth edition of *Systema Naturae* was published. (Redrawn from Steyskal, 1965.)

techniques for observing features. As mentioned earlier, biochemical, serological, physiological, behavioral, and so on, features are being used more and more.

Classification

Definition and Purpose. The thousands of species descriptions would be practically impossible to deal with if they were not organized around a standard. Obviously such organization is necessary because of the differences among organisms and at the same time is made possible by the similarities among them. By grouping organisms based on the degrees of similarity among them we can arrive at a system of classification. At this point in time an overall classification of organisms is pretty well established, and concern is primarily involved with modification in the light of new information and ideas and the addition of new species. It is generally at the most specific (least inclusive) levels of grouping, species and genus, where there is the most change, although such modifications do occur occasionally at higher levels.

We have pointed out that classification makes it possible to deal with the diversity and numbers of organisms. What exactly do we mean by "dealing with" organisms; in other words, what are the purposes of classification? Warburton (1967) has critically examined the writings of several taxonomists and has extracted a useful list of such purposes. Ideally the best classification would be one that satisfied all these purposes to the greatest extent. Warburton's list includes the following:

1. Classifications provide intellectual satisfaction for the taxonomists who make them.
2. They make possible the identification of species and higher groupings.
3. They ". . . provide a convenient, practical means by which zoologists may know what they are talking about and others may find out" (Warburton quoting G. G. Simpson).
4. They provide a system of information retrieval.
5. They may reflect phylogenetic relationships (to be discussed later).
6. They may serve as summarizing and predicting devises.

The first purpose requires no elaboration and is probably the primary motivation for the efforts of most scientific investigators. The level of identification made possible by a classification depends, of course, on the extent of that classification, but certainly the ability to place a given animal at some point in a classification facilitates further identification by ruling out a good part of the animal kingdom. For instance, if one can determine the class, order, or even family of a given animal by determining where it fits in an already established classification, he is several steps closer to a complete identification than he would otherwise have been.

A classification's value as a basis for communication is obvious. As mentioned earlier, an animal's name is the "file category" under which all information pertaining to that animal is stored. Within a classification one not only has the name of an animal but also the names of the successively more inclusive groups of which it is a part and hence the "file categories" of the literature of these groups. Many taxonomists feel or have felt that a scheme of classification should show the evolutionary relationships. We will elaborate on this topic later in this section. Regarding the last-mentioned purposes, a classification by the nature of the way it is constructed—a descending or ascending hierarchy of successively less or more inclusive groups—is a summarizing device, since all its categories are arrived at by the grouping of animals on the basis of similarities. Thus a great deal of information is summarized merely by recognizing that a given animal is an insect. Similarly, an animal's position in a classification enables one to predict certain aspects about that animal without previous examination. In addition, a classification by its nature enables one to make inductive generalizations regarding the animals included in the classification.

Features Used. Under description, potentially useful features and possible sources of variation were listed. Only certain features are useful in developing classifications—those which make a given organism or group of organisms different from other organisms or groups of organisms. These features are referred to as *taxonomic characters.* Blackwelder (1967) lists the following as attributes of "good" taxonomic characters:

(1) They are not subject to wide variation among the known specimens, (2) they do not show a high intrinsic genetic variability, (3) they are not readily modified by the environment, (4) they are consistently expressed, (5) they are available in the specimens which must be used, (6) they are visible with reasonable procedures, and (7) they can be effectively recorded . . . .

Certain taxonomic characters can stand alone as distinguishing features, singularly representing a species or higher category. Such singularly distinguishing features are referred to as *diagnostic* or *key characters.* It has sometimes been the practice to assume greater significance for certain characters to be used in a classification than others. This practice of *weighting* may introduce considerable bias into a classification unless the weighted features show a strong correlation with other features, which could support the classification as well as the weighted ones. Undoubtedly the safest procedure is to use several characters in the development of a classification.

Kinds of Classifications. Most systematists agree that a classification should be "natural," but definitions of the term "natural" vary somewhat. To some a natural classification is a phylogenetic one which reflects the evolutionary relationships among the groups that comprise it. Although this is an admirable goal for a classification, and there is certainly general agreement that the organisms which are classified are the contemporary products of eons of organic evolution, there is to date no absolutely known phylogenetic classification and only a very few likely ones (e.g., horses). This is simply because the fossil record is fragmentary at best, and comparative data gained by any approach, morphological or otherwise, cannot possibly provide a complete picture by itself.

Blackwelder (1967) offers a more realistic, attainable natural classification as follows: ". . . one in which the groups are recognized by having a maximum number of attributes in common, with their limits set by discontinuities in the diversity, and capable of yielding the maximum number of correct deductions about correlations of other features."

The numerical taxonomists also recognize the fact that a truly phylogenetic classification is unlikely to be attained and develop what they term *phenetic classifications.* The following quote from Sokal and Sneath (1963) adequately explains their position:

There may be confusion over the term "relationship." This may imply relationship by ancestry, or it may simply indicate the overall similarity as judged by the characters of the organisms without any implication as to their relationship by ancestry. For this meaning of overall similarity we have used the term "affinity," which was in common use in pre-Darwinian times.

We may also distinguish this sort of relationship by ancestry by calling it *phenetic* relationship . . . to indicate that it is judged from the phenotype of the organisms and not from its phylogeny.

These taxonomists believe that as many features as possible, or at least a very large number of features, should be used without any weighting in establishing a classification. Their approach is highly quantitative and the computer is one of their major research tools.

Blackwelder (1967) extends his concept of a natural classification into one which he calls an *omnispective classification*. This type of classification and the approach used to attain it seems to the present author to be by far the most realistic, pragmatic available. Blackwelder explains:

> In this system the extensive background of the experienced taxonomist enables him to pass over, almost without conscious thought, all the non-varying features of the organism, as well as the ones due to sex, age, pathology, and so on, and to use a workable number of features that are evidently correlated with many unmentioned ones. In this system, then, all features of the organisms are considered so far as they are available, and only those are employed which are necessary to show the groupings and distinctions that occur.

> This, then, is an all-seeing or all-considering system. It does not pretend to *use* all features, because it decides against the use of some. It knows of the probable existence of other features which it does not yet have access to. This has been called the omnispective system and is the one in current use by most taxonomists. It is at present the only workable system, and the reason why it has changed so little is that it was very largely adequate to the purposes for which it was designed.

Components of Biological Classification. Four components form the basis for most systems of biological classification: (1) hierarchy, (2) taxon, (3) category, and (4) rank. Simpson (1961) offers clear, concise, and useful definitions of each of these components. A *hierarchy* is "a systematic framework for zoological classification with a sequence of classes (or sets) at different levels in which each class except the lowest includes one or more subordinate classes." A *taxon* (pl. *taxa*) is "a group of real organisms recognized as a formal unit at any level of a hierarchic classification." A *category* is "a class, the members of which are all the taxa placed at a given level in a hierarchic classification." *Rank* is a category's "absolute position in a given hierarchic sequence of categories or its position relative to other categories." Thus Insecta (taxon) is a class (category) that in the generally accepted biological hierarchy (Fig. 10–12) is ranked between superclass and subclass.

Taxonomic Categories. Taxonomic categories can be conveniently examined in three parts as follows: (1) the species, (2) the subspecies, and (3) the higher categories.

The species forms the basic unit of a taxonomic classification and it refers directly to real populations of real animals in nature. The

Kingdom
 Phylum
  Subphylum
   Superclass
    Class
     Subclass
      Cohort
       Superorder
        Order
         Suborder
          Superfamily (*-oidea*)
           Family (*-idae*)
            Subfamily (*-inae*)
             Tribe (*-ini*)
              Genus
               Subgenus
                Species
                 Subspecies

Fig. 10–12

Hierarchy of generally accepted taxonomic categories.

subspecies is a subdivision of a species and refers to real entities. On the other hand, higher categories are all those categories in the biological hierarchy above the species level and are essentially subjective, their nature depending upon the professional opinions of the taxonomists that "recognize" them. They refer to real animals only indirectly in that they eventually include one to several species, depending on their position in the hierarchy.

Despite its objective reality, a number of different concepts revolve around the term "species." Early, pre-Darwinian biologists subscribed to the *type concept* of a species. This species was considered to have arisen as a result of special creation and was hence recognised as being composed of nonvarying individuals. With this idea of species a single specimen was all that was really necessary to describe a species and morphological traits and sometimes geographical data were generally the only ones used. It must be emphasized at this point that this concept of a species and the use of "types" in species description mentioned earlier are very different. Types in the latter sense serve a pragmatic purpose (as already explained), but the modern concern is with series or samples of individuals of a species, and great stress is placed on identifying the kind and degree of variation of any feature used.

With the advent of modern biology, with its understanding of organic evolution and heredity, a new concept of species, the *biological species*, developed. According to this concept, "Species are groups of actually (or potentially) interbreeding natural populations which are reproductively isolated from other such groups" (Mayr, Linsley, and Usinger, 1953). This concept of species has been purported to negate the earlier morphologically based concept of species.

Blackwelder (1967) takes exception with the idea that the "biological species" is the species of modern taxonomy. He does not deny the validity of this concept but explains that it in fact is the "species" of another area of biology—the "new science of speciation" —with little practical application in taxonomy as such. He argues that: "Morphogeographical species are those of the ordinary taxonomist, from Linneaus to modern times. Although data other than 'morphology' and geography have been increasingly used in recent years, these still remain the basic species of taxonomy." He does not deny that a taxonomist needs to be concerned with variation but merely recognizes that studies of speciation with their attendant terminologies and concepts do not negate the concepts of taxonomy, which continues to be a vital, important area of biological science.

Other definitions of species with less pervading meanings can be found in Blackwelder (1967) and Mayr, Linsley, and Usinger, (1953).

A *subspecies* may be defined as ". . . an aggregation of local populations of a species, inhabiting a geographical subdivision of the range of the species, and differing taxonomically from other populations of the species" (Mayr, 1963). In other words, subspecies are geographical races within the same species which are sufficiently different in some regard for them to be classified in this manner. Based on the biological species definition, subspecies should be able to interbreed if given the opportunity to do so. That two races of a species are "taxonomically different" is a decision that must be made only by a professional taxonomist. Many apply the 75 *percent rule*, which can be stated as follows: "population A can be considered subspecifically different from population B if 75 percent of the individuals of A are different from 'all' the individuals of population B" (Mayr, Linsley, and Usinger, 1953).

As mentioned earlier, the higher categories are subjective, depending on the opinions of taxonomists. This subjective nature is apparent when one realizes that a taxon that was once an order is now a family (or vice versa) or that a given taxon may be at the same time recognized as a family by some taxonomists and an order by others. Another piece of evidence for the subjective nature of higher categories is the fact that the taxa of one group of animals usually do not correspond with the taxa of another group in terms of the diversity shown in a given category. For example, the orders of birds are much less distinct than the orders of insects.

A higher category may appropriately be defined in relation to the category immediately below it in the hierarchy. For example, a genus may be defined as a group of species which have a sufficient number of features in common to warrant inclusion within a single group. In practical terms the decision regarding "a sufficient number of features" is up to the taxonomist making the decision. Also, inclusion of a given group of species in a genus may mean a monophyletic origin of this group to one taxonomist and to another may merely be indicative of a number of common features with no evolutionary overtones. Similar comments can be applied to the family, order, and so on, categories, and the generalization can be made that the higher the category, the greater the stability with regard to changes. In other words, genera change much more frequently than families or orders.

Future of Classification.   Classifications should be looked upon as organizational systems that are continually undergoing a sort of "evolutionary" development in light of new information about organisms and their relationships to one another. It is manifest that classification is destined to "evolve" for some time to come since as Blackwelder (1967) so appropriately points out, "The well-classified parts of the animal kingdom form no more than a hundredth part of the known fauna and probably only a thousandth part of what actually exists." Since the largest portion of the animal kingdom is held by insects it stands to reason that at least a major part of the job will have to be done by entomological taxonomists.

Nomenclature

Nomenclature is that area of taxonomy involved with the naming of organisms and groups of organisms and the rules and procedures to be followed in such naming. The rules of nomenclature have been developed and gradually changed and modified over the years, and the current rules are in the 1964 revision of the International Code of Zoological Nomenclature. This code is analyzed in considerable detail in Blackwelder (1964).

In this section we shall merely outline a few of the most basic rules.

Scientific Names. Every species and higher category has a Latinized scientific name. Regarding the species level, our system is commonly referred to as a *binomial system of nomeclature*, following the standard set by Carl Linne in the eighteenth century. In this system the scientific name of a species consists of two parts, the genus and the *specific epithet*, which comprise the *binomen*. An example would be *Aedes occidentalis,* in which *Aedes* is the genus and *occidentalis* the epithet. These two words together comprise the scientific name of a species of mosquito. The epithet is not the species name and should not be used as such. As a convention all species names are printed in italics or underlined when written or typed without italics and the generic name is capitalized while the epithet is not. Actually a complete species name can be considerably more complex than a simple binomen, although the binomen is still valid to use. For example, the complete scientific name of one of the subspecies of the mosquito species just mentioned is as follows: *Aedes (Finlaya) occidentalis occidentalis* (Skuse, 1889) *(Culex)*. The first and third terms are the binomen. The second term, in parentheses, is the subgenus and the fourth indicates the subspecies. Skuse is the taxonomist who first described this species and his description was published in 1889. The parentheses indicate that Skuse placed this species in a different genus than *Aedes*. The last name, *Culex*, in parentheses, is the genus in which Skuse placed this species. Sometimes a generic name is written followed by sp. or spp., for example *Aedes* sp. or *Aedes* spp. These abbreviations indicate a particular species or several species within a given genus, respectively.

The names of families are uninomials and end in -idae and name changes are strictly regulated by the code mentioned above. All family names are based on a type genus; for example, the family name of tiger beetles, Cicindelidae, is based on the generic name *Cicindela*.

The names of categories above family are all uninomials and are not extensively considered by the above code. They are used in the form of plural Latin nouns (e.g., Pterygota, Insecta) and no type categories are involved. The categories used in a given classification depend entirely upon the opinion of the taxonomist preparing the classification. Like family, several other categories have standardized endings (Fig. 10–12). A number of zoologists are supporting a system of uniform endings of higher category names for the entire animal kingdom (Honigberg et al., 1964): ". . . phyla, subphyla and superclasses end in '-a'; classes in '-ea'; subclasses in '-ia'; orders in '-ida'; and suborders in '-ina'."

Common Names. Many insects that are particularly common or outstanding in a given locality are often given vernacular names: For example, dragonflies have been variously referred to as "mosquito hawks," "snake feeders," and "devil's darners." Such names have little value other than in the context of local color. However, certain common names may be particularly useful when applied to agricultural or medical pests since they facilitate communication between the professional and layman. Most orders and families and many species have common names that over the years have become well established. Many of these names are included in the list entitled "Common Names of Insects Approved by the Entomological Society of America" published in the *Bulletin of the Entomological Society of America,* 11(4):287–320 (1965). Two particularly outstanding common names are "bugs" and "flies." Although these words are found in several ordinal and familial names (e.g., butterflies, mayflies, doodlebugs, lightning bugs, etc.), they apply specifically to members of the suborder Hemiptera (true bugs) and the order Diptera (true flies). When used in reference to members of these orders, "fly" and "bug" are written as separate words (e.g., house fly and conenose bug). However, if they are used for members of any other order, they are written with a modifier of some sort as a prefix (e.g., dragonfly or ladybug). When there is no common name, or when it is preferable not to use a common name, the scientific name may be used as a noun or modifier (e.g., mosquitoes may be called *culicids* derived from family Culicidae).

Selected References

EVOLUTION

Dodson (1960); Ehrlich and Holm (1963); Hennig (1966); Mayr (1963, 1970); Ross (1962b).

SYSTEMATICS

Blackwelder (1967); Ferris (1928); Mayr (1969); Mayr, Linsley and Usinger (1953); Simpson (1961); Sokal and Sneath (1963).

Survey of Class Insecta

The objective of this chapter is to provide a survey of the orders of insects. It is well beyond the intended scope of this text to provide anything approaching a comprehensive survey, and hence the coverage of each order will be comparatively brief.

The ordinal divisions of the class Insecta are a matter of contention and it is not easy to decide which divisions to use. The approach of this author has been to adopt the orders listed in the *Bulletin of the Entomological Society of America,* vol. 11, no. 4, 1965. However, it must be emphasized that there are many different ordinal groupings recognized by different systematists, and which one is "best" will no doubt remain a problem. Where appropriate, alternative names for a given order and other groupings will be given.

Instead of listing the number of species described in each order, a number that changes frequently, and to facilitate comparison of the sizes of the different orders to the extent that they are presently known, this information is presented in graphic form (Fig. 11–1). Figure 11–2 depicts the size standards used throughout this chapter.

Fig. 11–1

Relative sizes of the orders of Insecta. (Data from Borror and DeLong, 1971.)

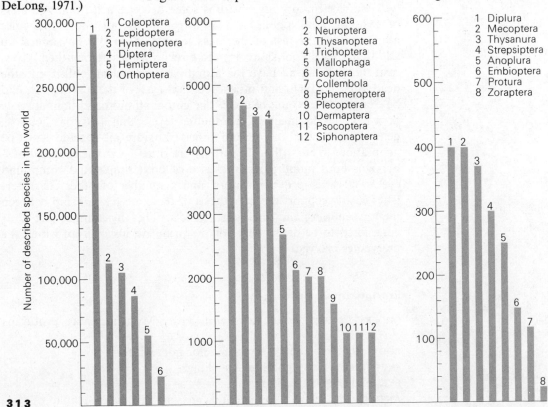

Fig. 11–2
Size standards used in this chapter. Mi,
minute; S, small; Me, medium; L, large;
VL, very large. (Size abbreviations in
following figure captions pertain to all
insects depicted, unless otherwise specified.)

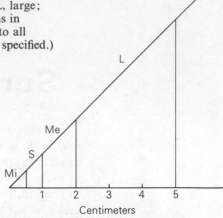

Apterygota

The members of the orders in this group of insects are all primitively
wingless (i.e., none of their ancestors possessed wings). Based on the
ordinal divisions recognized in this text, this group of insects includes
the following orders: Protura, Collembola, Entotrophi, and
Thysanura.

In the past the presence of six legs in the adult stage was used as a
characteristic that defined a given arthropod as an insect (Sharov,
1966). However, with more detailed studies, it has become fairly
well established, although there is some disagreement, that members
of Protura, Collembola, and Entotrophi do not share the close
affinities with the rest of the class Insecta as do the Thysanura. In
addition, they do not share any close relationships among themselves,
save the fact that all have the mouthparts somewhat pulled into the
head. Sharov (1966) and others feel that each of these groups should
be elevated to the rank of class. The entognathous condition, absence
or weak development of compound eyes, thin and pale cuticle,
general small size, and several other traits are all considered to be
adaptations to life within the soil. On the other hand, the exognathous-
hypognathous mouthparts, presence of both simple and compound
eyes, well-developed ovipositors, and a number of other characteristics clearly support the hypothesis of the close relationship between
the thysanurans and pterygote insects. The thysanurans are thus
considered to be descendents of the primitive insect from which the
pterygotes evolved.

Entognathous Apterygotes

ORDER Protura (*prot*, first; *ura*, tail); Myrientomata; proturans
  (Fig. 11–3A).
BODY CHARACTERISTICS minute, elongate, whitish.
MOUTHPARTS entognathous, sucking.
EYES and OCELLI lacking.
ANTENNAE lacking; "pseudoculi" may be vestigial antennae.

WINGS lacking.

LEGS first pair carried in elevated position suggestive of antennae; tarsi 1-segmented.

ABDOMEN short bilateral styli on first three segments; 12 segments; cerci lacking.

COMMENTS tracheae lacking; Malpighian tubules are small papillae.

As with other entognathous apterygotes, proturans are inhabitants of the soil and associated litter and require moist conditions. They apparently feed on decaying organic matter. Protura are worldwide in distribution and are common but are easily overlooked.

Proturan development is ametabolous (i.e., they do not undergo metamorphosis and hence the adults are essentially identical to the immatures except for the presence of fully developed gonads and 12 abdominal segments). First instar individuals have eight true abdominal segments and during the course of development three more are added (intercalated) between the telson and the eleventh segment. This type of growth is called *anamorphosis.*

ORDER Entotrophi (*ento*, within; *trophi*, food); Diplura, Aptera, Entognatha; campodeids and japygids (Fig. 11–3B).

BODY CHARACTERISTICS minute to small; slender; whitish.

MOUTHPARTS entognathous; chewing.

EYES and OCELLI lacking.

ANTENNAE long; filiform.

WINGS lacking.

LEGS 1-segmented tarsi.

ABDOMEN 10 visible segments; cerci forcepslike or long caudal filaments.

COMMENTS tracheae present; Malpighian tubules vestigial or absent.

Fig. 11–3
A. Proturan, *Acerentulus barberi barberi* Ewing. B. Entotrophi, *Iapyx diversiungus.* (Mi.) (A redrawn from Ewing, 1940; B redrawn from Essig, 1942.)

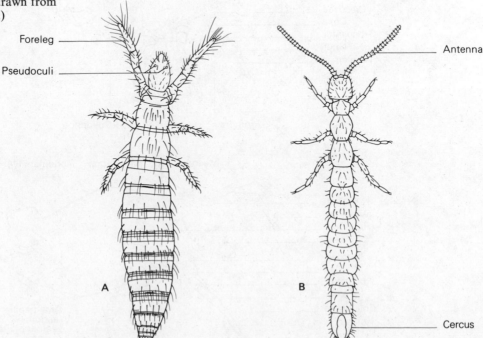

Foreleg

Pseudoculi

Antenna

A

B

Cercus

Telson

Entotrophi are found in damp situations in caves, under bark, in the soil, and in similar habitats. They undergo ametabolous development.

ORDER   Collembola (*coll*, glue; *embol*, a wedge); springtails (Fig. 11–4).

BODY CHARACTERISTICS   minute, somewhat tubular (suborder Arthropleona) or globose (suborder Symphyleona).

MOUTHPARTS   entognathous, chewing.

EYES and OCELLI   eye patches consist of 1 to several lateral ocelli; dorsal ocelli lacking.

ANTENNAE   short; usually 4-segmented.

WINGS   lacking.

LEGS   pair of claws (ungus and unguiculus) at distal end of tibia; tarsi fused to tibia (tibiotarsus) or 1-segmented.

ABDOMEN   6 segments; lobelike organ on venter of first segment, function unclear, once thought to be "adhesive" organ; forked structure, *furcula,* on venter of fourth segment; *tenaculum* on venter of third segment.

COMMENTS   postantennal organs in many species; furcula directed anteroventrally and secured by tenaculum; "springing" accomplished by sudden release of furcula from tenaculum; tracheae present or absent; Malpighian tubules lacking. Springtails are so named because of their habit of jumping by means of the furcula.

Fig. 11–4

Collembola, springtails.
A. Arthropleonid.
B. Symphypleonid. (Mi.)
(Redrawn from
Maynard, 1951.)

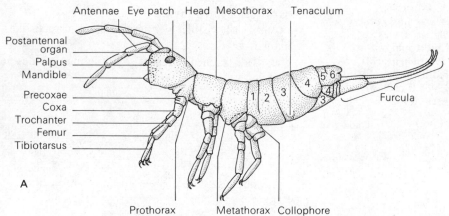

Antennae   Eye patch   Head   Mesothorax   Tenaculum

Postantennal organ
Palpus
Mandible
Precoxae
Coxa
Trochanter
Femur
Tibiotarsus

1 2 3 4 5 6

Furcula

A

Prothorax   Metathorax   Collophore

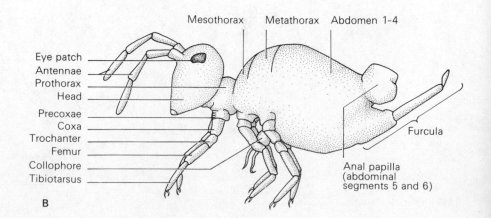

Mesothorax   Metathorax   Abdomen 1-4

Eye patch
Antennae
Prothorax
Head
Precoxae
Coxa
Trochanter
Femur
Collophore
Tibiotarsus

Furcula

Anal papilla
(abdominal
segments 5 and 6)

B

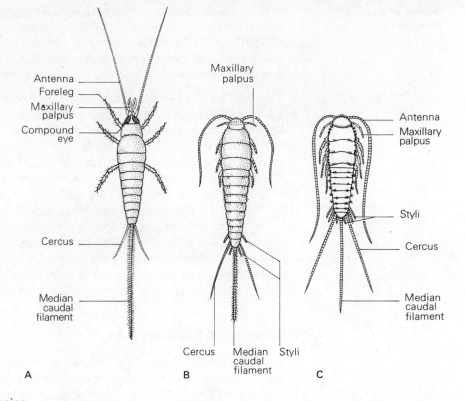

Antenna
Foreleg
Maxillary palpus
Compound eye

Maxillary palpus

Antenna
Maxillary palpus

Styli

Cercus

Cercus

Median caudal filament

Median caudal filament

Cercus   Median caudal filament   Styli

A   B   C

Fig. 11–5

Thysanura. A. Jumping bristletail, *Machilis* sp. B. Common silverfish, *Lepisma saccharina*. C. Firebrat, *Thermobia domestica*. (S to Me.) (A redrawn from Essig, 1942, after Lubbock; B and C courtesy of the U.S. Public Health Service (redrawn).)

Collembola are very widely distributed throughout the arctic, temperate, and tropical regions and are commonly found in very large populations (e.g., several million in a single acre). They are found in diverse, but always moist, situations ranging from intertidal zones at the seashore to the surface of snow. They are common in soil, leaf litter, and other decaying organic matter and may be found on the surfaces of bodies of water. Some species inhabit ant nests and others are found in caves. They apparently feed on decaying organic matter, which is abundant in the habitats mentioned. These insects are ametabolous.

Although they are of no medical importance, certain species of Collembola have been identified as pests of mushrooms, truck garden crops, forage and cereal crops, sugar cane, and in households (Maynard, 1951). The most outstanding is *Sminthurus viridis*, the "Lucerne flea," a serious introduced pest of alfalfa in Australia.

Ectognathous Apterygotes

ORDER Thysanura (*thysan*, fringe; *ura*, tail); Ectognatha, Ectotrophi; bristletails (Fig. 11–5).
BODY CHARACTERISTICS small; usually covered with scales.
MOUTHPARTS ectognathous, chewing.
EYES and OCELLI compound eyes usually present; 0 or 3 ocelli.
ANTENNAE elongate, filiform.
WINGS lacking.
LEGS 2- to 4-segmented tarsi.
ABDOMEN small bilateral styli on venter of several segments; females

have elongate jointed ovipositor; cerci elongate with many segments; elongate median caudal filament present.

COMMENT   some (e.g., *Machilis*) good jumpers; styli interpreted as vestiges of locomotor appendages inherited from myriapodlike ancestors; open tracheal system; Malpighian tubules present.

Most species of Thysanura (e.g., *Machilis*, Fig. 11–5A) live in soil, leaf litter, rotting wood, and similar damp habitats.

Bristletails undergo ametabolous development. These insects are different from the pterygote forms in that they continue to undergo periodic moltings after they have reached sexual maturity.

Some thysanuran species [e.g., the common silverfish, *Lepisma saccharina* (Fig. 11–5B), and the firebrat, *Thermobia domestica* (Fig. 11–5C)] are domesticated and share human habitations. Silverfish are found in cool, damp locations; firebrats frequent warmer locations and are generally found around steam pipes, furnaces, and similar heat-producing objects. Both species can become pests because they feed upon such things as book bindings, starched clothing, and cloth of various sorts. They have no medical significance.

Paleopterous Exopterygota

As explained in Chapter 10, the insects in this group have wings in the adult stage, but the wings are primitive in that they cannot be flexed and laid down over the abdomen and hence must, at rest, either be held laterally or together above the thorax and abdomen. All other winged members of Pterygota have a wing-flexion mechanism or are descendants of insects that possessed such a mechanism [e.g., Lepidoptera (butterflies and moths)]. Although the fossil record indicates the past existence of other orders of paleopterous insects, members of Odonata (dragonflies and damselflies) and Ephemeroptera (mayflies) are the sole survivors.

Paleopterous insects undergo simple metamorphosis during their development from egg to adult (i.e., they are hemimetabolous).

ORDER   Ephemeroptera (*ephemero*, for a day; *ptera*, wings); Ephemerida, Plectoptera; mayflies (Fig. 11–6).

BODY CHARACTERISTICS   small to medium; fragile; soft-bodied.

MOUTHPARTS   nymphs: chewing; adults: vestigial.

EYES and OCELLI   compound eyes present; 3 dorsal ocelli.

ANTENNAE   short; setaceous.

WINGS   adults and subimagoes: 2 pairs; membranous; forewings larger than hindwings; few species with only mesothoracic wings; held vertically at rest; typically many veins.

LEGS   3- to 5-segmented tarsi.

ABDOMEN   nymphs: bilateral ventilatory gills on the first 4 to 7 segments; adults: pair of long, filamentous cerci; some with additional long median caudal filament.

COMMENTS   nymphs: closed tracheal system, gill ventilators; adults: open tracheal system.

The members of this order are aquatic in the nymphal stages and terrestrial-aerial as adults.

Mayflies are unique in the class Insecta in passing through a winged, subadult (subimaginal) stage which is essentially identical to the adult stage. The subimaginal and adult stages live for a brief period of time (a few hours to a few days) and their apparent sole function is to survive long enough to copulate and deposit their eggs, which are often simply dropped into the water and subsequently sink to the bottom. Some species attach their eggs to partially submerged rocks or aquatic vegetation. The nymphal period is generally fairly long, a year or more, although some species may develop in a few months.

Mayflies commonly emerge from lakes, streams, and rivers in fantastically large numbers and in areas near towns or cities the shed subimaginal cuticles and dead adults may literally accumulate in piles on streets and sidewalks. This was once a common occurrence in towns along the shores of Lake Erie until pollution took its toll.

Nymphal mayflies (Fig. 11–6B) are solely responsible for the intake of nutrients since the adults do not feed. Their food consists of aquatic vegetation and small aquatic organisms.

Mayfly nymphs are important food for aquatic predators, particularly fish, and provide a periodic feast for insectivorous birds and dragonflies when they emerge in vast numbers. They have been likened to terrestrial herbivores in that they are close to the bottom of several food chains, converting plant material into animal material. They also

Fig. 11–6
Ephemeroptera, mayflies.
A. Adult. B. Nymphs.
(S to Me.) (Redrawn from Burks, 1953).

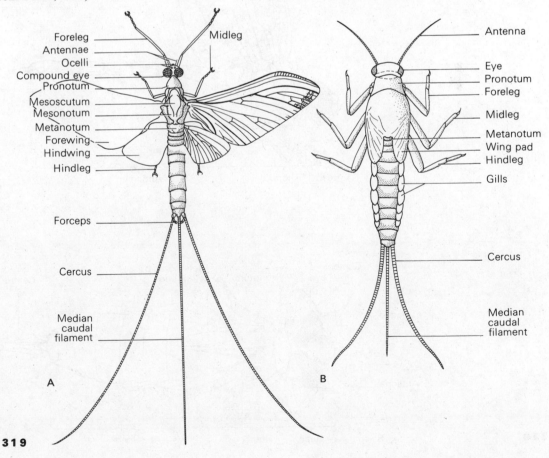

provide excellent models for many of the common "tied flies"
cherished by avid fly fisherman.

In areas where large populations of adult mayflies invade towns and
cities, there have been reports of rather severe hypersensitive responses
(asthma, conjunctivitis, and so on) to airborne detritus composed of
the broken casts of subimagoes and dried bodies of adults.

ORDER   Odonata (*odon*, a tooth); dragonflies and damselflies (Figs.
11–7 and 11–8).

BODY CHARACTERISTICS   medium to very large; elongate.

MOUTHPARTS   chewing; nymph with prehensile labium.

EYES and OCELLI   large compound eyes; 3 dorsal ocelli.

ANTENNAE   short; bristlelike.

Fig. 11–7
Odonata, Anisoptera,
dragonflies. A. Adult, *Anax
junius.* B. Nymph (VL.)
(Redrawn from Garman, 1927.)

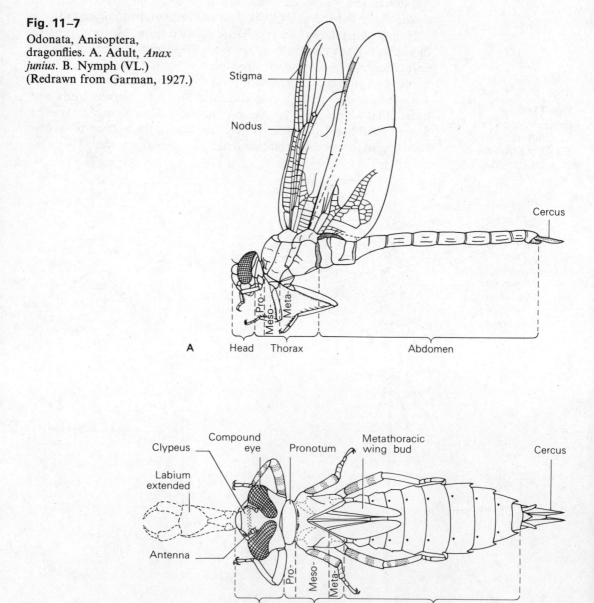

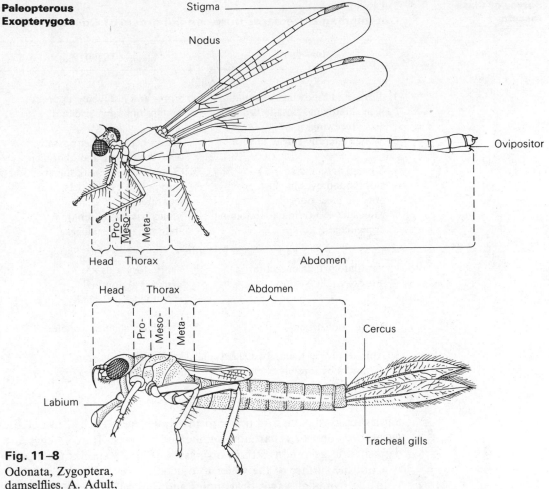

Fig. 11–8
Odonata, Zygoptera, damselflies. A. Adult, *Lestes vigilax*.
B. Nymph. (VL.) (Redrawn from Garman, 1927.)

WINGS 2 pairs; membranous; netlike venation; characteristic pigmented cell immediately posterior to the costal vein near the apex of each wing, the *stigma*; well-developed cross vein near the middle of the leading edge of each wing, the *nodus*.

LEGS adults: basketlike arrangement, adaptation for prey capture; 3-segmented tarsi.

ABDOMEN elongate; adult males with gonopores on ninth segment and complex penis on venter of second segment; 1-segmented cerci in males serve as claspers during copulation.

COMMENTS nymphs: closed tracheal system, gill ventilators; adults: open tracheal system.

Odonata can be divided into two suborders, Anisoptera, the dragonflies (Fig. 11–7), and Zygoptera, the damselflies (Fig. 11–8). Members of these two suborders are easily distinguished from each other both in the nymphal and adult stages. Table 11–1 lists some of these differences.

Immature dragonflies and damselflies are aquatic, whereas the adults are terrestrial–aerial but are generally found in the vicinity of water. Nymphs usually reach the adult stage in a single season; however, some species require two or three years to develop into

Table 11–1
Outstanding Differences Between Suborders of Odonata

Anisoptera	Zygoptera
Adults	
1. Fore- and hindwings unequal in size; hindwings basally broader than forewings.	1. Fore- and hindwings approximately the same size and shape.
2. Wings laterally spread at rest	2. Wings held together dorsally over the thorax and abdomen.
3. Strong, agile fliers.	3. Comparatively weak fliers.
4. Compound eyes in close proximity or meet dorsally.	4. Compound eyes widely separated.
5. Males have 3 terminal abdominal appendages.	5. Males have 4 terminal abdominal appendages.
Nymphs	
1. Ventilatory gills in rectum, not externally visible.	1. Ventilatory gills are 3 terminal abdominal appendages, externally visible.
2. Stout, robust body.	2. Relatively slender, fragile body.
3. Able to propel themselves short distances by forcibly ejecting water from the rectum.	3. Lack "jet-propulsion" mechanism.

adults. Eggs may be dropped into the water and attached to aquatic vegetation or other partially submerged objects. Rarely, eggs are deposited in gelatinous strings or masses. In a few species females go beneath the surface of the water to oviposit.

In the nymphal stages dragonflies and damselflies prey on a wide variety of aquatic organisms, including tadpoles and fish for some of the larger species. Adults generally prey upon other flying insects (e.g., mosquitoes and midges, moths, bees and other dragonflies). They are able to catch, hold, and devour prey in flight.

Although Smith and Pritchard (1956) list these insects as occasional pests in apiaries and mention that they occasionally prey upon trout fry, these insects are generally considered to be quite beneficial and many species are aesthetically very pleasing. They are harmless to man.

Neopterous Exopterygota

Orthopteroid Orders

The following orders form a fairly distinct group within the neopterous exopterygotes: Plecoptera, Orthoptera, Dermaptera, Isoptera, Embioptera, and Zoraptera. The affinities between members of these orders which support the validity of the orthopteroid grouping include (1) primitive, "generalized" mandibulate mouthparts, (2) wings with highly complex venation and the hindwings with large anal lobes, (3) the presence of many Malpighian tubules, (4) the presence of cerci, and (5) ventral chain ganglia distributed segmentally in the thorax and several abdominal segments (Richards and Davies, 1957; Oldroyd, 1968). Some authors do not include the stoneflies (Plecoptera) with the orthopteroids, but recognize their close affinity with them; other authors do include them with the orthopteroids. For convenience, we shall follow suit with the latter authors.

Like the paleopterous forms, orthopteroid insects are hemimetabolous.

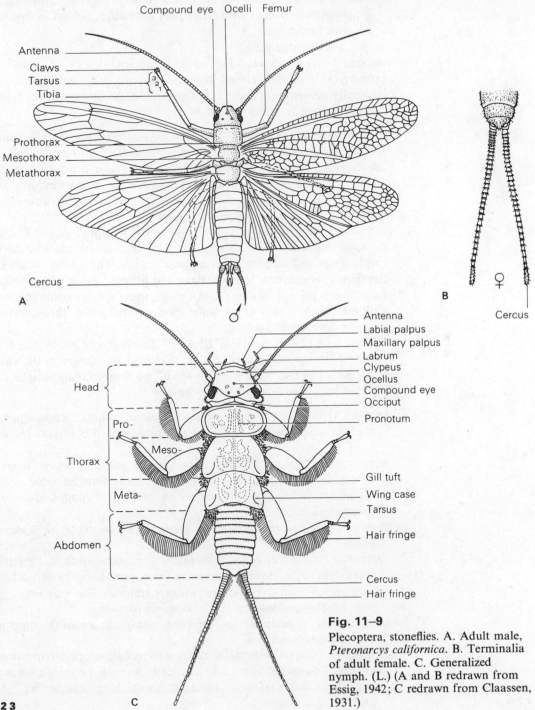

Compound eye Ocelli Femur

Antenna

Claws
Tarsus
Tibia

Prothorax
Mesothorax
Metathorax

Cercus

A

B

Cercus

Head

Pro-
Meso-

Thorax

Meta-

Abdomen

Antenna
Labial palpus
Maxillary palpus
Labrum
Clypeus
Ocellus
Compound eye
Occiput
Pronotum

Gill tuft
Wing case
Tarsus

Hair fringe

Cercus
Hair fringe

C

Fig. 11–9

Plecoptera, stoneflies. A. Adult male, *Pteronarcys californica*. B. Terminalia of adult female. C. Generalized nymph. (L.) (A and B redrawn from Essig, 1942; C redrawn from Claassen, 1931.)

ORDER   Plecoptera (*pleco*, twist; *ptera*, wing); stoneflies (Fig. 11–9).

BODY CHARACTERISTICS   small to medium; soft-bodied; elongate, flattened; body nearly parallel-sided.

MOUTHPARTS   chewing; commonly vestigial in adults.

EYES and OCELLI   well-developed compound eyes; usually 3 dorsal ocelli, rarely 2.

ANTENNAE   long and tapering.

WINGS   2 pairs; membranous; males of some species apterous or brachypterous (short wings); at rest, hindwings folded in "plaits" beneath forewings.

LEGS   3-segmented tarsi.

ABDOMEN   ovipositor lacking; many-segmented cerci.

COMMENTS   nymphs: closed tracheal system, gill ventilators, gills usually on venter of thorax, but sometimes on other parts of body; adults: open tracheal system.

Stonefly nymphs (Fig. 11–9C) are aquatic and are usually found around and beneath stones in fast-moving, well-aerated water, and the adults (Fig. 11–9A, B) are terrestrial–aerial and are found in the vicinity of such water courses. The nymphs are campodeiform and resemble the adults in general appearance except for the absence of wings and the presence of tracheal gills.

Stoneflies are mostly phytophagous in the nymphal stages, feeding on various forms of small aquatic plant life; a few are said to be carnivorous and feed on other aquatic insects. The adults with well-developed mouthparts feed on algae and lichens; those with vestigial mouthparts do not feed at all. Although most species complete their nymphal life in a year or so, some species require two, three, or four years to develop into adults.

Comstock (1940) mentions adults of *Taeniopteryx pacifica* as fruit tree pests in the Wenatchee Valley, Washington. However, the value of stoneflies as fish food and fish "bait" no doubt far outweighs this isolated instance of pestiferous behavior.

ORDER   Orthoptera (*ortho*, straight; *ptera*, wing); grasshoppers, crickets, cockroaches, and related forms (Figs. 11–10, 11–11, and 11–12).

BODY CHARACTERISTICS   minute to very large; the largest is over 25 cm; among the largest of all insects; body form variable.

MOUTHPARTS   chewing; all gradations between hypognathous and opisthognathous.

EYES and OCELLI   well-developed compound eyes; 0, 2, or 3 dorsal ocelli.

ANTENNAE   variable; filiform in many (e.g., cockroaches, katydids, and crickets); somewhat shorter in many grasshoppers and others.

WINGS   most with 2 pairs; forewings parchmentlike *(tegmina)*; at rest, hindwings folded pleatlike beneath forewings.

LEGS   variable; adapted for jumping, grasping, running, digging; 1- to 5-segmented tarsi.

ABDOMEN   females in several families with well-developed ovipositors; in some, ovipositor concealed beneath seventh or eighth sternite (e.g., cockroaches); cerci variable; short, long, clasperlike, segmented, unsegmented.

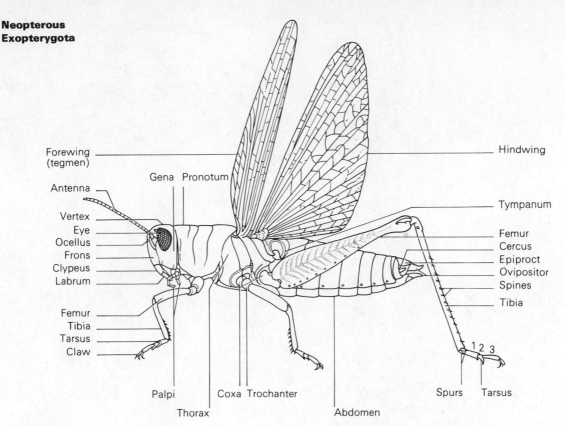

Forewing (tegmen) — Hindwing
Gena Pronotum
Antenna
Tympanum
Vertex
Eye
Ocellus
Frons
Clypeus
Labrum
Femur
Cercus
Epiproct
Ovipositor
Spines
Tibia
Femur
Tibia
Tarsus
Claw
1 2 3
Palpi
Coxa Trochanter
Spurs Tarsus
Thorax
Abdomen

Fig. 11–10
Orthoptera, a short-horned grasshopper (Acrididae), showing major external structures. (L.) (Redrawn from Essig, 1942.)

The order Orthoptera taken in its narrowest sense includes only the grasshoppers, crickets, and close allies and not the cockroaches, mantids, and phasmids (stick and leaf insects). The latter three groups of insects have been variously classified: cockroaches, mantids, and phasmids (order Dictyoptera or Cursoria); cockroaches and mantids (order Dictyoptera); phasmids (order Phasmida); and cockroaches (order Blattaria). On the other hand, some authors group these insects as Orthoptera, and this is the classification that will be followed here. Since Orthoptera is one of the largest orders and is composed of very diverse forms, it is appropriate to at least comment upon the various families. This order can be conveniently divided into the saltatorial (jumping) Orthoptera (Fig. 11–11) and the cursorial (running) Orthoptera (Figs. 11–10 and 11–12). In the saltatorial Orthoptera, the hind femora are enlarged, accommodating the muscles that produce the jumping force. This adaptation is lacking in the cursorial Orthoptera.

Members of this order are typically terrestrial and are found in trees, bushes, and other vegetation; on the surface of or burrowing into the ground; and some in caves. A few species are aquatic or semiaquatic. Orthoptera includes general scavengers, many voracious herbivores, and some carnivorous species.

There are generally five or more nymphal instars in an orthopteran life cycle. All form eggs, laid either singly, in masses, or encased in "egg packets" or oothecae (see Fig. 8–11). Some species retain the eggs in their bodies until they hatch into first instar nymphs.

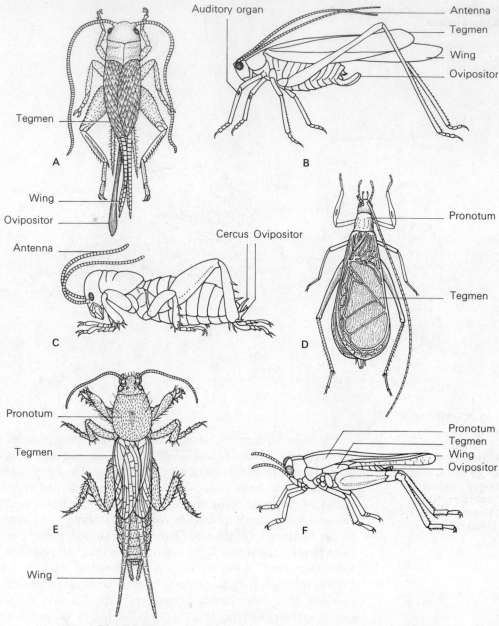

Fig. 11–11

Representative saltatory Orthoptera. A. Field cricket, *Gryllus assimilis* (Gryllidae). B. Bush katydid, *Scudderia furcata* (Tettigoniidae). C. Jerusalem cricket, *Stenopelmatus longispina* (Gryllacrididae). D. Tree cricket, *Oecanthus niveus* (Gryllidae). E. Mole cricket, *Gryllotalpa hexadactyla* (Gryllidae). F. Pygmy grasshopper, *Telmatettix hesperus* (Tetrigidae). (A, D, and F, Me; B, C, and E, L.). (Redrawn from Essig, 1942.).

The ability to produce and perceive sound is found among many orthopterans. Although sound production is effected in a variety of ways, most, save the spiracular hissing of certain species of cockroaches (e.g., *Gromphadorhina*) involve the rubbing of one part of the body against another. The females of some species are capable of producing sound, but it is typically the males that are the accomplished singers. The songs are most commonly involved with bringing members of the opposite sexes together. The snowy tree cricket, *Oecanthus fultoni*, is purported to be a "living thermometer" because adding 40 to the number of chirps in 15 seconds approximates the temperature in degrees Fahrenheit. This, of course, assumes that one is able to recognize the song of the snowy tree cricket. A phonograph record (Alexander and Borror, 1956) is available which includes the songs of many orthopterans, including the snowy tree cricket.

Orthoptera includes a large number of economically important species. Members of the family Acrididae, the short-horned grasshoppers, are serious pests of field crops, particularly small grains. Some of the long-horned grasshoppers (family Tettigoniidae) are also

Fig. 11–12

Representative cursorial Orthoptera. A. Chinese mantid, *Tenodera aridifolia* (Mantidae). B. Walking stick, *Diapheromera femorata* (Phasmidae). C. German cockroach, *Blatella germanica* (Blattidae). D. Grylloblattid, *Grylloblatta campodeiformis*. (A and B, VL; C, Me; D, L.) (A and D redrawn from Essig, 1942; B redrawn from Hebard, 1934; C courtesy of U.S. Public Health Service) (redrawn).)

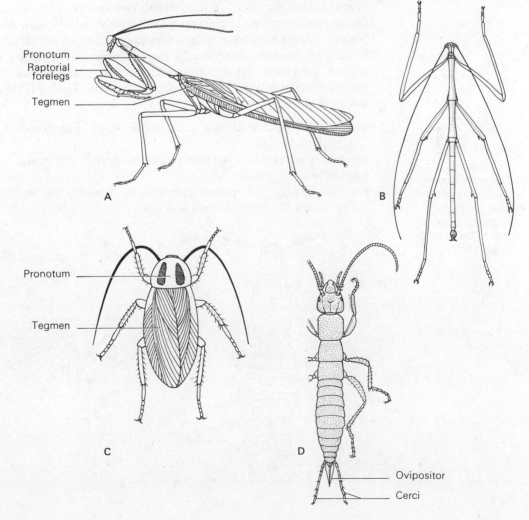

serious pests of field crops. For example, the Mormon cricket, *Anabrus simplex*, attacks several crops in the West. In Salt Lake City, Utah, a statue commemorating the sea gulls that saved early Mormon settlers from destruction of their crops has been erected in Mormon Square. The coulee cricket, *Peranabrus scabricollis*, is a common pest of field crops in the northwestern United States. A common species of stick insect, *Diapheromera femorata*, may defoliate forest trees when it occurs in large numbers.

The cockroaches (family Blattidae; Fig. 11–12C) are the only Orthoptera of major medical, as well as economic, importance. They are common household pests and feed on a wide variety of household goods, but the major indictment against them is that they are dirty, distasteful, and odoriferous creatures and are attracted to such materials as garbage, feces, and foodstuffs consumed by humans. The most common household-invading species (domiciliary) in the United States are (1) *Periplaneta americana*, the American cockroach; (2) *Blattella germanica*, the German cockroach; and (3) *Blatta orientalis*, the oriental cockroach. Although there is little conclusive evidence that points to cockroaches as disseminators of pathogenic organisms, the circumstantial evidence is strong and it has been suggested that they may, in fact, rival house flies in their capacity for disease transmission. Their nocturnal, secretive habits have perhaps been the reason they have been somewhat overlooked in this capacity in the past. Readers interested in more information regarding cockroaches, both their biology and medical-economic significance, should consult Roth and Willis (1957, 1960), Guthrie and Tindall (1968), and Cornwell (1968).

ORDER   Dermaptera (*derma*, skin; *ptera*, wing); Euplexoptera; earwigs (Fig. 11–13).

BODY CHARACTERISTICS   small to medium; narrow; elongate.

MOUTHPARTS   chewing.

EYES and OCELLI   compound eyes present in most, but vestigial or absent in some; dorsal ocelli lacking.

Fig. 11–13
Dermaptera. A.
European earwig,
Forficula auricularia,
male. B. Right
hindwing unfolded.
(Me.) (Redrawn from
Essig, 1942.)

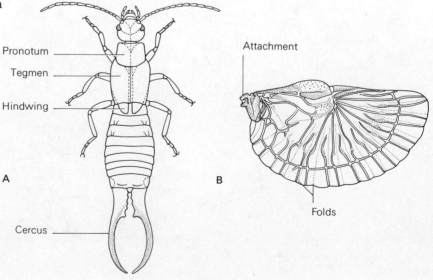

Pronotum

Tegmen

Hindwing

Attachment

A

B

Folds

Cercus

ANTENNAE long; slender.

WINGS forewings short, leathery tegmina; hindwings semicircular in shape and membranous with radially arranged veins; hindwings folded fanlike beneath forewings; wingless species common.

LEGS 3-segmented tarsi.

ABDOMEN forcepslike, unsegmented cerci; not covered by wings.

COMMENTS cerci apparently used in prey capture, defense, folding and unfolding wings, and possibly during copulation.

Most species of earwigs are nocturnal and prefer damp situations (e.g., in the soil, under bark, under stones, among vegetation, and near bodies of water). They are mostly omnivorous and feed upon such materials as small living or dead insects, decaying plant material, and tender parts of living plants. Some earwigs are parasitic (e.g., *Hermimerus* sp., which inhabits the fur of African rats).

Earwigs are remarkable in that they demonstrate the phenomenon of parental care of offspring, which is rather uncommon among the members of most insect orders and may represent an early stage in the evolution of social behavior. Eggs are deposited in the soil and the females literally "roost" on them until they hatch and subsequently care for the newly hatched young.

When disturbed, earwigs commonly raise the abdomen with its imposing cerci, somewhat resembling the behavior of a scorpion. An additional defensive adaptation is their ability to squirt for a distance of several inches a repugnatorial substance produced by glands in the dorsum of the second or third abdominal segments.

Some species (e.g., *Forficula auricularia*), when in abundance may damage flowers and other tender foliage. The common name "earwig" apparently came about because these insects were purported to enter the ears of sleeping persons, but this has little basis in fact and they are considered to be harmless. Richards and Davies (1957) point out that their common name may be a corruption of "earwing," in reference to the hindwing resembling a human ear.

ORDER Embioptera (*embio*, lively; *ptera*, wing); Embiidina; web-spinners (Fig. 11–14).

BODY CHARACTERISTICS minute to small; brown to yellow; elongate.

MOUTHPARTS chewing.

EYES and OCELLI compound eyes present; no dorsal ocelli.

ANTENNAE filiform; shorter than body.

WINGS some males with 2 pairs of membranous wings of nearly equal size and shape; "smokey" in appearance; females and some males wingless.

LEGS basal tarsal segment of foreleg enlarged and contains silk-producing glands; 3-segmented tarsi.

ABDOMEN pair of cerci present and of unequal size in males.

Embioptera is composed of a small group of uncommon insects which are mostly inhabitants of the tropics. However, they are also found in temperate zones. Webspinners are so named because they live in a network of silken tunnels beneath stones, bark, and so on. The tunnels are constructed by both sexes and immatures from the secretions of glands in the tarsi of the forelegs. These insects are

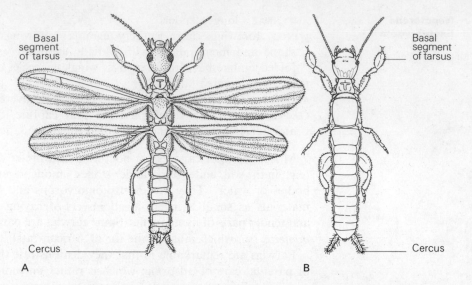

Basal
segment
of tarsus

Basal
segment
of tarsus

Cercus

Cercus

A

B

Fig. 11–14
Embioptera,
webspinners. A. Male,
Oligotoma saundersii.
B. Female. (Mi.)
(Redrawn from Essig,
1942.)

gregarious and a network of tunnels may contain many individuals.
It is interesting to note that they are able to run forward and backward
within these tunnels. In addition to serving as a habitation, the tunnels
apparently serve as protection against potentially predaceous insects
and other arthropods and may play a role in the maintenance of a
humid atmosphere. Webspinners feed mainly on decaying plant
matter.

These insects reproduce within their silken tunnels and the eggs are
deposited in scattered groups. They are like the Dermaptera in that
the females actively attend the eggs. At least one species is parthenogenetic.

ORDER   Isoptera (*iso,* equal; *ptera,* wings); termites (Figs. 11–15 and
11–16).

BODY CHARACTERISTICS   minute to large.

MOUTHPARTS   chewing.

EYES and OCELLI   compound eyes present in all winged forms, present
or lacking in apterous forms; 0 or 2 dorsal ocelli.

ANTENNAE   short, moniliform or filiform.

WINGS   when present, 2 identical pairs; membranous; shed by
breakage along a basal fracture line.

LEGS   4- to 5-segmented tarsi.

ABDOMEN   genitalia lacking or weakly developed; pair of short cerci.

COMMENTS   characteristic depression on dorsum of head called
fontanelle.

The termites are a very primitive order of orthopteroid insects, being
most closely related to the cockroaches. They are mainly tropical and
subtropical, but they also occur in temperate regions. Termites in
certain families live in dry wood (e.g., that frequently used in the construction of buildings and furniture). Others are subterranean and
require very humid conditions. The subterranean forms probably
play an ecological role similar to earthworms in that they aerate soil
and add nutriment accumulated in wood to it. In addition to wood,
termites feed on a wide variety of cellulose-containing materials, fungi,
and dried animal remains. In many termites the digestion of cellulose

is carried out by flagellate protozoans which are mutualistic inhabitants of the gut.

Termites are commonly confused with ants. This is probably the
reason that termites are sometimes referred to as "white ants." However, there are several distinct differences between these two groups of
insects. Some of the more obvious of these differences include the following. Comparing winged forms, in ants the hindwings are smaller
and have fewer veins than the forewings and at rest are held vertically
over the abdomen, whereas in termites the wings are similar in size and
venation and flexed horizontally over the abdomen at rest. Comparing
wingless forms, ants have relatively dark, sclerotized bodies, whereas
termites are pale and soft-bodied. In both winged and apterous forms
the abdomen of ants is basally constricted, but this is not the case in
termites.

All termites are social insects. A colony's population is initiated and
and maintained by a queen that may live for many years, perhaps as
long as 50 years in some species. Two to several castes may be present,
depending on the species, and all castes are composed of both males
and females. Conveniently, castes can be divided into reproductives

Fig. 11–15
Isoptera, termites. A. Winged form, *Zootermopsis angusticollis*. B. Eggs.
C. Third instar nymph. D. Last instar nymph. E. Soldier. F. Head of
Nasutitermes sp. (A and E, Me.) (A–E redrawn from Essig, 1942.)

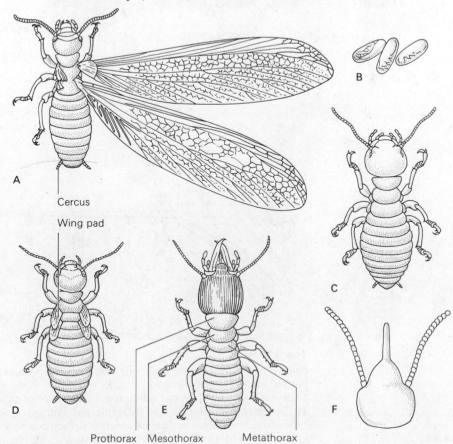

A

Cercus

Wing pad

B

C

D

E

F

Prothorax   Mesothorax   Metathorax

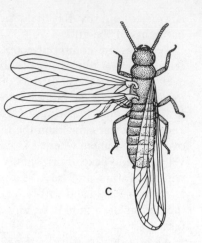

A

B

C

D

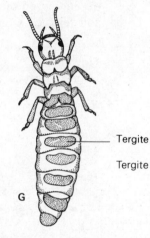

E

F

G — Tergite

H — Tergite

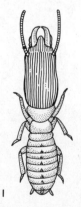
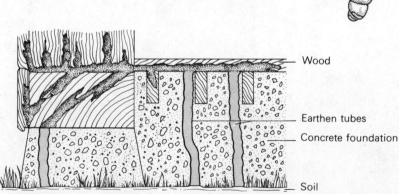

Wood

Earthen tubes

Concrete foundation

Soil

I

J

Fig. 11–16

Subterranean termite, *Reticulitermes*. A. Eggs. B. Nymph. C. Female winged form. D. Male winged form. E. worker. F. King. G. Supplementary reproductive (abdomen distended with enlarged ovaries). H. Queen (abdomen distended with enlarged ovaries). I. Soldier. J. Earthen tubes from soil across surface of concrete foundation to wooden structures. (Mi to S.) (Redrawn from CCM General Biological, Inc., key card.)

and nonreproductives. There may be two kinds of reproductives present in a colony of a given species: primary reproductives and secondary reproductives. The primary reproductives are the queen (Fig. 11–16 C, H) and king (Fig. 11–16 D, F) and are thought to comprise the original caste in isopteran phylogeny. They typically have dark, sclerotized (at least compared to other castes) bodies with completely developed wings and compound eyes. Secondary reproductives (Fig. 11–16 G) occur in a variety of forms (e.g., ones with shorter wings, less pigmentation, and smaller compound eyes than the primary reproductives or ones with no wings and pale bodies resembling workers).

The nonreproductive castes (Fig. 11–16E, I) are the workers and soldiers. The workers (Fig. 11–16E) are typically wingless, unpigmented, soft-bodied forms without compound eyes or well-developed mandibles and function as the labor force of the colony, maintaining and adding to the structure of the nest, tending fungus gardens, and feeding members of other castes and immatures. Soldiers (Figs. 11–15E and 11–16I), on the other hand, may or may not have compound eyes and generally have enlarged, sclerotized heads with well-developed mandibles. They function as defenders of the colony either with their mandibles or in some species by plugging up holes with their heads. In the soldier caste (called nasuti; Fig. 11–15F) of *Nasutitermes* spp., the duct from a gland in the head opens at the end of a long, narrow snout through which a sticky secretion from the gland can be forced. This secretion is squirted as a means of defense. In the simplest colony composition, characteristic of certain primitive termite species, there are reproductives and soldiers and the immatures of both these castes function as workers (Fig. 11–15A–E).

The control of caste differentiation has long been a topic of interest and investigation and is still a long way from being understood completely. There is strong evidence that complex interactions of pheromones are involved (Lüscher, 1961b; Engelmann, 1970).

New colonies are formed when, at certain times of the year, winged primary reproductives appear and swarm from the nest. Males and females pair off during this swarming, but mating usually occurs on the ground. Subsequently, their wings drop off at the fracture lines and the pair sets out to find an appropriate site for nest construction. Another method of nest foundation is the *sociotomy* (Lindauer, 1965), the process by which colonies divide by the separation of immatures and secondary reproductives from the parent colony or by division of a migrating colony into two daughter colonies. Colony size may vary from a few individuals to over 100,000.

The nests of termites vary from simple cavities in soil or wood to vast subterranean complexes or elaborate edifices that project well above the ground. Richards and Davies (1957) cite an investigator who found one nest of a species in northern Australia, *Nasutitermes triodiae,* to be 20 feet high with a diameter of 12 feet at its base! Exceedingly elaborate ventilatory systems, designs that favour maintenance of a constant temperature, canopies that deflect rainwater, and other structural adaptations of nests have been described in various termite species. The means by which the behavior of individual members of a colony are coordinated to produce such complex structures has long been a source of amazement and will undoubtedly occupy the time of ardent investigators for many years to come. It has been suggested that a termite colony might quite appropriately be looked upon as a "superorganism" which grows, differentiates, attains a steady state, and reproduces.

Termites are very significant structural pests, causing considerable damage to wooden structures (e.g., furniture, building timbers, and wooden floors). According to Helfer (1963) they do approximately $40,000,000 worth of damage annually in the United States alone. Some of the most destructive termites in the United States are subterranean and in the genus *Reticulitermes*. They can construct earthen tubes on concrete and hence are able to invade a structure even though it is not in direct contact with the soil. The presence of these earthen tubes (Fig. 11–16J) is one of the characteristics used to "diagnose" a termite infestation. Not all destructive termite species are subterranean; for example, colonies of *Kalotermes* (dry-wood termites) and *Zootermopsis* (damp-wood termites) exist entirely in wood. These termites are found along the Pacific Coast and in the southern United States.

ORDER Zoraptera (*zor,* pure; *ptera,* wings); zorapterans (Fig. 11–17).
BODY CHARACTERISTICS minute.
MOUTHPARTS chewing.
EYES and OCELLI compound eyes and dorsal ocelli in winged forms; compound eyes and dorsal ocelli lacking in wingless forms.
ANTENNAE 9-segmented; moniliform or filiform.
WINGS 2 pairs when present; membranous; forewings larger than hindwings; shed by breakage along a basal fracture line; both sexes of a given species may have winged and wingless forms; wingless forms more common.
LEGS 2-segmented tarsi.
ABDOMEN short, 1-segmented cerci.

This is the smallest order in the class Insecta. Only slightly more than 20 species have been described in the world. All described species are included in the family Zorotypidae.

Members of this order have been found in such places as decaying wood, humus, sawdust piles, termite nests, and under bark and stones. They tend to be gregarious but apparently lack a definitive social structure. These insects are most likely scavengers, feeding on such materials as fungal spores and the remains of other arthropods. Zorapterans occur in the southeastern United States but are uncommon.

They are significant in that they are considered by some to provide an evolutionary link between the orthopteroid orders and Psocoptera and hence the hemipteroid orders (Oldroyd, 1968).

Hemipteroid Orders

The orders included under this heading are Psocoptera, Mallophaga, Anoplura, Thysanoptera, and Hemiptera. Members of these orders share certain features, including the following; (1) specialized

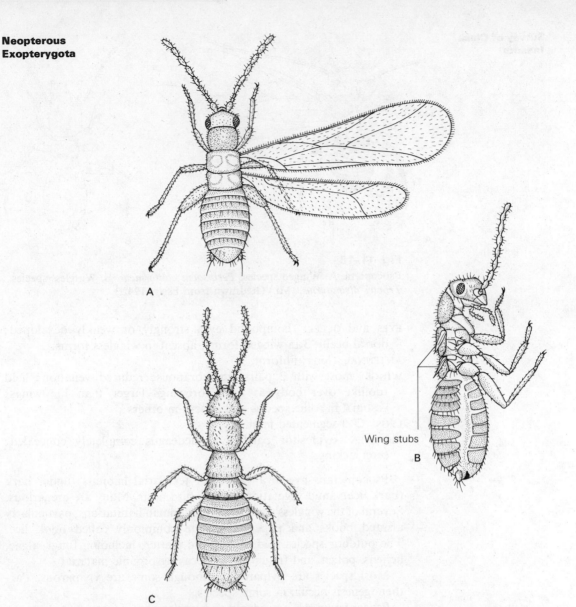

Fig. 11–17

Zoraptera, *Zorotypus hubbardi*. A. Alated (winged) form of adult female.
B. Dealated form of adult female. C. Adult female of apterous form. (Mi.)
(Redrawn from Caudell, 1918.)

mandibulate or haustellate mouthparts, (2) hindwings lacking a
large anal lobe, (3) cerci lacking, (4) few Malpighian tubules, and
(5) a tendency toward concentration of the ventral chain ganglia
(Richards and Davies, 1957).

Development is hemimetabolous in this group of insects.

ORDER Psocoptera (*psoco,* rub small; *ptera,* wing); Corrodentia;
   book lice and relatives (Fig. 11–18).

BODY CHARACTERISTICS minute; head capsule large compared to
   rest of body.

MOUTHPARTS chewing; labial silk glands present.

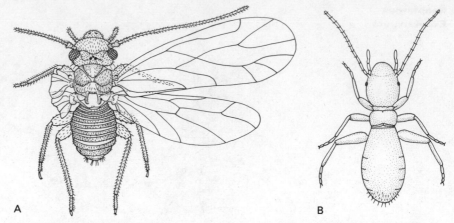

Fig. 11–18

Psocoptera. A. Winged species, *Peripsocus californicus*. B. Wingless species, *Troctes divinatorius*. (Mi.) (Redrawn from Essig, 1942.)

EYES and OCELLI   compound eyes strongly or weakly developed; dorsal ocelli: 3 in winged forms, absent in wingless forms.

ANTENNAE   long; filiform.

WINGS   most with 2 pairs; membranous; reduced venation; held rooflike over body at rest; forewings larger than hindwings; vestigial in some species and absent in others.

LEGS   2- 3-segmented tarsi.

ABDOMEN   ovipositor partly and aedeagus completely concealed; cerci lacking.

Psocopterans live in a variety of terrestrial habitats: under bark (bark lice), amid vegetation, and in bird nests. Many are gregarious. Several of the wingless species occur in human habitations, particularly around books and papers, and are commonly called book lice. The outdoor species feed on organic matter, including fungi, algae, lichens, pollen, and fragments of decaying organic material.

Most species are oviparous, although some are viviparous. Parthenogenesis occurs in some species.

Psocopterans are considered to be somewhat intermediate between the zorapterans (an orthopteroid group) and the remaining hemipteroids.

Psocoptera may reach large enough populations among books, papers, stored cereal grains, insect collections, and other materials to constitute a pest situation, but they are generally of little economic importance.

ORDER   Mallophaga (*mallo*, wool; *phaga*, eat); chewing lice (Fig. 11–19).

BODY CHARACTERISTICS   minute; dorsoventrally flattened; triangular head broader than thorax.

MOUTHPARTS   chewing.

EYES and OCELLI   reduced compound eyes; dorsal ocelli lacking.

ANTENNAE   3- to 5-segmented; usually capitate or filiform; when capitate, concealed beneath the head.

WINGS   lacking.

LEGS   tarsi modified for grasping hairs in species with mammalian hosts; 1- to 2-segmented tarsi.

ABDOMEN   cerci lacking.

Most Mallophaga are ectoparasites of birds, but some (members of the families Trichodectidae and Gyropidae) have mammalian hosts. They feed on various organic fragments (e.g., from feathers and skin) and epidermal secretions, and certain species [e.g., *Menacanthus stramineus* (Fig. 11–19B)] are known to gnaw through the skin and obtain blood. Some actively obtain sebum and perhaps serum by attacking hair follicles.

Some authors include the Mallophaga with the closely related order Anoplura (the sucking lice) in the order Phthiraptera. Other than the Anoplura, the Mallophaga are apparently related to the Psocoptera.

Chewing lice pass through three nymphal instars. The eggs are attached to host feathers or hairs and generation after generation is spent on the same host. Individuals are transferred from one host to another by direct contact and they soon die if separated from an acceptable host. Different species sometimes show different preferences as to the part of the host body attacked. Since these insects are wingless, they must be dispersed by contact between two hosts.

Although probably all domestic animals are attacked by one or more species of chewing lice, they usually do not constitute major problems. However, if an infestation is large, some damage may occur, particularly to poultry, which may suffer restlessness, decrease in egg production, and loss of feathers. Three common chewing lice that attack poultry are *Menopon pallidum,* the common chicken louse; *Menacanthus stramineus* (Fig. 11–19B), the chicken body louse; and *Menopon gallinae,* the shaft louse. Cattle, horses, sheep, goats, dogs, and cats are among the domestic animals attacked by species of chewing lice. Humans are not attacked by these lice. Man may occasionally become infested by direct contact with infested animals, but the lice do not survive long on a human host. Indirectly,

Fig. 11–19
Mallophaga, chewing lice. A. Dog-biting louse, *Trichodectes canis*. B. Chicken body louse, *Menacanthus stramineus*. (Mi.) (A courtesy of U.S. Public Health Service (redrawn); B redrawn from CCM General Biological, Inc., key card.)

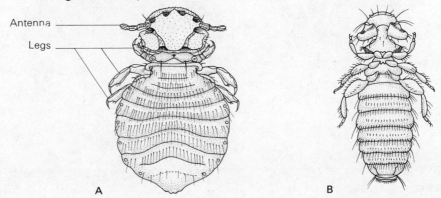

Antenna

Legs

A   B

however, the common dog louse, *Trichodectes canis* (Fig. 11–19A), does at times constitute a threat to human health because it serves as an intermediate host for the double-pored dog tapeworm, *Dipylidium caninum*. If an infected louse is inadvertently ingested by a human playing with an infected dog, the human can become infected with the tapeworm.

ORDER   Anoplura (*anopl,* unarmed; *ura,* tail); Siphunculata; sucking lice (Fig. 11–20).

BODY CHARACTERISTICS   minute; dorsoventrally flattened; head narrower than thorax.

MOUTHPARTS   piercing-sucking; retracted into head when not feeding.

EYES and OCELLI   compound eyes weakly developed or lacking; dorsal ocelli lacking.

ANTENNAE   short; 3 to 5 segments.

WINGS   lacking.

LEGS   1-segmented tarsi; tarsi usually adapted for grasping hairs.

ABDOMEN   cerci lacking.

Sucking lice are all blood feeders. They attack a wide variety of mammals, even marine forms (e.g., seals), and demonstrate a high degree of host specificity.

The major pests of domestic animals in this order are in the families

Fig. 11–20

Anoplura, sucking lice. A. Crab louse, *Phthirus pubis.* B. Human louse. *Pediculus humanus.* C. Egg or nit of crab louse attached to host hair. D. Egg of human head louse attached to host hair. (Mi.) (A and B courtesy of U.S. Public Health Service (redrawn); C and D redrawn from CCM General Biological, Inc., key card.)

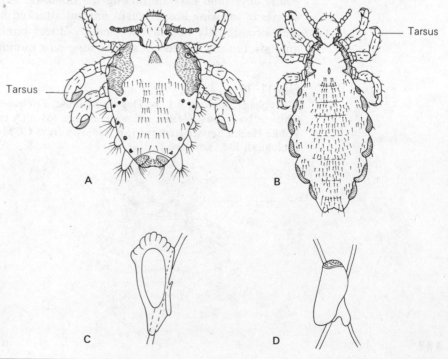

Tarsus

Tarsus

A   B   C   D

Haematopinidae, (e.g., the hog louse, *Haematopinus suis*) and Linognathidae (e.g., *Linognathus vituli,* the long-nosed ox louse).

Two species of sucking lice attack man, *Phthirus pubis* (Fig. 11–20A) and *Pediculus humanus* (Fig. 11–20B). Some systematists include these two species in a single family, Pediculidae; others split them into two families, *Pediculus* in Pediculidae and *Phthirus* in Phthiriidae.

Phthirus pubis is the human pubic louse or crab louse and infests the pubic and armpit areas, although it may be found in other areas of the body with coarse hair. This louse causes intense itching, and an infestation may result from contact with an already infested individual via toilet seats, blankets, and sexual intercourse. The life cycle ordinarily takes about 1 month to complete and the eggs, or *nits,* are attached to hairs on the host (Fig. 11–20C). Pubic lice are not known to transmit any human pathogens.

Pediculus humanus can be divided into two subspecies, *P. humanus capitis,* the head louse, and *P. humanus corporis,* the body louse. The first infests particularly the head region, but it has been found on other hairy areas of the body. The body louse infests regions that come into frequent or continuous contact with clothing (e.g., armpits, neck and crotch). The eggs are usually attached to clothing, which provides an ideal vehicle for transmission to a new host. Head lice, on the other hand, attach their eggs to hairs (Fig. 11–20D) and a new infestation may result from contact with a stray hair bearing an egg. Both head and body lice are spread via direct contact. They develop quite rapidly, a complete generation occurring in about 3 weeks.

Infestation with lice is sometimes referred to as *pediculosis.* The attendant discomforts range from intense itching to anemia and pathologic changes in the skin resulting from the chronic feeding of lice and the blood loss associated with it. In addition to the problems caused directly by their feeding, body lice are involved in the transmission of a number of human pathogens, the most important of which are epidemic relapsing fever, caused by a spirochaete, *Borellia recurrentis;* epidemic typhus, caused by a rickettsia, *Rickettsia prowazeki;* and murine typhus fever, caused by *Rickettsia mooseri.* All three of these diseases have occurred in large epidemics and in the case of epidemic typhus have claimed thousands of lives, particularly during wartime when populations become highly concentrated and sanitary practices are lax. Several other pathogenic microbes may be transmitted by lice.

ORDER Thysanoptera (*thysano*, a fringe; *ptera*, wing); Physapoda; thrips (Fig. 11–21).

BODY CHARACTERISTICS minute to small; slender; usually darkly colored.

MOUTHPARTS rasping-sucking; asymmetrical, the right mandible lacking or reduced.

EYES and OCELLI small compound eyes; 3 dorsal ocelli in winged forms; dorsal ocelli lacking in wingless forms.

ANTENNAE short; 6 to 10 segments.

WINGS 2 pairs; narrow with fringe of hairs; coupled by basal hooks.

LEGS 1- to 2-segmented tarsi are bladderlike at the tip.

ABDOMEN cerci lacking.

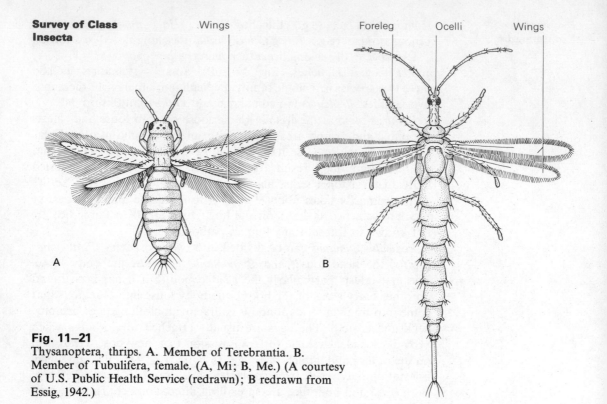

Wings   Foreleg   Ocelli   Wings

A   B

Fig. 11–21
Thysanoptera, thrips. A. Member of Terebrantia. B.
Member of Tubulifera, female. (A, Mi; B, Me.) (A courtesy
of U.S. Public Health Service (redrawn); B redrawn from
Essig, 1942.)

Thysanoptera is divided into two suborders, based mainly on the
presence or absence of an ovipositor. Females in the suborder Terebrantia (Fig. 11–21A) have a well-developed ovipositor which is used
to deposit eggs in plant tissues and the wings, if present, have veins.
Females in the suborder Tubulifera (Fig. 11–21B) have either a weakly
developed ovipositor or it is lacking altogether and the wings, when
present, are nearly or completely veinless. The posterior extremity of
the abdomen is tubular and eggs are deposited in crevices.

Although Thysanoptera is classified as a hemimetabolous group of
insects, the life cycle is strongly suggestive of holometabolism. The
early instar nymphs are active and wingless but otherwise resemble
the adults. However, during the last two or three nymphal stadia,
they become quiescent, sometimes in a cocoon or earthen cell during
which time wings, if they are going to be present, develop. They have
a relatively short generation time and there are usually several generations per year. Parthenogenesis is common in this group.

Some species of thrips are predaceous (all in the suborder Tubulifera), feeding on aphids and various other small insects or mites, but
most are plant feeders, usually feeding on sap. They are widely distributed among all sorts of vegetation and although generally weak
fliers, some species are dispersed over many miles via the wind.

Several species are pests of certain crops, particularly truck and
floricultural, although they may become pests of field crops (e.g.,
tobacco). They not only do considerable damage by their feeding
activities but also may transmit plant diseases.

They can cause human discomfort by attempting to pierce the skin
and may constitute a serious annoyance if present in large numbers.

**Table 11–2
Outstanding Differences Between Suborders of Hemiptera**

Heteroptera	Homoptera
1. Labial sheath arises from the anterior part of the head.	1. Labial sheath arises more posteriorly.
2. Forewings are hemelytra; i.e., basal portion leathery and apical portion membranous.	2. Forewings not hemelytra; both pairs of wings essentially uniform in texture.
3. Wings held horizontally over abdomen at rest.	3. Wings held rooflike over abdomen at rest.

ORDER   Hemiptera (*hemi*, one-half; *ptera*, wing); true bugs, aphids, leafhoppers, and relatives (Figs. 11–22 to 11–27).

BODY CHARACTERISTICS   minute to very large; extremely variable.

MOUTHPARTS   piercing-sucking; degenerate palpi.

EYES and OCELLI   variable.

ANTENNAE   variable; concealed in some.

WINGS   usually 2 pairs; many wingless forms.

LEGS   variable; 2- to 3-segmented tarsi.

ABDOMEN   cerci lacking.

This is the largest of the hemimetabolous orders. Its members form a very diverse, heterogeneous group both in structure and habits. This order is divided into two large suborders (Table 11–2): Heteroptera (true bugs) and Homoptera (aphids, leafhoppers, and relatives). Some authors place each of these suborders in separate orders. This is apparently logical in terms of North American species, but not so when world fauna are considered (Fox and Fox, 1964). Members of the family Peloridiidae found in Australia, New Zealand, and South America are transitional, displaying both heteropteran and homopteran traits (Oldroyd, 1968).

Heteroptera.   These insects (Figs. 11–22, 11–23, and 11–24) usually have compound eyes and there may be two ocelli or none. The four- or five-segmented antennae are comparatively long and in many are concealed from immediate view. These insects usually have a pronounced pronotum with a distinct triangular scutellum on the mesothorax between the bases of the forewings. Although most are winged, there are many wingless and brachypterous forms.

Many adult Heteroptera possess repugnatorial or scent glands which are located in the region of the metathoracic coxae. In some nymphs, similar glands are located on the dorsum of the abdomen. Many species give off a very unpleasant odor when handled. It is said that bed bugs (family Cimicidae) give off a very characteristic odor and that a relatively large infestation can be identified by an experienced person by the smell alone.

A large number of species are able to produce sounds (Richards and Davies, 1957) and stridulatory organs have been identified in the prosternal region (Reduviidae and Phymatidae), abdominal sterna (some Pentatomidae), forelegs and clypeus (Corixidae), coxae (certain Nepidae), and dorsum of the abdomen in association with the wings

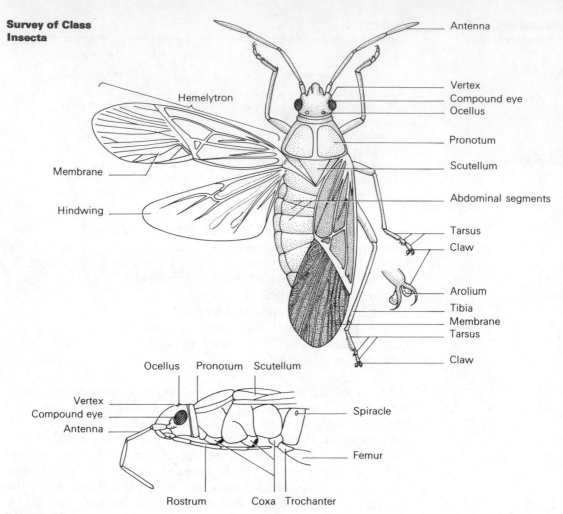

Hemelytron

Vertex
Compound eye
Ocellus

Pronotum

Scutellum

Membrane

Abdominal segments

Hindwing

Tarsus
Claw

Arolium
Tibia
Membrane
Tarsus

Claw

Ocellus   Pronotum   Scutellum

Vertex
Compound eye
Antenna

Spiracle

Femur

Rostrum   Coxa   Trochanter

Fig. 11–22
Hemiptera,
Heteroptera, true bugs.
Major external features
of the boxelder bug,
Leptocoris trivittatus
(Coreidae), and lateral
view showing
mouthparts. (Me.).
(Redrawn from Essig,
1942.)

(some Pentatomidae). The habitats in which heteropterans are found
are highly diversified, terrestrial, aquatic, semiaquatic, and ectoparasitic forms being known. Their feeding habits are likewise diverse,
some being phytophagous (plant juices), others carnivorous, and some
hematophagous.

The Heteroptera can be divided into two groups of families based
on whether they have short antennae concealed in grooves on the
venter of the head (Cryptocerata = "hidden antennae") or long
antennae which are not so concealed and are freely movable (Gymnocerata = "naked antennae").

The Cryptocerata (Fig. 11–23) are all aquatic or semiaquatic
and the most common families are Corixidae, waterboatmen;
Notonectidae, backswimmers; Nepidae, water scorpions; Belostomatidae, giant water bugs; Naucoridae, creeping water bugs; Gelastocoridae, toad bugs; and Gerridae, water striders. The vast majority of
these insects are predatory, and certain ones, such as backswimmers
and giant water bugs, can inflict a painful bite if carelessly handled.
Most are capable of flight and many appear in the vicinity of artificial
lights at night. These families show many interesting variations of
mechanisms involved in ventilation in an aquatic situation.

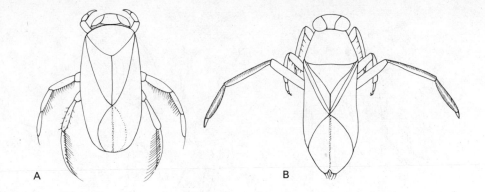

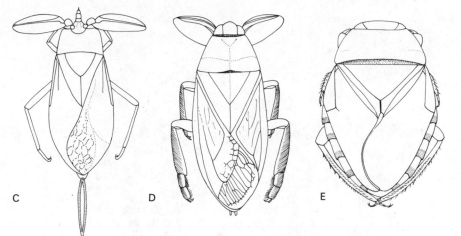

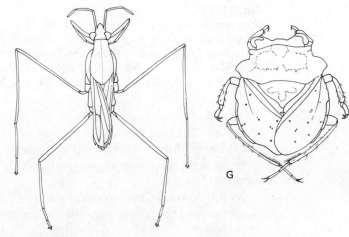

Fig. 11–23

Representative Cryptocerata. A. Waterboatman (Corixidae).
B. Backswimmer, *Notonecta irrorata* (Notonectidae). C. Water scorpion,
Nepa sp. (Nepidae). D. Giant water bug, *Lethocerus americanus*
(Belostomatidae). E. Creeping water bug, *Pelocoris femoratus* (Naucoridae).
F. Water strider, *Gerris buenoi* (Gerridae). G. Toad bug, *Gelastocoris
barberi* (Gelastocoridae). (A, S to Me; B, C, E, and F, Me; D, L; G, S.)
(B–G redrawn from Britton, 1923.)

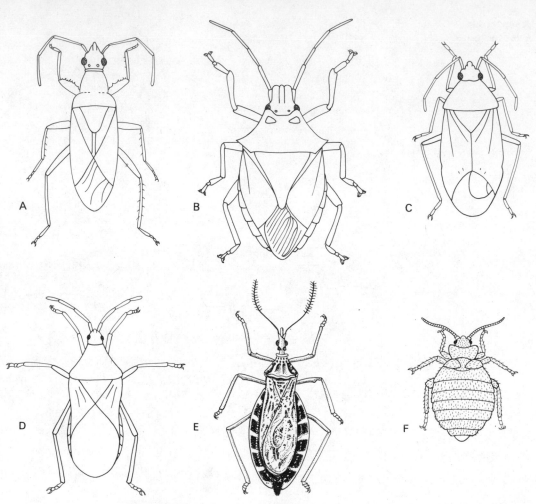

Fig. 11–24
Representative Gymnocerata. A. Seed bug, *Ligyrocoris diffusus*
(Lygaeidae). B. Stink bug, *Apateticus cynicus* (Pentatomidae). C. Tarnished
plant bug, *Lygus lineolaris* (Miridae). D. Squash bug, *Anasa tristis*
(Coreidae). E. Conenose bug (Reduviidae). F. Common bedbug, *Cimex
lectularius* (Cimicidae). (A, C, and F, S; B, S to Me; D, Me; E, Me to L.)
(B–G redrawn from Britton, 1923.)

Among the Gymnocerata (Fig. 11–24) are found terrestrial, aquatic,
and parasitic forms. This group contains a much larger number of
families than the Cryptocerata and usually is divided into several
superfamilies.

Gymnocerata contains numerous species of economic and medical
significance. The chinch bug, *Blissus leucopterus* (family Lygaeidae)
attacks corn, sorghum, and small grains in the United States. Pfadt
(1962). reports: "In 1934, another year of serious infestation, the
chinch bug caused an estimated loss of $27,500,000 to the corn crop
and $28,000,000 to wheat, barley, rye, and oats." Many species in the
family Miridae, plant bugs, are serious pests of leguminous forage
crops, particularly alfalfa. The families Pyrrhocoridae, Coreidae,
Pentatomidae (stink bugs), and others also contain important agricultural pests.

Medically important Gymnocerata are in the families Cimicidae (bed bugs and relatives) and Reduviidae (assassin bugs and relatives). Members of the family Cimicidae are wingless ectoparasites of birds and mammals (bats, various rodents, and others). Three species attack man (James and Harwood, 1969): *Cimex lectularius* (Fig. 11–24F), *Cimex hemipterus* (tropical America, Africa, Asia, East Indies, and certain Pacific islands), and *Leptocimex boueti*. Although experimentally bed bugs can transmit pathogens that cause a number of human diseases, there is little evidence that they are the natural vectors of any pathogen. The reduviids (Fig. 11–24E) are mostly predaceous, feeding on other insects (many can inflict painful bites), but members of the subfamily Triatominae (kissing bugs and relatives) are exclusively ectoparasites of vertebrates. Several species (e.g., *Rhodnius prolixus* and *Triatoma* spp.) are vectors of Chagas disease, caused by the protozoan *Trypanosoma cruzi*. On the positive side, *Rhodnius prolixus* has been the subject of much significant biological research, particularly that carried out by a well-known insect physiologist, Sir Vincent Wigglesworth.

Homoptera. The majority of these insects (Figs. 11–25, 11–26, and 11–27) are small, with a few exceptions (e.g., some of the fulgorid plant-hoppers and cicadas). Compound eyes and ocelli are usually present and antennae vary from filiform to bristlelike. The prothorax is usually quite inconspicuous except in the family Membracidae, where it takes on a variety of bizarre shapes. Members of several groups have wax glands which commonly secrete a powdery material. The females of some have distinct ovipositors. Cicadas and certain leafhoppers possess the ability to produce sound by means of specialized structures on the dorsolateral portion of the base of the abdomen. Many insects in this suborder (e.g., aphids, whiteflies, and others) produce honeydew, a liquid material excreted from the anus.

The Homoptera differ from the Heteroptera in being exclusively terrestrial herbivores.

As with all the Heteroptera, most homopterans undergo hemi-

Fig. 11–25
Homoptera. Major external structures of a leafhopper, *Paraphlepsius irroratus*. A. Dorsal view. B. Anterior view. (S.) (Redrawn with modifications from Borror and DeLong, 1971.)

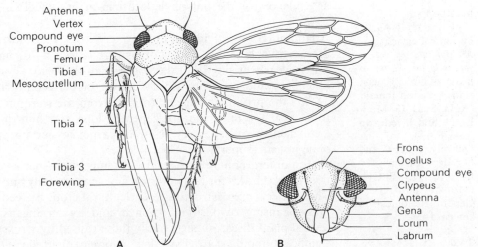

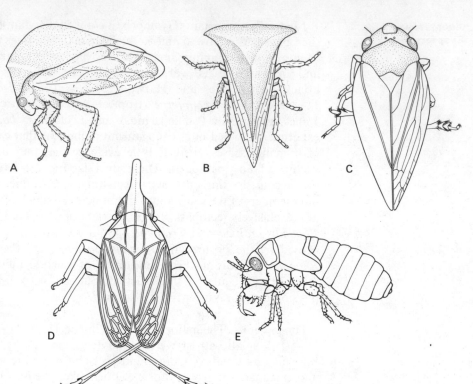

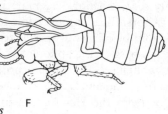
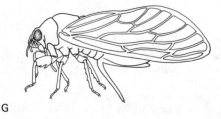

Fig. 11–26
Homoptera,
Auchenorrhyncha.
A. Treehopper,
Stictocephala bubalus
(Membracidae).
B. Same species as A,
dorsal view. C.
Spittlebug (Cercopidae).
D. Fulgorid
planthopper
(Fulgoridae). E–G.
Periodical cicada,
*Magicicada
septendecim:* E, last
instar nymph; F, cast
skin; G, adult female.
(A–D, S; D–G, Me to
L.) (A and B redrawn
from Borror and
DeLong, 1971; C and
D redrawn from
Britton, 1923; E–G
redrawn from CCM
General Biological, Inc.,
key card.)

346

metabolous development. However, certain scale insects, whiteflies, and phylloxerans are essentially holometabolous. This fact is viewed as evidence for the polyphyletic origin of holometabolism (Hinton, 1948).

Homoptera is divided into two groups: Auchenorrhyncha and Sternorrhyncha. The Auchenorrhyncha (Fig. 11–26) have the proboscis clearly arising from the posterior of the head; short, bristlelike antennae; and are usually very active. In the Sternorrhyncha (Fig. 11–27) the proboscis appears to arise from the sternum between the forecoxae, the antennae are generally long and filiform, and many lead inactive or sedentary lives, in particular certain stages of the scale insects (superfamily Coccoidea).

Homoptera contains no medically important species, but it more than compensates for this in the area of economically important pests. These insects are destructive to agricultural crops by feeding on plant juices, causing oviposition damage, and by serving as vectors for several plant diseases, particularly those caused by viruses. Especially significant families are Psyllidae, jumping plant lice, Aleyrodidae,

Fig. 11–27
Homoptera, Sternorrhyncha.
A. Aphid, *Aphis pomi* (Alphididae), wingless viviparous female.
B. Same species as A, winged viviparous female. C–F, San Jose scale, *Aspidiotus perniciosus* (Coccoidea); C, scales on a twig; D, nymph; E, adult male; F, adult female with scale removed. (Mi.) (Redrawn from CCM General Biological, Inc., key card.)

whiteflies, Aphididae, aphids or plant lice (Fig. 11–27A, B), Coccidae and related families, the scale insects (Fig. 11–27C–F), and Cicadellidae, leafhoppers (Fig. 11–25). Members of these families cause feeding damage and may also produce a toxic effect on their hosts. Aphids and leafhoppers are extensively involved in plant disease transmission. Aphids are well known for their complex life cycles, which involve various combinations of parthenogenesis, sexual generations, wingless and winged forms, and alternation of plant hosts. They feed on plant juices extracted from nearly every part of a plant, roots included.

Although many homopterans are common, the cicadas (Fig. 11–26E–G) are particularly well known. These insects range in length from about 12 mm to approximately 10 cm, and certain species are the largest homopterans. The nymphs are subterranean and feed on the roots of plants; the adults live above the ground in trees. The periodical cicadas, *Magicicada* spp., are a particularly well known group which spend either 13 or 17 years in the soil, depending on the species. At the end of their time in the soil, they emerge as

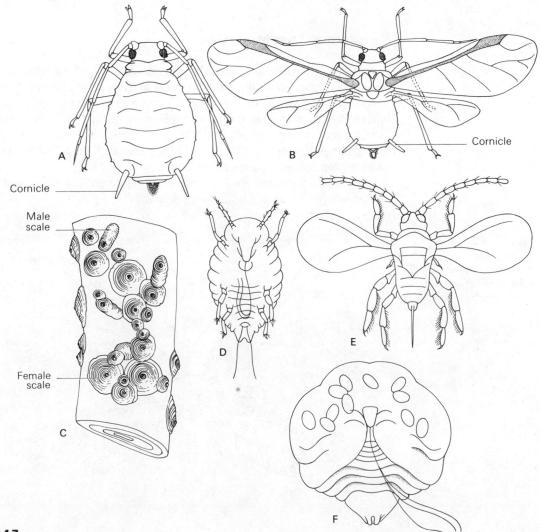

adults, mate, and the females lay their eggs in slits in the twigs of trees, sometimes causing considerable damage. The emergence of a large brood may cause large patches of forest to appear brown due to dead twigs and leaves. In a few weeks the newly emerged nymphs drop to the ground and burrow in, commencing another 13 or 17 years of feeding. There are several broods in the United States and they emerge in different years. Another group of cicadas (e.g., *Tibicen* spp.) are referred to as "dog-day" cicadas and emerge from their subterranean home annually in May. Cicadas are chronic noisemakers, only the males being able to produce sound.

Another commonly encountered and interesting group of homopterous insects is the spittlebugs (family Cercopidae; Fig. 11–26C). Spittlebugs are so called because the nymphs surround themselves in a frothy spittlelike substance. The function of this substance is apparently twofold, protection from potential predators and parasites and prevention of desiccation (Richards and Davies, 1957). The material is released through the anus. These are small insects, less than 12 mm in length, and the adults are active jumpers.

Neopterous Endopterygota

This group of orders is generally considered to be of monophyletic origin. They are sometimes referred to collectively as the neuropteroid orders, but among them the orders Neuroptera, Mecoptera, Trichoptera, Lepidoptera, Diptera, and Siphonaptera have been classified as being members of the panorpid orders (superorder Panorpoidea or the "panorpid complex"; Hinton, 1958) because of the affinities they show with panorpid (scorpionfly) larvae.

The insects included in this group all undergo complete metamorphosis (i.e., they are holometabolous).

Fig. 11–28
Neuroptera. Green lacewing, *Chrysopa perla* (Chrysopidae). (Me.) (Redrawn from Essig, 1942.)

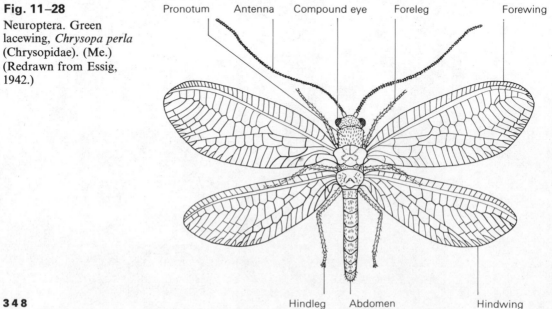

Pronotum   Antenna   Compound eye   Foreleg   Forewing

Hindleg   Abdomen   Hindwing

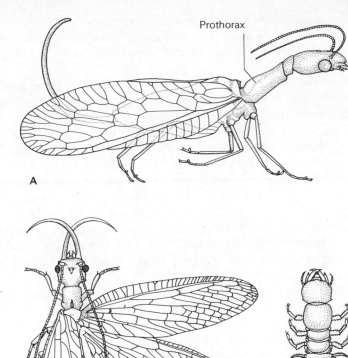

Prothorax

A

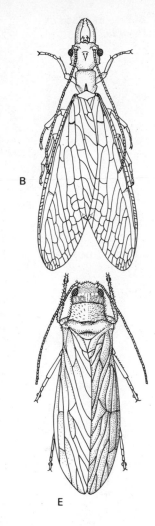

B

C

D

E

Fig. 11–29

Neuroptera. A.
Snakefly, *Agulla bractea*
(Raphidiodea;
Raphidiidae). B, C, and
D. Dobsonfly, *Corydalis
cornutus*, female, male,
and larva, respectively
(Megaloptera;
Corydalidae). E.
Alderfly, *Sialis mohri*
(Megaloptera; Sialidae).
(A, Me; B–D, VL;
E, Me.) (A redrawn
from Wolglum and
McGregor, 1958; B–D
redrawn from CCM
General Biological, Inc.,
key card; E redrawn
from Frison, 1937.)

349

ORDER   Neuroptera (*neuro*, nerve; *ptera-* wing); lacewings, dobsonflies, and related forms (Figs. 11–28, 11–29, and 11–30).

BODY CHARACTERISTICS   small to very large; soft-bodied; larvae:
campodeiform; pupae: exarate in silken cocoon.

MOUTHPARTS   chewing; larvae: grasping-sucking.

EYES and OCELLI   compound eyes present; dorsal ocelli present or
absent.

ANTENNAE   usually filiform.

WINGS   2 pairs; similar in size and appearance; usually held rooflike
over body at rest.

LEGS   5-segmented tarsi.

ABDOMEN   cerci lacking.

COMMENTS   mostly weak fliers.

This order is recognized as the most primitive of the Endopterygota.

Neuropterans are found about vegetation and often near bodies of
water since many have aquatic larvae. They feed on fluid materials and
various soft-bodied insects. The larvae are carnivorous, capturing and
holding prey in their well-developed mandibles.

Some neuropteran species undergo hypermetamorphosis.

Members of this order have no medical or negative economic significance and many, because of their predatory habits, are recognized as beneficial.

Neuroptera is divided by some authorities into three suborders: Megaloptera (Fig. 11–29B–E), Raphidiodea (Fig. 11–29A) and Planipennia (Fig. 11–30). Megaloptera are characterized by having aquatic larvae with chewing mouthparts, hindwings basally slightly broader than forewings, with the anal region of the hindwings folding at rest. Members of the suborder Raphidiodea, the snakeflies (Fig. 11–29A), have a very distinctive appearance in that the prothorax is quite elongate, with the legs attached to the posterior end, giving the impression of a very long neck. Snakefly larvae are terrestrial. Planipennia is the largest neuropteran suborder and its members are characterized by having usually terrestrial larvae with mandibulate sucking mouthparts and wings of similar shape and size with a non-folding anal region on the hindwing.

Megaloptera contains two families, Corydalidae (dobsonflies and fishflies) and Sialidae (alderflies). Dobsonflies (Fig. 11–29B–D) are comparatively large and their aquatic larvae, or hellgrammites (Fig. 11–29D), are well known to fisherman as fish bait. Fishflies are smaller than dobsonflies. Alderflies (Fig. 11–29E) are also smaller than dobsonflies, also have aquatic larvae, and tend to be darkish in color.

The suborder Planipennia can be grouped into two superfamilies, Hemerobioidea and Myrmeleontoidea.

Among the families in Hemerobioidea are included Mantispidae (mantidflies; Fig. 11–30A), Chrysopidae (common lacewings; Fig. 11–28), Hemerobiidae (brown lacewings; Fig. 11–30B), and Sisyridae (spongillaflies; Fig. 11–30C). Mantidflies (Fig. 11–30A) bear a superficial resemblance to snake flies but have raptorial forelegs which attach anteriorly to the elongate prothorax. Their larvae are terrestrial and attack the spiderlings in the nests of ground spiders. Both brown and green lacewings are terrestrial groups. The larvae of both families live on vegetation and prey on aphids, mites, and other smaller soft-bodied insects. The common lacewings (Fig. 11–28) tend to be greenish in color and often have yellowish metallic-appearing eyes. They are often 2 or 3 cm long and their eggs are deposited at the ends of stalks which are anchored to leaves. The larvae often carry a layer of debris about on their dorsum, which provides them with camouflage. The brown lacewings (Fig. 11–30B) are usually brownish in color and are smaller than the common lacewings. They do not deposit their eggs at the ends of stalks. Spongillaflies (Fig. 11–30C) are small insects and feed on freshwater sponges during the larval stage.

Myrmeleontoidea contains two families, Myrmeleontidae (antlions) and Ascalaphidae (owlflies). Antlion adults (Fig. 11–30D) superficially resemble damselflies and are similar in size. However, they are weaker fliers, have a much different pattern of wing venation, have long, distinctly clubbed antennae, and are soft-bodied. The larvae (Fig. 11–30E) are terrestrial and live at the bottom of funnel-shaped pits in sand. When a potential prey insect or other small arthropod passes close to the pit, the antlion creates a miniature landslide, bringing the prey down within reach of its prehensile, sucking mandibles and subsequently dines.

Owlflies (Fig. 11–30F) bear a resemblance to dragonflies but differ from them in the same features that the antlion adult differed from the damselflies. In addition, their clubbed antennae are even longer than those of the adult antlions. Their larvae are predatory and are often concealed by a layer of debris on their bodies.

Fig. 11–30
Neuroptera, Planipennia. A. Mantispid, *Mantispa brunnea occidentis* (Mantispidae). B. Brown lacewing, *Sympherobius angustatus* (Hemerobiidae). C. Spongillafly, *Climacia areolaris* (Sisyridae). D. Antlion adult, *Hesperoleon abdominalis* (Myrmeleontidae). E. Antlion larva, *Myrmeleon sp.* F. Owlfly (Ascalaphidae). (A, D and F, L; B and C, S; E, Me.). (A, B, and F redrawn from Essig, 1942; C redrawn from Froeschner, 1947; E redrawn from Peterson, 1951.)

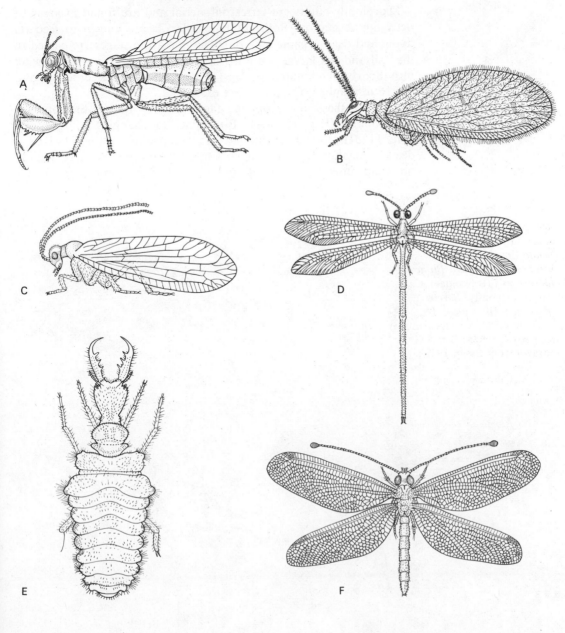

ORDER Mecoptera (*meco,* long; *ptera,* wing); scorpionflies (Fig. 11–31).

BODY CHARACTERISTICS small to medium; fragile; larvae: eruciform; pupae: exarate.

MOUTHPARTS chewing; commonly at apex of rostrum formed by elongation of head capsule; larvae: chewing.

EYES and OCELLI compound eyes present; 0 or 3 dorsal ocelli.

ANTENNAE long and filiform.

WINGS most species with 2 pairs; membranous; commonly with dark spots; some brachypterous and apterous (wingless) forms.

LEGS long and slender; 5-segmented tarsi.

ABDOMEN in some long and slender and genitalia prominent; last segment in male carried upright suggestive of a scorpion; short cerci present.

Scorpionfly adults are terrestrial–aerial and are found in areas of decaying vegetation. Both adults and larvae are known to feed on living and dead organic material. The immature stages are passed in the soil and the larvae are eruciform or scarabeiform with chewing mouthparts. Abdominal prolegs, when present, lack crochets and are 16 in number. The pupae are exarate.

Three of the four mecopteran families—Panorpidae, the common scorpionflies (Fig. 11–31A); Bittacidae, the hanging scorpionflies (Fig. 11–31B); and Boreidae, the snow scorpionflies (Fig. 11–31C)—deserve mention in this context. Common scorpionflies feed mainly on dead insects and plant materials. The hanging scorpionflies

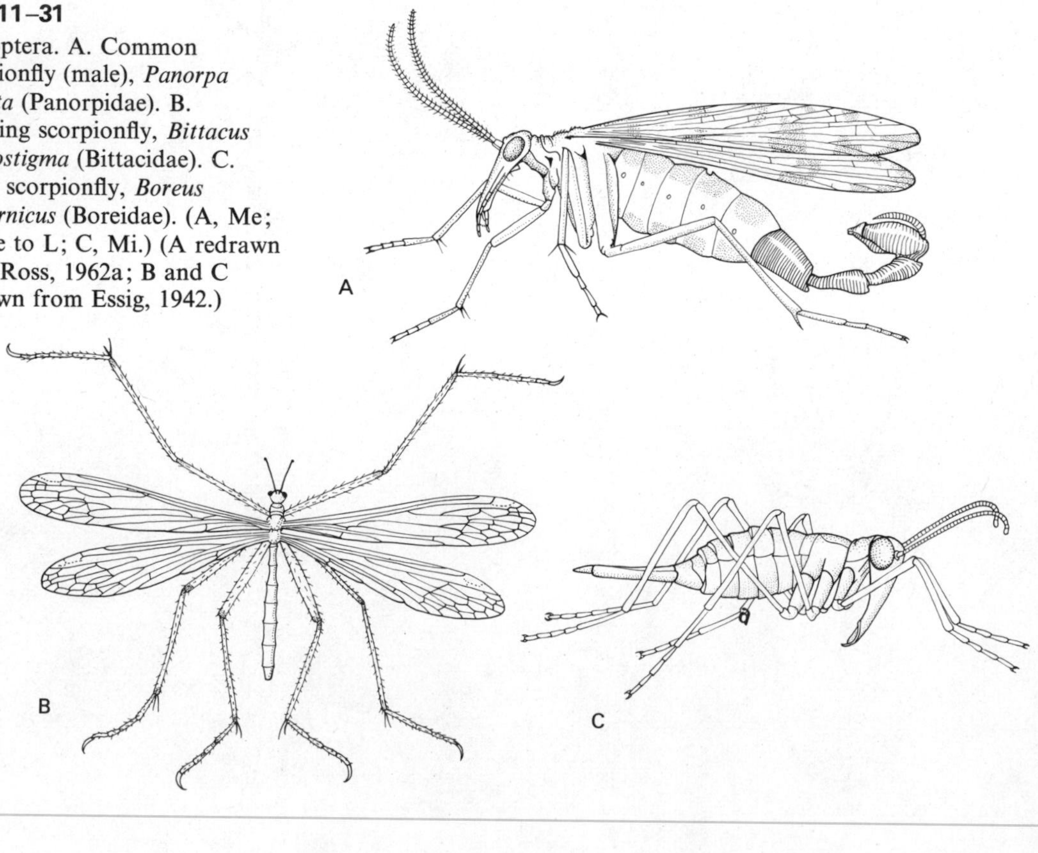

Fig. 11–31
Mecoptera. A. Common scorpionfly (male), *Panorpa chelata* (Panorpidae). B. Hanging scorpionfly, *Bittacus chlorostigma* (Bittacidae). C. Snow scorpionfly, *Boreus californicus* (Boreidae). (A, Me; B, Me to L; C, Mi.) (A redrawn from Ross, 1962a; B and C redrawn from Essig, 1942.)

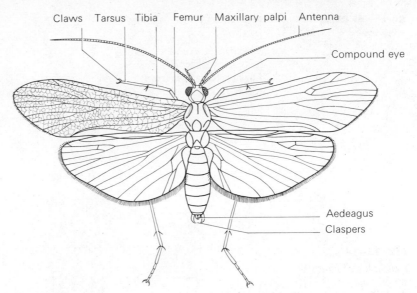

Claws Tarsus Tibia Femur Maxillary palpi Antenna

Compound eye

Aedeagus
Claspers

Fig. 11–32
Trichoptera, a caddisfly. (S to Me.) (Redrawn from Essig, 1942.)

superficially resemble crane flies and hang suspended from vegetation by their forelegs. They capture small flies with their raptorial tarsi. Snow scorpionflies inhabit moss and damp areas under stones and sometimes are found on the surface of snow. They feed on plant material.

ORDER Trichoptera (*tricho,* hair: *ptera,* wing); caddisflies (Figs. 11–32, 11–33, and 11–34).

BODY CHARACTERISTICS small to medium; generally somber colored, usually brownish; larvae: eruciform; pupae: exarate.

MOUTHPARTS some nonfeeding in adult stage; adapted for imbibing fluid; mandibles weakly developed or lacking; larvae: chewing.

EYES and OCELLI compound eyes present; 0 or 3 dorsal ocelli.

ANTENNAE range from setaceous to filiform.

WINGS 2 pairs; membranous; hindwings broader than forewings; covered with modified hairlike setae; held rooflike over body at rest.

Caddisflies are aquatic in the egg and larval stages (a few live in brackish water) and adults are terrestrial–aerial and may be found some distance from water. Eggs are deposited in the form of strings or masses near or in the water, on aquatic vegetation, beneath stones, and other similar locations. The larvae (Figs. 11–33B and 11–34) of most groups construct cases with a silken foundation into which a variety of environmental materials are incorporated. Exactly which environmental materials and the method by which they are incorporated is a family or sometimes a generic characteristic. The larvae that inhabit cases have hooked caudal appendages used to maintain a hold within the cases. Larval and pupal ventilation is by means of abdominal gills. The larvae feed on small aquatic plants and animals and some are important predators of black flies.

Caddisflies are important as fish food.

Fig. 11–33
Caddisfly, *Rhyacophila fenestra*. A. Adult in resting posture. B. Larva. (Redrawn from Ross, 1938.)

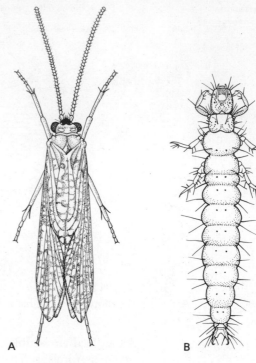

A

B

Fig. 11–34
Examples of caddisfly larval cases.
A. *Leptocella*.
B. *Triaenodes*.
C. *Agrypnia*.
D. *Polycentropus*.
(Redrawn from Cummins, Miller, Smith, and Fox, 1965.)

A

B

Young

Mature

C

D

ORDER   Lepidoptera (*lepido,* scale; *ptera,* wing); butterflies, moths, and skippers (Figs. 11–35 and 11–36).

BODY CHARACTERISTICS   small to very large; usually covered with scales; much diversity in color and color pattern; larvae: eruciform; pupae: obtect.

MOUTHPARTS   usually long, sucking proboscis; vestigial in some; larvae: strong mandibles, chewing.

EYES and OCELLI   large compound eyes; 0 or 2 dorsal ocelli; larvae: clusters of lateral ocelli (stemmata).

ANTENNAE   always prominent; vary in form.

WINGS   2 pairs; membranous; covered with scales; large in proportion to body; few cross veins; a few wingless; forewings larger than hindwings.

LEGS   5-segmented tarsi; larvae: 3 pairs.

ABDOMEN   cerci lacking; larvae usually with 5 pairs of prolegs on segments 3 through 6 and 10; apex of each proleg with tiny hooks (crochets).

COMMENTS   members in several families have organs of hearing.

Lepidoptera is the second largest order of insects. It can be divided into suborders on the basis of wing-coupling mechanisms. These are the suborders Frenatae (or Heteroneura) and Jugatae (or Homoneura). In the Frenatae the fore- and hindwings are coupled by a spine or a group of spines, the *frenulum* (see Fig. 2–40B), at the anterior base of the hindwing or by an enlarged area along the anterior base of the hindwing (humeral angle). The hindwings are smaller and have fewer veins than the forewings, hence the alternative, Heteroneura ("mixed nerves"). The wings of the Jugatae are coupled by the *jugum* (see Fig. 2–40C), a projection from the base of the posterior edge of the forewing. The venation is similar in both fore- and hindwings, hence the alternative name Homoneura ("same nerves"). The vast majority of Lepidoptera are in the suborder Frenatae, the Jugatae being a small group of uncommon insects.

The suborder Frenatae is further divided into two major groups, the Macrolepidoptera and the Microlepidoptera. The Macrolepidoptera are a diverse group of butterflies (Fig. 11–36A), skippers (Fig. 11–36B), and moths (Fig. 11–36C) of variable size, but usually with a wingspan of more than 2.5 cm. The Microlepidoptera are a large group of moths, most of which are much smaller than Macrolepidoptera, usually with a wingspan of 20 mm or less. Wing venation and other characters are used in further subdivision of these two groups.

Fig. 11–35

Stages in the life cycle of the monarch butterfly, *Danaus plexippus*. A. Egg. B. Larva. C. Pupa (chrysalis). D. Adult. (L.) (Redrawn from CCM General Biological, Inc., key card.)

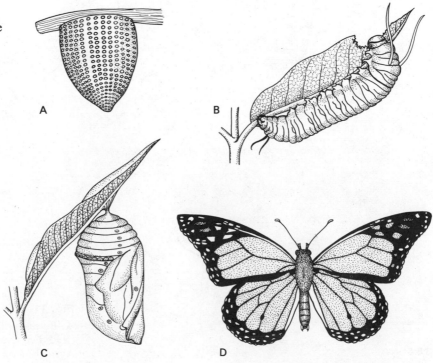

Some authors divide the Lepidoptera differently than above—into the suborders Rhopalocera (butterflies and skippers) and Heterocera (moths). The Rhopalocera are then characterized by (1) being diurnally active; (2) holding the wings vertically at rest; and (3) having variable antennae, often feathery, but usually not clubbed; and (4) usually having a frenulum.

Lepidoptera includes a large number of families with diverse habits, the treatment of which would be inappropriate in this brief survey.

Many lepidopterous larvae or caterpillars have repugnatorial glands of various sorts. The lobelike, eversible *osmeteria,* which are located on the dorsum of the prothorax of papilionid (swallowtail butterflies) larvae, are examples of such glands. Caterpillars all have the ability to produce silk. This substance is secreted by large labial silk glands which open via a common duct into the highly modified labium.

Lepidopterous pupae (Fig. 11–35C) are generally obtect, although some are exarate, and are usually encased in some sort of external protective structure constructed by the last instar larva. This protective structure may, for example, be a silken cocoon, with or without various kinds of detritus incorporated into it, detritus held together by a sticky secretion that hardens into a foundation matrix, a cell in the soil, or a tentlike structure made by "tying" edges of a leaf together. In many, no protective structure is constructed and they must depend upon various kinds of protective coloration and patterning. Such is the case with most butterflies that attach themselves to twigs or other objects by means of a silken thread and/or a silken patch at the tip of a projection from the posterior part of the

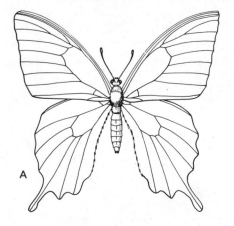

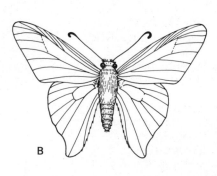

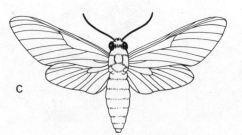

Fig. 11–36
A. Butterfly. B. Skipper. C. Moth.
(Redrawn from U.S. Deparment of
Agriculture, 1952.)

body, the *cremaster*. These naked pupae are commonly referred to as *chrysalids*.

Although the vast majority of Lepidoptera are terrestrial, certain members of several families (e.g., Arctiidae, tiger and footman moths; Pyralidae, snout moths and relatives; Sphingidae, sphinx moths; and several others) have been found in aquatic environments, both as larvae and sometimes as adults.

Most Lepidoptera feed on flowering plants and are particularly voracious in the larval stages, most adults being capable of obtaining only fluid meals from flowers and other sources and being unable to masticate plant material. Their plant-feeding habits make several species among the most important agricultural pests. Many attack stored grain and some attack fiber in clothing. Most pest species are moths. The common names of several of the major lepidopteran pests will give you an idea of the various crops they attack and in some instances the part of the plant: European corn borer, corn earworm, beet armyworm, garden webworm, alfalfa webworm, bollworm, imported cabbageworm, oriental fruit moth, cherry fruit- worm, grape root borer, strawberry crown moth, and rose budworm. A few lepidopterous species, particularly the silkworm moth, *Bombyx mori* (see Chapter 12), are highly valued insects.

A small number of species have some medical and veterinary importance (James and Harwood, 1969). Certain noctuid moths in Africa feed on the lachrymal secretions of cattle and may be the vectors of the etiological agent of infectious keratitis, a disease that may cause blindness. Some species have been observed to feed in the eyes of humans. The caterpillars of certain members of at least 10 families possess urticating hairs which may contain a toxin and which can cause severe blistering and even blindness if they get into the eyes.

ORDER Diptera (*di,* two; *ptera,* wing); true flies (Figs. 11–37 to 11–40).

BODY CHARACTERISTICS minute to very large; larvae: apodous (leg- less), some with well-defined head and thorax and others wormlike (vermiform); pupae: obtect or coarctate.

MOUTHPARTS various modifications of sucking: larvae: various modifications of mandibulate mouthparts.

EYES and OCELLI compound eyes present; most species with 3 ocelli; ocelli lacking in some.

ANTENNAE variable.

WINGS 1 pair; membranous; metathoracic halteres; many apterous species.

LEGS 5-segmented tarsi.

ABDOMEN cerci present or absent.

The order Diptera is the fourth largest order of insects and its members are found in a wide diversity of habitats. Some are adapted to what would have to be considered biologically adverse environ- ments; for example, *Psilopa petrolei* (family Ephydridae) spends its immature stages in crude oil, and the larvae of other ephydrid flies are found in the Great Salt Lake. A survey of the feeding habits of the order could well apply to the entire class Insecta. Members of several families are phytophagous in the larval stages. Some are mycetophagous, "fungus eaters." Many flies are saprophagous and play a vital role in the breakdown of decaying organic material. A large number are predaceous and are to be counted among the beneficial forms. Many are adapted to feed on the blood of vertebrates or are external or internal parasites. Such associations have resulted in their involvement with human and animal diseases.

The Diptera are generally divided into three suborders: Nematocera, Brachycera, and Cyclorrhapha. Members of these suborders are separated on the basis of such features as antennal characteristics, the shape of the exit hole from the pupal case, the nature of the pupa, and larval characteristics.

The Nematocera (Figs. 11–37 and 11–38) are so called because, in the adult stage, they have comparatively long antennae (Nematocera = "thread horn"). The larvae (see Fig. 3–14) have well-developed heads with mandibles that move in a horizontal plane. The pupae are obtect.

Flies in the suborder Brachycera (Fig. 11–39) are characterized by having antennae which are shorter than the thorax (Brachycera = "short horn"). The pupae of these flies are exarate. The larvae have weakly developed heads which are generally retractile, and the mandibles move in the vertical plane. As with Nematocera, Brachycera contains a large number of families. However, many of them are composed of uncommon species.

Members of the suborder Cyclorrhapha (Fig. 11–40) have coarctate pupae. Cyclorrhapha ("circular seam") refers to the circular shape of

Fig. 11–37
Diptera. A. mosquito, *Anopheles maculipennis* Meigen (female).
(Redrawn from James and Harwood, 1969.)

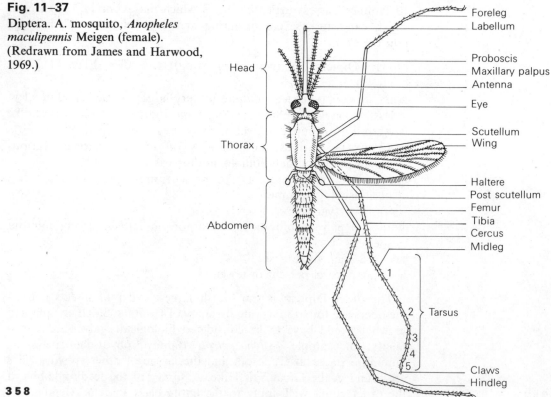

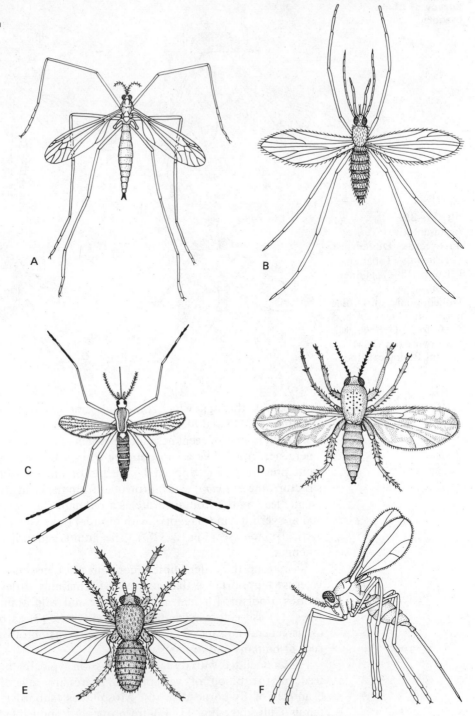

Fig. 11–38

Representative nematocerous Diptera. A. Crane fly (Tipulidae). B. Sand fly,
Phlebotomus (Psychodidae). C. Mosquito, *Aedes aegypti* (Culicidae).
D. Biting midge, *Culicoides furens* (Ceratopogonidae). E. Black fly,
Simulium venustum (Simuliidae). F. Gall midge (Cecidomyiidae). (A, Me to
L; B, C, and E, S; D and F, Mi.) [A and F redrawn from CCM General
Biological, Inc., key card; B–E courtesy of U.S. Public Health Service
(redrawn).]

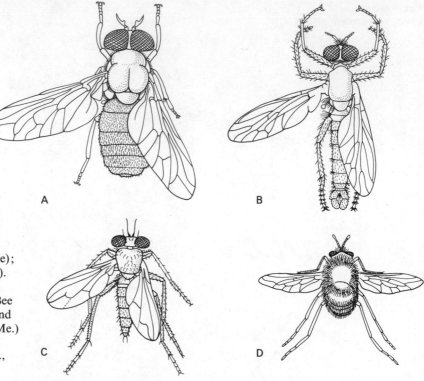

Fig. 11–39
Representative
brachycerous Diptera.
A. Horse fly (Tabanidae);
B. Robber fly (Asilidae).
C. Long-legged fly
(Dolichopodidae). D. Bee
fly (Bombyliidae). (A and
B, Me to L; C, S; D, Me.)
(Redrawn from CCM
General Biological, Inc.,
key card.)

the opening through which the adult emerges from the puparium
(see Figs. 8–21D and 8–23C–D). Near one end of the puparium there
is a circular line of weakness along which the cuticle splits at adult
emergence, much like a cap popping from a bottle. The split results
from pressure being applied by means of an eversible bladderlike
structure, the *ptilinum,* which protrudes from the head after emergence,
usually leaving the frontal suture as a remnant. The larvae lack heads
and are vermiform (maggots), with mouth hooks which operate in the
vertical plane (see Fig. 2–32D). The adults generally have aristate
antennae.

As a group the order Diptera contains more insects of medical and
veterinary importance than any of the remaining orders. Flies cause
serious problems for man and domestic and wild animals in several
ways: (1) as vectors of many serious diseases, (2) as blood feeders,
(3) as general nuisances, and (4) by direct invasion of organs and tissues of humans and animals.

Flies associated with disease transmission typically, but not always,
feed on the blood of vertebrates. There are several outstanding
examples of fly-borne diseases. Mosquitoes (suborder Nematocera,
family Culicidae; Fig. 11–37) are probably responsible for the transmission of more diseases of man than any other group of insects.
They are the vectors of malaria (caused by four species of protozoa
in the genus *Plasmodium*); filariasis (caused by the nematode
Wuchereria bancrofti, and others); and several viral diseases (yellow
fever, dengue, and the encephalitides). Two other nematoceran
families also include disease-carrying members: Simuliidae (black
flies; Fig. 11–38E) and Psychodidae (moth flies and phlebotomus

flies; Fig. 11–38B). Black flies are vectors of onchocerciasis, a disease caused by a filarial worm, *Onchocerca volvulus*, which is typified by the development of painful nodules in the skin and often results in blindness. Psychodid flies in the genus *Phlebotomus* transmit the leishmanias, diseases caused by protozoans in the genus *Leishmania*. The tsetse flies, *Glossina* spp. (suborder Cyclorrhapha, family Muscidae), serve as vectors of the trypanosomes that cause African sleeping sickness in humans (*Trypanosoma gambiae* and *T. rhodesiense*) and nagana (*T. brucei*), a serious disease of cattle in Africa.

Other biting flies capable of transmitting certain diseases but more often involved as nuisances through their blood-feeding activities are the biting midges or no-see-ums (family Ceratopogonidae; Fig. 11–38D); horse flies and deer flies (family Tabanidae); certain snipe flies (family Rhagionidae); and the stable fly, horn fly, and other biting muscoids (family Muscidae). An interesting group of biting flies, which usually do not attack man but are intermittent blood-feeding parasites of sheep, goats, and various wild birds and mammals, are the louse flies and bat flies. The larvae of these flies are retained in the female parent until nearly ready to pupate, at which time they are released. The sheep ked (Fig. 11–40E), *Melophagus ovinus*, is a common wingless ectoparasite of sheep and goats.

Nonbiting flies that pass their larval stages in garbage, excrement, and other decaying organic matter may also be involved in the transmission of disease, generally through the contamination of food. The flies in this category are largely in the families Muscidae (house fly, Fig. 11–40C, and relatives), Calliphoridae (blow flies; Fig. 11–40F), and Sarcophagidae (flesh flies). Flies in these groups may transmit various gastrointestinal diseases, both of bacterial (dysentery,

Fig. 11–40
Representative cyclorrhaphous Diptera. A. Fruit fly (Drosophilidae). B. Picture-winged fly (Otitidae). C. House fly (Muscidae). D. Syrphid fly (Syrphidae). E. Sheep ked, *Melophagus ovinus* (Hippoboscidae). F. Green bottle fly, *Phaenicia sericata*. (A, Mi; B–F, S.) (A–D redrawn from CCM General Biological, Inc., key card; E and F courtesy of U.S. Public Health Service (redrawn).)

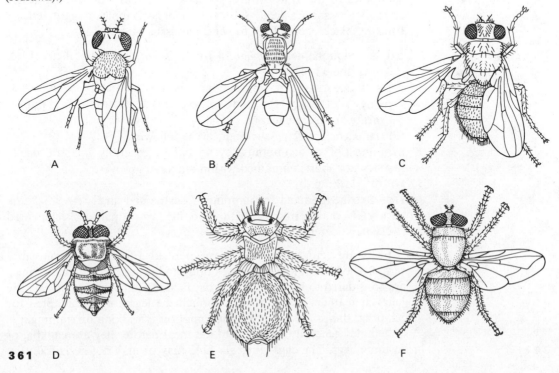

A   B   C

D   E   F

typhoid fever, and others) and protozoan (*Entamoeba*) etiology, cholera, yaws, poliomyelitis, and eye infections. These flies are also vectors of various diseases of domestic animals, including several caused by helminths (tapeworms and nematodes or roundworms). Chloropidae (frit flies and relatives) is another family of nonbiting flies, certain species of which transmit disease—those in the genus *Hippelates*. These are commonly called eye gnats and have been identified as probable vectors of conjunctivitis ("pinkeye"), yaws, and bovine mastitis. These flies are attracted to body secretions of various sorts, which results in their tendency to cluster around the mouth, eyes, nose, and genital and anal openings as well as sores exuding pus.

The maggots of many families of flies may invade the organs and tissues of humans and animals, sometimes causing severe injury and even death. Such invasions are referred to as *myiasis*. Invasion may be brought about accidentally (e.g., accidental ingestion of maggots in food) or may be due to adult female flies being attracted to an open wound or body opening and depositing eggs. Several flies (e.g., the screwworm fly, *Cochliomyia hominivorax*), require an animal host in the larval stages. This situation is called *obligate myiasis*. Other examples of obligate myiasis occur in the bot flies in the families Gasterophilidae (horse bot flies) and Oestridae (warble flies and bot flies).

Surprisingly few true flies are pests of agricultural significance. Among those that are, the Hessian fly, *Phytophaga destructor* (family Cecidomyiidae), is outstanding. This fly is the most destructive of the insects that attack wheat in the United States. It was apparently introduced from Europe into the United States during the Revolutionary War in the straw bedding used by Hessian soldiers. One can gain some idea of the kinds of crops attacked by flies from the names of several pest species (e.g., seed corn maggot, apple maggot, cherry fruit fly, black cherry fruit fly, and narcissus bulb fly).

ORDER Siphonaptera (*siphon*, a tube; *aptera*, wingless); fleas (Figs. 11–41 and 11–42).

BODY CHARACTERISTICS minute; hard-bodied; bilaterally compressed; many posteriorly directed spines; larvae: vermiform; pupae: exarate, in silken cocoons.

MOUTHPARTS piercing-sucking; larvae: chewing.

EYES and OCELLI compound eyes lacking; generally 2 lateral ocelli.

ANTENNAE short; contained within antennal grooves.

WINGS lacking.

LEGS coxae modified for jumping; 5-segmented tarsi.

ABDOMEN male terminalia highly modified with 1 pair of 2-segmented claspers; cerci lacking.

Fleas are ectoparasites of warm-blooded animals (mammals and birds). The bilaterally compressed body and posteriorly directed spines are considered to be adaptations that facilitate movement between the hairs or feathers of the host. Although the legs are well adapted for jumping, the primary mode of locomotion is walking or running.

Females generally require a blood meal before they are capable of forming eggs. The eggs (Fig. 11–42A) may or may not be laid on the

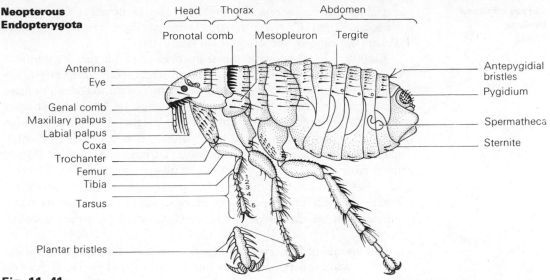

Head   Thorax   Abdomen

Pronotal comb   Mesopleuron   Tergite

Antenna
Eye

Genal comb
Maxillary palpus
Labial palpus
Coxa
Trochanter
Femur
Tibia

Tarsus

Plantar bristles

Antepygidial
bristles
Pygidium

Spermatheca

Sternite

Fig. 11–41

Siphonaptera, a flea.
(Mi.) (Courtesy U.S.
Public Health Service
(redrawn).)

host, but ultimately they fall to the resting place of the host. The
vermiform larvae (Fig. 11–42B) feed on organic detritus and some on
the partially digested blood in the feces of adults. They are whitish
in appearance, quite active, and pupate within silken cocoons (Fig.
11–42C). The pupae (Fig. 11–42D) are exarate. Adults are able to
remain quiescent in the cocoon for long periods of time and exit in
response to vibrations of the substrate and other stimuli associated
with a potential host. It is a common occurrence for persons with
pets to return from a vacation and be deluged with fleas which have
been quiescent within their pupal cocoons. This behavior is an
obvious adaptation which decreases the probability of an adult flea
becoming active when no potential hosts are available. In addition, the
adults can survive prolonged periods of starvation. The life cycle
varies from 3 to several weeks, depending on the species.

Although fleas are highly specialized insects, they show definite
affinities with members of the orders Diptera and Mecoptera.

Fleas are problems from several standpoints. Through their bloodsucking activities they are sources of irritation to man as well as
many wild and domestic animals. Although fleas tend to show host
preferences, they will feed on a variety of available hosts. *Pulex irri-
tans*, for example, is often referred to as the human flea, but it has
many hosts in addition to man. Other well-known fleas include

Fig. 11–42

Life stages of a flea. A.
Egg. B. Larva. C. Pupa.
D. Adult. (Courtesy
U.S. Public Health
Service (redrawn).)

Cocoon

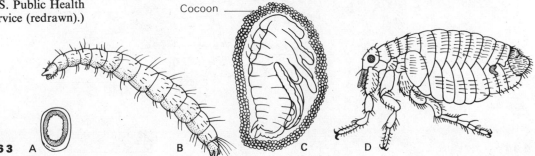

A   B   C   D

Echidnophaga gallinacea, the sticktight flea of poultry; *Ctenocephalides canis* and *C. felis*, the dog and cat fleas, respectively; and *Xenopsylla cheopis*, the oriental rat flea.

Fleas are involved with the transmission of plague. Plague is a disease of wild rodents, the etiological agent, *Yersinia pestis (Pasteurella pestis)*, of which is transmitted from rodent to rodent by fleas. Under certain conditions human beings may be bitten by infected fleas and develop the symptoms of the plague. In regions of high population concentration and poor sanitary conditions, the potential may exist for epidemics of this disease. Throughout recorded history there have been sporadic outbreaks of plague, or the "Black Death," as it has been called at times. One of the aspects of plague which makes it especially prone to becoming epidemic is that in its pneumonic form, in the lungs, it can be transmitted by airborne cough droplets.

One species of flea, *Tunga penetrans*, the chigoe, or any of a variety of other names, is purported to burrow into the skin. These fleas do not actually burrow into the skin, but apparently induce tissue changes, which result in their being partially enveloped by host tissues. Only the females "burrow," and they generally attack in the region of the toes and may open the way for secondary infection. In some instances, autoamputation of toes has directly been attributed to this species. Chigoes are found in the tropical and subtropical regions of North and South America, the West Indies, and Africa.

ORDER Coleoptera (*coleo*, sheath; *ptera*, wing); beetles (Figs. 11–43 to 11–45).

BODY CHARACTERISTICS minute to very large; hard-bodied; larvae: campodeiform or eruciform; pupae: exarate.

MOUTHPARTS chewing, typically prognathous; some hypognathous; larvae: chewing.

EYES and OCELLI compound eyes present or absent; ocelli generally

Fig. 11–43
Coleoptera, ground beetle, *Calosoma semilaeve* with left elytron and hindwing spread. (Redrawn from Essig, 1942.)

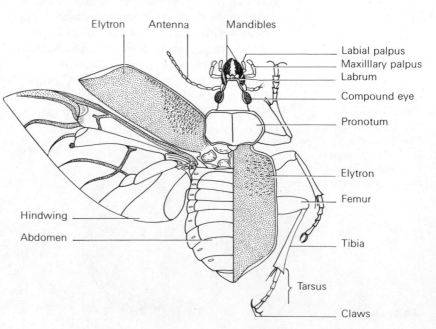

Elytron Antenna Mandibles

Labial palpus
Maxilllary palpus
Labrum
Compound eye
Pronotum
Elytron
Femur
Hindwing
Abdomen
Tibia
Tarsus
Claws

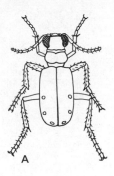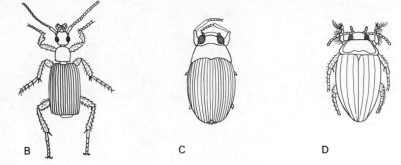

Fig. 11-44

Representative members of the suborder Adephaga. A. Tiger beetle (Cicindellidae). B. Ground beetle (Carabidae). C. Whirligig beetle Gyrinidae). D. Predaceous diving beetle (Dytiscidae). (S to L.) (Redrawn from CCM General Biological, Inc., key card)

lacking, but 1 or 2 in certain groups; larvae: compound eyes lacking, but lateral ocelli usually present.

ANTENNAE   variable; small in larvae.

WINGS   forewings sclerotized elytra which protect the membranous hindwings at rest; some flightless with fused elytra; a few species are wingless.

LEGS   typically cursorial; some adapted for swimming, jumping, or digging; number of tarsal segments variable and taxonomically useful; larvae: commonly 6 legs, some legless.

ABDOMEN   Larvae: some with posterior pair of prolegs (no crochets); usually a pair of spiracles on each of the first 8 abdominal segments plus 1 or 2 pairs on the thorax.

COMMENTS   pronotum = single, conspicuous sclerite; larvae are often called grubs.

This is the largest order of insects and therefore the largest order in the animal kingdom. Beetles comprise approximately 40% of all species of insects.

As a group, beetles are found in nearly as wide a diversity of habitats as the entire class of insects. A similar statement can be made relative to their feeding habits. However, surprisingly few are parasitic. The largest numbers are phytophagous or predatory, while somewhat fewer are scavengers or feed on fungi. The predatory and scavenging species are, of course, beneficial in that they either destroy other insects or take part in the breakdown of organic material.

Coleoptera is divided into three suborders: Archostemmata, Adephaga (Fig. 11–44), and Polyphaga (Fig. 11–45). The first is small and composed of obscure groups.

The majority of beetles are oviparous, but a few are viviparous and a few are parthenogenetic. Hypermetamorphosis occurs in some. The pupae are exarate. They often reside in earthen cells, within host plants, in cocoons, some of which are composed of a secretion from the Malpighian tubules, or they may be protected by the cast of the last larval instar.

The beetles rival the moths in their economic importance. This order contains a large number of very destructive pests of agricultural crops;

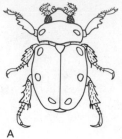 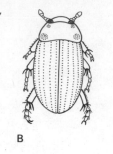 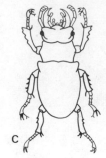  

A B C D E

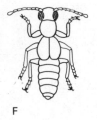 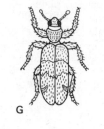   

F G H I J

   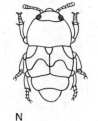 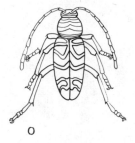

K L M N O

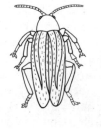 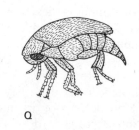 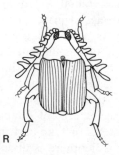 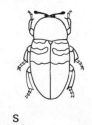

P Q R S

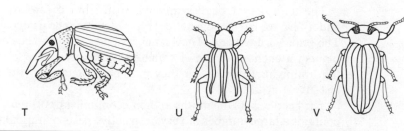

T U V

Fig. 11-45

Representative members of the suborder Polyphaga. A. Scarab beetle (Scarabaeidae). B. Water scavenger beetle (Hydrophilidae). C. Stag beetle (Lucanidae). D. Click beetle (Elateridae). E. Carrion beetle (Silphidae). F. Rove beetle (Staphylinidae). G. Checkered beetle (Cleridae). H. Cucujid beetle (Cucujidae). I. Blister beetle (Meloidae). J. Brentid beetle (Brentidae). K. Darkling beetle (Tenebrionidae). L. Languriid beetle (Languriidae). M. Lady beetle (Coccinellidae). N. Sap beetle (Nitidulidae). O. Long-horned beetle (Cerambycidae). P. Firefly (Lampyridae). Q. Tumbling flower beetle (Mordellidae). R. Seed beetle (Bruchidae). S. Carpet beetle (Dermestidae). T. Snout beetle (Curculionidae). U. Leaf beetle (Chrysomelidae). V. Metallic wood borer (Buprestidae). (Redrawn from CCM General Biological, Inc., key card.)

stored grains, seeds, and grain products; other stored products, such as tobacco, nuts, and chocolate; and shade trees and shrubs. Several species of beetles serve as vectors of plant disease. One of the best known is the European bark beetle, which transmits the Dutch elm disease. This disease is caused by a fungus, and the spores are introduced into the cambium of a host tree by the boring activities of the beetles. Following are the common names of some examples of pest beetle species: white grub (June beetle larva), wireworm (click beetle larva), Japanese beetle, Colorado potato beetle, boll weevil, corn rootworm, bronze birch borer, alfalfa weevil, cane borer, flat-headed apple borer, powder post beetle, drug store beetle, granary weevil, rice weevil, mealworm, cadelle, confused and red flour beetles, sawtoothed grain beetle, pea weevil, and bean weevil.

With the exception of the occasional accidental ingestion of a beetle larva (canthariasis), the more common beetle "pinch" with the large mandibles of some species, and the vesicating (blister-forming) properties of body secretions of the "blister" beetles (family Meloidae and others), beetles are of no particular medical significance.

ORDER Strepsiptera (*strepsi,* turning or twisting; *ptera,* wing); stylopids or twisted-wing parasites (Fig. 11-46).

BODY CHARACTERISTICS minute.

MOUTHPARTS reduced mandibulate.

EYES and OCELLI compound eyes present in males and free-living females.

ANTENNAE males: flabellate; females: lacking.

WINGS males: mesothoracic, clublike; metathoracic, membranous and fan-shaped; females: lacking.

LEGS larvae and adult females mostly legless; 2- to 5-segmented tarsi.

ABDOMEN 3 to 5 genital pores in adult females.

This is a very small order of mostly endoparasitic insects which are sometimes included in the order Coleoptera. They attack several groups in the orders Hemiptera and Hymenoptera. The males are endoparasitic as larvae but emerge as free-living adults. The females of parasitic species are parasitic throughout their lives and are quite reduced in structure and legless. Strepsipteran larvae undergo a hypermetamorphic development. The first instar (triungulin) is free-living and possesses legs. When this larva comes into contact with a host, it burrows through the host's cuticle and becomes endoparasitic. The second and remaining larval instars are vermiform.

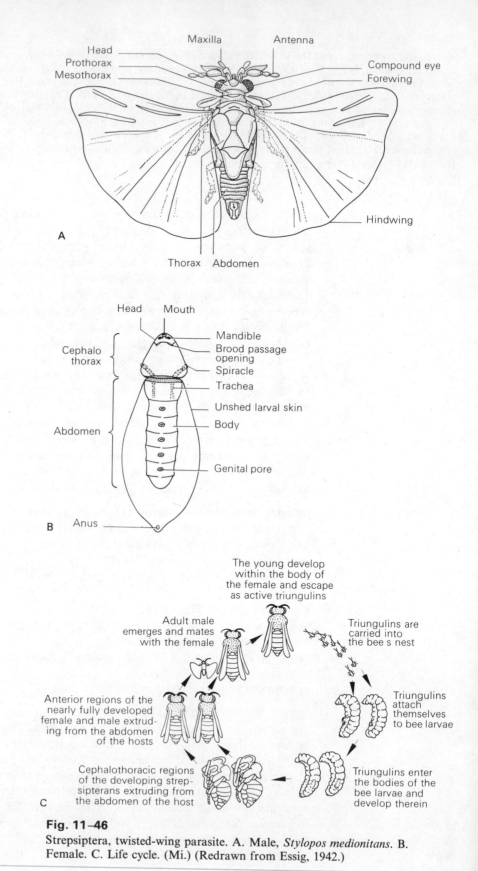

Fig. 11–46

Strepsiptera, twisted-wing parasite. A. Male, *Stylopos medionitans*. B. Female. C. Life cycle. (Mi.) (Redrawn from Essig, 1942.)

ORDER Hymenoptera (*hymen*, membrane; *ptera*, wing); ants, bees, wasps, and relatives (Figs. 11–47, 11–48, and 11–49).

BODY CHARACTERISTICS small to very large; larvae: usually legless with distinct head, rarely eruciform with legs and prolegs; pupae: exarate, usually within a cocoon.

MOUTHPARTS variable; chewing to lapping or sucking.

EYES and OCELLI compound eyes usually well developed; commonly 3 dorsal ocelli, but lacking in some.

ANTENNAE commonly long and filiform or geniculate with elbowed scape and distal segments clubbed.

WINGS most with 2 pairs of membranous wings; hindwings smaller than forewings; many wingless.

LEGS usually 5-segmented tarsi.

ABDOMEN in most, the first abdominal segment *(propodeum)* is fused with the metathorax and is constricted posteriorly, forming the *petiole* between the thorax and abdomen; females with ovipositor or homolog modified for piercing plant tissues, sawing, or stinging.

Using complexity and diversity of behavior as criteria, the Hymenoptera are generally recognized as the most advanced group of insects.

Hymenoptera can be divided into two suborders, Symphyta (Chalastogastra) and Apocrita (Clistogastra). The most obvious characteristic separating these two suborders is the relationship between the thorax and abdomen. The Symphyta (Fig. 11–48) lack the petiole described earlier, which is characteristic of the Apocrita (Fig. 11–49). Other differences include the following. The Symphyta are less behaviorally sophisticated than the Apocrita, parasitic forms are almost nonexistent, the ovipositor is usually fitted for sawing or piercing plant tissues, and the larvae are eruciform. Among the

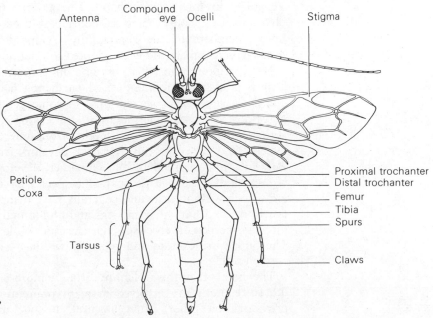

Fig. 11–47
External features of Hymenoptera. An ichneumon wasp (Ichneunomidae). (Redrawn from Essig, 1958.)

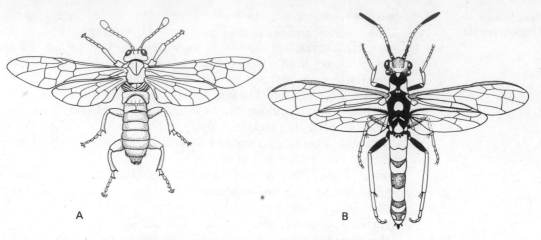

Fig. 11–48

Hymenoptera, Symphyta. A. Sawfly (Tenthredinoidea). B. Raspberry horntail, *Hartigia cressoni* (Siricoidea). (Me.) (A. redrawn from U.S. Department of Agriculture, 1952; B redrawn from Essig, 1958.)

Apocrita are many parasitic forms, the ovipositor is specialized for piercing in parasitic groups and stinging in others, the larvae are legless, and a distinct head is lacking in some of the parasitic groups.

The habits of Hymenoptera are very diverse and they have representatives adapted to nearly every mode of insectan life. Several groups [the ants (family Formicidae), vespid wasps (family Vespidae), and bees in the family Apidae] have elaborate patterns of social behavior. These insects, as well as many other Hymenoptera, are pollen feeders and are important plant pollinators. A large number are parasitic on other insects: for example, all members of the super-families Ichneumonoidea and Proctotrupoidea, and most members of the Chalcidoidea. Members of the superfamilies Sphecoidea and Vespoidea are predatory in habit. The majority of the Symphyta are phytophagous, some being of considerable economic importance. Some members of the superfamily Cynipoidea are gall-formers. A few Hymenoptera are even adapted to aquatic environments.

Hypermetamorphosis is common in certain families. Parthenogenesis is also very common and parthenogenetic generations may alternate with sexual generations. In some groups parthenogenesis plays a role in sex determination; for example, in the honey bee, *Apis mellifera,* males (drones) develop from unfertilized eggs, whereas queens and workers (all females) develop from fertilized eggs. *Polyembryony,* more than one individual developing from one egg, is known to exist in a few parasitic species. Hymenopterous larvae are usually without legs and generally have distinct heads. Some are eruciform and have thoracic legs and abdominal prolegs which lack the tiny hooks or crochets characteristic of lepidopterous larvae. The pupae are exarate and typically within a cocoon or waxen or earthen cell.

The members of this order are certainly more beneficial than harmful, since many are involved in plant pollination, are parasitic or predatory on other, often harmful, insects, and produce useful products (e.g., honey and beeswax). However, there are several species which are destructive pests. Most of these are sawflies in the families Diprionidae (conifer sawflies), Tenthredinidae (common sawflies), and Cephidae (stem sawflies). The larvae of common sawflies feed on foliage of trees and can be very damaging. The larvae of stem sawflies bore into the stems of grasses and berries, and two species are important pests of wheat. Ants (family Formicidae) may constitute a problem when they invade the household and infest foodstuffs.

Fig. 11–49

Hymenoptera, Apocrita. A. Ichneumon wasp (Ichneumonoidea; Ichneumonidae). B. Chalcid wasp (Chalcidoidea; Chalcidae). C. California gallfly, *Andricus californicus* (Cynipoidea). D. Velvet ant (Scolioidea; Mutillidae). E. Thief ant, *Solenopsis molesta* (Scolioidea; Formicidae). F. Golden polistes, *Polistes aurifer* (Vespoidea; Vespidae). G. Thread waisted wasp (Sphecoidea; Specidae). H. Honey bee, *Apis mellifera* (Apoidea; Apidae). (A, F, and G., Me; B, C, and E, Mi; D, S to Me.) (A, B, G, and H redrawn from U.S. Department of Agriculture, 1952; C and F redrawn from Essig, 1958; D and E courtesy U.S. Public Health Service (redrawn).)

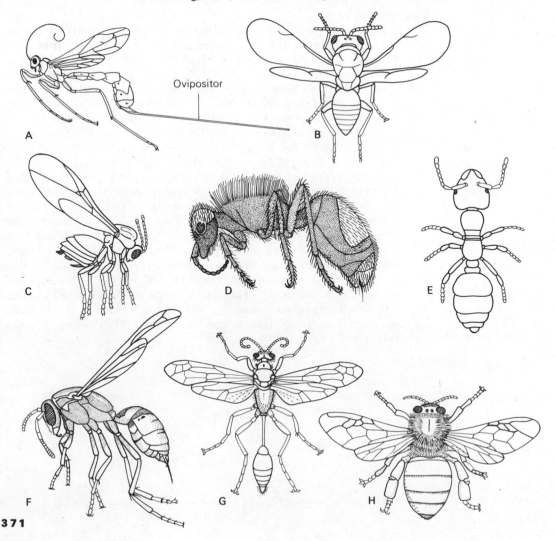

Ovipositor

A

B

C

D

E

F

G

H

Although no hymenopterous insects serve as disease vectors, many are capable of inflicting a painful sting which can lead to serious consequences if the person stung happens to be hypersensitive to the injected venom. Stinging hymenoptera are in the following seven superfamilies (James and Harwood, 1969): (1) Chrysidoidea, cuckoo wasps and relatives; (2) Bethyloidea, bethyloid wasps; (3) Scolioidea, scoliid wasps, velvet ants, and relatives; (4) Formicoidea, ants; (5) Vespoidea, hornets, yellow jackets, spider wasps, and relatives; (6) Sphecoidea, sphecoid wasps; and (7) Apoidea, bees.

Selected References

GENERAL

Borror and DeLong (1971); Borror and White (1970); Brues, Melander, and Carpenter (1954); Chu (1949); Commonwealth Scientific and Industrial Research Organisation (1970); Comstock (1940); Edmondson (1959); Essig (1942, 1958); Jacques (1947); Metcalf and Metcalf (1928); Oldroyd (1968); Pennak (1953); Peterson (1948, 1951); Richards and Davies (1957); Ross (1965); Swain (1957); Tietz (1963); Usinger (1956a).

SPECIFIC GROUPS

PROTURA: Ewing (1940), Tuxen (1964). ENTOTROPHI: Smith (1960). COLLEMBOLA: Christiansen (1964), Maynard (1951), Scott (1961). THYSANURA: Slabaugh (1940). EPHEMEROPTERA: Day (1956), Edmunds (1959), Edmunds and Allen (1963), Needham, Traver, and Hsu (1935). ODONATA: Corbet (1963), Gloyd and Wright (1959), Needham and Heywood (1929), Needham and Westfall (1955), Smith and Pritchard (1956). PLECOPTERA: Jewett (1956), Needham and Claassen (1925). ORTHOPTERA: Blatchley (1920), Cornwell (1968), Guthrie and Tindall (1968), Helfer (1963), Rehn and Grant (1961), Roth and Willis (1957, 1960), Uvarov (1966). DERMAPTERA: Popham (1965). EMBIOPTERA: Ross (1970). ISOPTERA: Howse (1970), Krishna and Weesner (1969, 1970), Snyder (1954). ZORAPTERA: Gurney (1938). PSOCOPTERA: Mockford (1951). MALLOPHAGA: Emerson (1964). ANOPLURA: Ferris (1951). THYSANOPTERA: Stannard (1968). HETEROPTERA: Britton (1923), Hungerford (1959), Lawson (1959), Metcalf (1954–1963, 1962–1967), Miller (1956), Usinger (1956b, 1966). NEUROPTERA: Chandler (1956), Gurney and Parfin (1959). MECOPTERA: Carpenter (1931). TRICHOPTERA: Denning (1956), Ross (1959). LEPIDOPTERA: Ehrlich and Ehrlich (1961), Holland (1903), Klots (1951), Mitchell and Zim (1962). DIPTERA: Cole (1970), Curran (1965), Demerec (1950), Oldroyd (1964), Stone (1965). SIPHONAPTERA: Ewing and Fox (1943). COLEOPTERA: Arnett (1968), Dillon and Dillon (1961). STREPSIPTERA: Ulrich (1966). HYMENOPTERA: Andrewes (1971), Evans and Eberhard (1970).

PART THREE
Applied Aspects of Entomology

Applied Entomology

In this chapter we shall deal with the area of applied entomology, which concerns itself with the pragmatic aspects of the science of entomology. This is the point at which the results of both basic and applied research are brought to bear on specific problems. Our approach will be to first consider how insects are problems relative to man, then the attempted solutions to these problems, and finally a brief discussion of insects that are directly or indirectly beneficial. A number of excellent applied entomology texts are available and should be consulted for more detailed information.

The Problems

Fortunately, the vast majority of insects are neutral or indirectly beneficial relative to man, but those comparatively few which are pests have done and will continue to do all too effective a job of taxing his ingenuity to its fullest. According to the Subcommittee on Insect Pests (1969): "It is estimated that in the United States 150 to 200 species or complexes of related species frequently cause serious damage. From time to time, 400 to 500 additional species are pests and may cause serious damage. Approximately another 6,000 species of insects are pests at times, but seldom cause severe damage . . . ." These figures represent but a small fraction of the approximately 84,557 described species of insects in North America north of Mexico (Borror and DeLong, 1964).

There is nothing, biologically speaking, which clearly defines an insect as a pest. Two similar insects may have very nearly the same biological patterns and be in the same family and one be considered a pest and another not, because one attacks man or something valued by man and the other does not. Rolston and McCoy (1966) illustrate this point with three chrysomelid beetles:

The Colorado potato beetle effectively defoliates the potato, and the dock beetle, *Gastrophysa cyanea* Melsheimer, defoliates its host with equal regularity. That the former is a pest and the latter is not reflects the value we place on the potato and our disinterest in dock. We may even encourage the destruction of some plants by insects. St. John's wort, or Klamath weed, has been cleared from thousands of acres of rangeland in western states by a group of insects, but principally by a leaf beetle, *Chrysolina quadrigemina* (Suffrian), introduced specifically to control this weed.

A variety of man-influenced factors have contributed to the creation of pest situations. These may be usefully grouped into three categories: (1) concentration and ecosystem simplification, (2) transportation, and (3) man's attitudes and demands. Man through his various activities has modified the environment, sometimes to his advantage, often to his disadvantage. As a result of this modification, which takes many forms, he has simplified many ecosystems (i.e., reduced the number of components). It is a fundamental principle of ecology that the simpler the system of interacting organisms, the less its inherent stability and the greater the likelihood of large fluctuations of populations of the component species. Thus it is not difficult to understand why there are a multitude of problems associated with the great monocultures of organisms, particularly plants, which man has developed. These "agroecosystems" are much less complex than the "natural ecosystems" that preceded them and hence are prone to produce pest situations.

A good example of a vast monoculture is the corn grown yearly in the United States. This crop occupies approximately 85,000,000 acres of land and represents about one fifth of the total cropland (Pfadt, 1962). About $900,000,000 worth of damage is caused annually by the pests that compete with man for this $5 billion crop. Another concentration process brought about by man has been the storage of of vast quantities of materials that are palatable to certain insects. Among these stored materials are millions of bushels of grains, fresh vegetable, various other food products, fiber, and so on. Such "environments" furnish optimal conditions for certain species. Man has also concentrated himself and often under other than sanitary conditions. These concentrations of human populations have afforded the ideal mileau for insect-carried pestilences, particularly louse-borne typhus and the plague. With the advent of the many forms of modern transportation, insects are no longer limited in their dispersal capabilities by natural geographic barriers such as mountains and oceans, but are carried about as hitchhikers.

The introduction of an exotic insect into a new region away from its natural enemies may have disastrous results. For example, the European corn borer is a major pest of corn in the United States, but it is only a minor pest in its original home (Fronk, 1962b). Conversely, plants introduced into a new region may prove to be ideal food for indigenous insects, which then constitute a pest problem. For example, the cucumber originated in the East Indies, while one of its major pest insects, the striped cucumber beetle, originated in North America (Fronk, 1962b). Man's attitudes and demands may create a pest problem or greatly exaggerate what might, in fact, be a minor one. For instance, any slight evidence of insect damage on a vegetable or fruit may decrease or destroy its economic value to a producer, even though its food value may be the same if undamaged, for the simple and unfortunate reason that the average consumer is very particular about such matters and demands blemish-free produce. This commonly forces producers to employ costly and ecologically undesirable means of insect control. Many insects are considered to be pests not because they attack man or destroy any of his goods, but simply because he finds them distasteful. It was a sad experience to once watch a young woman destroy a Luna moth with a golf club merely because it had the misfortune to fly by.

The conflict between man and insects may be resolved into three general categories in terms of what the insects attack: (1) growing plants; (2) stored products, household, and structural materials; and (3) man and wild and domestic animals valued by man. Specific examples of many of the insects involved in the conflict with man are included in Chapter 11.

Growing Plants

Virtually every growing plant valued by man is shared, to a greater or lesser extent, with one or usually more species of insect. For example, approximately ". . . 400 species of insects infest the apple, though only 25 of these are of economic importance" (Johansen, 1962). Almost all the damage these insects cause is associated with their feeding activities. Depending on the species and life stage, they remove chunks of tissue by chewing, suck sap from, and/or bore through virtually every part of a plant. Not only are all parts of a plant susceptible to insect attack but all life stages as well. The feeding of insects not only causes the removal of tissue and sap but may poison a plant by the injection of toxic saliva. As a result of their feeding activities several species of insects are responsible for the introduction, transmission, and dissemination of several plant diseases caused by viruses, primarily, but also by bacteria, fungi, and a few protozoans. Pfadt (1962), describing the various kinds of injury incurred by small grains (e.g., wheat, barley, and oats) and the guilty insects, illustrates the above points quite graphically:

> Insects injure small grains in the field from the time the seed is planted until the grain is harvested. Both wireworms and false wireworms feed on the planted seed. In dry soil they may even destroy a crop before rains stimulate germination and growth. After the grain germinates, these pests devour the tender sprouts just as they push out of the seeds.

> Wireworms also kill seedlings by boring into and shredding the underground portion of stems. Wireworms and white grubs feed on roots and sever them from the plant. Cutworms, white grubs, and false wireworms cut off young plants near soil level. In addition, the wounds left by soil pests allow rot pathogens to enter the plant.

> Grasshoppers, Mormon crickets, and armyworms may devour young plants completely; they may strip the leaves from older plants, feed on maturing heads, or cut through the stems below the heads.

> By extracting juices from the stem for food, larvae of Hessian flies retard or kill seedlings and reduce the yields of older plants. Weakened stems of older plants are likely to cause the crop to lodge. Wheat stem sawfly, wheat jointworm, and wheat straw-worm bore within culms and obstruct the flow of sap. This damage reduces the number and weight of kernels. Boring insects also cause grain to lodge.

Insects such as the chinch bug and various aphids impoverish plants by sucking juices from leaves or stems. Moreover, by injecting toxic saliva they produce fatal necroses. Small grain pests may transmit serious plant diseases, such as wheat streak mosaic by the wheat curl mite, striate by the painted leafhopper, and barley yellow dwarf by several species of aphids.

Barley yellow dwarf, a new and widespread virus disease of cereals characterized by leaves rapidly turning light green and yellow beginning at the tips, is causing much concern in the United States. Transmission of the disease is solely by aphids. Six species have been incriminated, the English grain aphid, apple grain aphid, corn leaf aphid, greenbug, rose grass aphid, and bluegrass aphid. Recent research has demonstrated that three different strains of the virus are present and that they are usually carried by different species of aphids.

Another way by which insects cause damage to growing plants is through their egg-laying activities. Several kinds of insects have well-developed ovipositors with which they are able to penetrate rather hard surfaces like the bark of woody stems. The periodical cicada provides a good example of an insect that can cause considerable oviposition damage. The larvae of these insects spend 13 or 17 years feeding on plant roots and then emerge from the ground for a brief period of time to mate and lay eggs. The mated females deposit their eggs on young stems of shrubs and trees, causing severe damage and often necrosis of the affected area.

Stored Products, Household, and Structural Materials

The wide variety of materials accumulated by man and stored in containers ranging from large bins to small boxes present ideal environmental conditions for certain insects to thrive, essentially free of any natural enemies. The presence of these insects may go undetected for long periods of time and hence the potential extent of damage is great. Stored grains are particularly susceptible to attack and probably incur the most damage of any stored material. Stored grain pests not only consume grain but also render large quantities useless by contaminating it with fecal material, webbing, odors, shed exoskeletons, and whole or fragmented dead individuals. These insects may also, through their activities, cause heating of grain. This heating causes moisture-laden warmed air to rise to the surface, where it is cooled, resulting in the condensation of the accumulated moisture on the surface of the grain. This in turn causes caking of the grain and affords an ideal situation for the growth of molds and encourages spoilage (Wilbur, 1962).

The major stored grain pests include the rice and granary weevils, the flour moths, and the Angoumois grain moth. The damage done by stored grain insects is estimated to be over 5% of the world's production, and much higher in certain areas. Wilbur (1962) points out that "Destruction of food by stored grain insects is a major factor responsible for the low levels of subsistence in many tropical countries. If these losses could be prevented, it would alleviate much of the food shortages in the famine areas of the world." Packaged food materials may be attacked by insects at any point from the processing plant to the consumer's home. What constitutes food for an insect is not necessarily food for man. Certain insects, such as the cigarette beetle and drug-store beetle, both members of the family Anobiidae, find a wide variety of nonhuman food items quite appetizing, including such delicacies as tobacco and several drugs.

Several species of insects have become common cohabitants with man. These include cockroaches, silverfish and firebrats, and a large variety of ants, beetles, and moths. The stored-products insects also fall into this category. Even when these insects do little or no damage, their presence in a modern home is deemed highly undesirable. Several household items are susceptible to attack, including rugs, furniture, clothing, books, paper, food, and insect collections. Davidson and Peairs (1966) cite an estimated $200 to $500 million annual loss due to the destructive activities of these pests.

Any structures made out of wood may be attacked by certain wood-infesting insects, and extensive and serious damage can be the result. The structural timbers of houses, buildings, and bridges, and wood constructs such as fences, railroad ties, and telephone poles are all susceptible to attack. Subterranean termites are the major pests involved, although insects such as powder post beetles may also cause trouble.

Man and Animal

The ways in which insects attack man and animals valued by man are essentially the same and hence will be treated together. The terms *medical* and *veterinary entomology* are commonly used in reference to this area of applied entomology. As with the preceding two categories of the man–insect conflict, this one also has strong economic overtones. Animals valued by man, wild or domestic, may be killed, weakened, and/or decreased in value as a result of being attacked by insects. Similarly, man may be killed or weakened and literally millions of man-hours of productive labor lost. Insect-borne diseases such as malaria and trypanosomiasis are particularly responsible for the latter. Vast areas of the world have been made uninhabitable or nearly so due to the presence of insects and the diseases they carry. For example, the tsetse flies, *Glossina* spp., because they transmit the trypanosomes which cause nagana in cattle and two types of African sleeping sickness in man, have prevented the development of millions of square miles of tropical Africa.

Insects affect the lives of man and other animals in two basic ways: (1) directly as the causative agents of disease and discomfort, and (2) indirectly as the transmitters (vectors) of the causative agents of disease (bacteria, viruses, spirochaetes, protozoans, helminths). Many of the diseases of man and animal transmitted by insects have been discussed in Chapter 11.

Insects as the Causative Agents of Disease and Discomfort. The ways insects accomplish the role as causative agents include various combinations of feeding activities, physical injury, secretions, invasion and infestation, and psychological disturbances.

The most important feeding activity that affects man and animals is the tendency of several species of insects, such as mosquitoes, black flies, other biting flies, bed bugs, and conenose bugs, to suck blood. These insects can make life miserable and may in some instances cause serious illness or even death. Illness due to blood loss in human beings is usually not significant since they can generally defend themselves or escape. However, the pain of being bitten and other complications to be explained below make insect blood feeding of major significance to both man and animal. On the other hand, blood loss in wild and domestic animals may be considerable, and when the population of the offending insects is large, blood loss coupled with interference with feeding causes death.

One of the most dramatic examples of this has been recounted by Oldroyd (1964): "An extreme instance is the oft- quoted census made in Rumania, Bulgaria, and Yugoslavia in the year 1923, where nearly 20,000 domestic animals, horses, cattle, sheep, and goats are said to have been killed, along with many wild animals, by the Golubatz fly, *Simulium columbaschense*." Some non-blood-feeding insects are attracted to animals and man and may feed on sebaceous and lachrymal secretions of perspiration and in this way cause considerable disturbance. Certain species of noctuid moths in Africa have been observed to feed on lachrymal secretions of cattle and may be involved in the transmission of a disease that results in blindness. Some Asian noctuid moths have been known to feed on lachrymal secretions in man (James and Harwood, 1969). Flies in the genus *Hippelates* (family Chloropidae) are attracted to lachrymal secretions as well as to mucous and sebaceous secretions, pus, and blood (James and Harwood, 1969).

Physical injury may occur if an insect flies into the eye or ear, or as a result of a defensive reaction of an insect that is carelessly handled, or because of inadvertent contact with an insect. Many insects equipped with powerful jaws or legs may pinch, bite, or jab. For example, many of the predaceous Hemiptera, giant water bugs and backswimmers, may inflict a very painful bite. As with anything which causes a break in the skin of man or animal, secondary infection is always a possibility.

Insect secretions that are harmful to man and animal may be divided into two groups: those which are inherently toxic, *venoms*, and those whose effect depends on the physiological response of the victim, *allergens*. Introduction of a venom usually follows a predictable course, causing pathological conditions; the response to the introduction of an allergen may vary from no reaction to anaphylactic shock, which may result in death. To become allergic or hypersensitive requires an initial introduction (injection) of allergen, at which time there is no response except that antibodies are produced which will couple in some way with the allergen when subsequently injected and cause the allergic response. A given secretion may act both as a venom and as an allergen (e.g., bee venom usually causes pain and a local reddening of the skin and if introduced into an allergic or hypersensitive individual may cause serious complications).

James and Harwood (1969) describe four ways by which venoms and allergens may be introduced: (1) bite, (2) sting, (3) contact, and

(4) active projection. Blood-feeding insects typically inject an amount of saliva into their host before they begin to remove blood. This saliva may act as an allergen, as has been demonstrated with mosquitoes, bed bugs, and others. Thrips, certain phytophagous Hemiptera, and certain predatory Hemiptera have also been observed to inflict bites on man, and their salivary secretions may act as venoms or allergens or both. Several members of the order Hymenoptera (i.e., several ants, bees, and wasps) are equipped with stinging apparatus and venom glands on the posterior part of the abdomen. These structures serve for defensive purposes in some and defense and prey paralysis in others. The venoms injected by these insects can act as allergens. Direct contact with secretions from the bodies of blister beetles in the family Meloidae can cause blistering of the skin. Some lepidopterous larvae in the families Saturniidae (e.g., the io moth caterpillar), Lymantriidae (e.g., the brown-tail moth caterpillar), Megalopygidae (e.g., certain flannel-moth caterpillars), and others possess urticating (stinging) hairs with which are associated one or more poison glands. These hairs can cause trouble either by contact with the caterpillar or may become airborne as a result of the death and drying up of the caterpillar and eventually come into contact with the skin or be inhaled. The effects of these urticating hairs is much the same as stinging nettles. Butterfly scales and the dried exuviae and fragmented bodies of dried, dead mayflies and other insects have been shown or are suspected to cause allergic reactions in susceptible individuals. Some insects, such as certain ants and predatory Hemiptera and others, are able to actively project or spray venom/allergen from their bodies. This ability is associated with the defense mechanisms of insects and has been discussed in Chapter 7.

Several species of insects spend all or at least a portion of their existence on man or other animals. Some of these are accidental invaders while others are definitely attracted to man or animal and must live on or in them to survive. Several insects fall into the facultative category; that is, they can complete their life cycles without invading or infesting man or animal but can take advantage of such hosts should they become available. Among the larvae of true flies there are several that will invade the various cavities or open wounds of animals. This type of invasion is referred to as *myiasis*. Students interested in pursuing this topic further should consult Zumpt (1965). Other animal-infesting insects include the chewing and sucking lice.

In some instances there may be psychological disturbances associated with the presence of insects, both in human and in wild and domestic animals. These disturbances may be totally unrelated to the fact that the offending insects are harmless or not. In humans, Pomeranz (1959) recognizes two categories of psychological disturbances in man caused by insects: (1) arthropod phobia (entomophobia when referring specifically to insects)—"the irrational, persistent fear of recurrent arthropod infestation"; and (2) hallucination of arthropod infestation—"a condition in which the subject imagines he is being molested by small and difficult-to-locate forms which reach and localize on the body despite all sorts of extraordinary preventive measures." In wild and domestic animals, the persistent buzzing and biting or oviposition attempts by flying insects may cause considerable behavioral disturbance and interfere severely with grazing. For example, horses show extreme nervousness in response to the buzzing and oviposition attempts of the adult female bot flies.

Solutions

Man has responded to the multitude of problems posed by insects in diverse ways, some of which have been quite successful, others of which have backfired on him. More and more it is being realized that in attempting to control or eradicate a given insect species consideration must be given to the "life system" (see Chapter 9) of the species and to what the immediate and long-term ecological effects of any measures applied are going to be. When the life system of an insect is understood, a combination of techniques can be applied at appropriate times to achieve maximum control with minimum disturbance of the environment. Thus the term *pest management* is most appropriate in the modern context. In this section we shall review the major approaches that have been used in attempting to solve insect problems and some of those which show promise for the future.

Biological Control

Under natural conditions, insect populations are kept in check by the influences of a multitude of environmental factors. Together these factors bring about the "natural control" of a given population. Among these natural control factors are parasites, predators, pathogenic microbes, and competing species. Biological control has generally been defined as the purposeful use of members of the first three groups, and more recently the fourth, in the destruction or suppression of undesirable insects (and other animals and plants). In the past two decades techniques have been developed that make it possible to make an insect species literally its own enemy. These techniques include the use of sterilized insects and a number of different applications of genetic manipulations. In most instances these techniques are still in the experimental–developmental stages but based on a few outstanding successes have great promise for the future, at least for certain pest situations. Since all these techniques ultimately involve the use of living organisms against living organisms, it seems justifiable to include them as categories of biological control. Another approach using living organisms against living organisms and one that is essentially a defensive tactic, instead of offensive as the ones above, is the development of plant and animal hosts which are in some way resistant to the attacks of pest insects. Sweetman (1936a) quite logically includes this approach in his concept of biological control.

Incorporating all the above categories into a single concept leads to the following definition: Biological control is the destruction, suppression, or repulsion of pest species by the use of living organisms of the same or different species.

Biological control offers certain advantages that many other methods of insect control do not. It is inherently safe both in that there is usually little or no danger involved during its application and that no toxic residues contaminate the environment and destroy beneficial animals as well as harmful insects. Once started it promises to provide long-lasting control and hence although the initial outlay of money for research and development may be comparatively large, it is quite economical in the long term. Also, the pest is usually not likely to develop resistance to the controlling agent.

Parasites and Predators. There are three basic ways parasites and predators have been manipulated for use in biological control: introduction, conservation, and augmentation.

The introduction of exotic species with hopes of controlling pest species has been by far the most successful of the three methods. There are two situations in which this method is appropriate: (1) when there are "unoccupied niches in the life system of the pest, which could be filled by an introduced species"; and (2) when "a certain niche is occupied by an organism that is inherently inefficient as a regulator and that might be displaced by a more efficient exotic regulator" (Subcommittee on Insect Pests, 1969). Both situations exist particularly when a given pest species has been accidentally introduced from another area. A large percentage of the pest insect species in the United States fit into this category. These introduced species find themselves in an environment in which they are free to multiply in the absence of their natural enemies, and hence if they happen to feed on something which is valued by man, they can quickly become problems.

Although it has long been known that pest insects are attacked by various natural enemies, the purposeful use of their natural enemies has occurred only in the last couple of centuries. According to Sweetman (1936a), the first written record of the use of beneficial insects describes the protection of date palms in Arabia from phytophagous ant species by the introduction of another species of ant that was predatory on the phytophagous forms. The first noteworthy example of the application of this type of control in the United States and probably the most significant early demonstration of this method as a valid approach was in California in 1887–1888 (Subcommittee on Insect Pests, 1969). At that time the citrus industry was severely threatened by the cottony-cushion scale, *Icerya purchasi*, and the insecticides available were ineffectual in controlling it. Since this pest species was a native of Australia, an appropriate control species was sought there, resulting in the discovery of the vedalia beetle, *Rodolia cardinalis*. Within a year following the introduction of this predatory beetle, effective control had been attained and is still in effect. Since that time over 100 pest species throughout the world have been effectively controlled by the use of parasites and predators.

Conservation methods include the application of pesticides at times when they are likely to do the least harm to beneficial species, provision of a place to live for a beneficial species, possible reduction of populations of natural enemies of beneficial insects, and other similar approaches. Augmentation (or inundation) involves the mass rearing and subsequent dissemination of parasites or predators in order to supplement the actions of the natural populations of these same species. An example of this method is the use of parasitic wasps (e.g., *Trichogramma* spp.), which attack the eggs of several lepidopterous species. Neither of these latter two approaches, conservation and augmentation, has been as effective as the introduction of exotic species.

A fourth possible approach is the artificial selection of highly resistant strains of beneficial insects which could be used in association with insecticides (integrated control) without themselves being harmed. Insecticide resistance has, in fact, been observed to develop in certain beneficial species.

Many vertebrates (e.g., birds, reptiles and amphibians, fish, and some mammals) are predatory on insects and many are of importance in the natural control of insect populations. Some of these have been used in biological control and a number may have untapped potential. One particularly outstanding example of the use of a vertebrate in insect control has been the use of certain cyprinodont fishes, especially *Gambusia affinis*, against mosquito larvae. Several traits of *Gambusia* which make it useful in this way are its broad tolerances for salinity and organic pollution, viviparous reproduction, high fecundity, small adult size, top feeding habits, ability to penetrate regions where mosquito larvae breed, and ease of transport (Sweetman, 1936a; Bay, 1967). The giant toad, *Bufo marinus*, is another example of a vertebrate that has been useful as an insect control agent (Sweetman, 1936a). It has been introduced into the West Indies, the Hawaiian Islands, the Philippines, Australia, and Louisiana, and has provided control of sugar cane white grubs in Puerto Rico.

Pathogenic Microbes. Although man has been long aware of the "natural" control of insect populations by microbes, the first record of the idea of using them for insect control was in the eighteenth century, and the study of insect microbiology and pathology did not become a serious topic of study until the early 1900s. Even to the present, there have been only a handful of successful attempts at insect control with microbes. The microbes that may be associated with insects (i.e., bacteria, fungi, viruses, rickettsiae, and protozoans) were discussed in Chapter 9. Our concern here is how man has tried to direct the effects of certain of these microbes on insects to his advantage.

The use of microbes in control has great promise in spite of numerous drawbacks. Many of the microbes can be cultured in large numbers economically, they are generally fairly specific in action and usually not harmful to other plants and animals, and they leave no toxic residues. Disadvantages include (1) dependency on certain weather conditions (e.g., most fungi pathogenic for insects are effective only under warm, moist conditions); (2) difficulties in mass producing certain microbes (e.g., viruses); (3) slow kill of hosts (e.g., rickettsiae); and (4) potential hazard to other animals (e.g., rickettsiae and certain protozoans).

Particularly outstanding examples of the successful use of microbes include the treatment of soil with spores of the bacterium *Bacillus thuringiensis* to control the Japanese beetle, *Popillia japonica*. This same bacterium has also been applied successfully against the alfalfa caterpillar, European corn borer, and the pink bollworm. Brown, red, and yellow fungi which attack whiteflies have been effective against these citrus pests in Florida. There are numerous examples where certain microbes are known to be very effective in natural control but have not yet been developed as control agents.

Competitors. The use of competitors as agents of biological control is a comparatively new idea and is based on the ecological principle of competitive displacement. This process occurs when one of these two species of animals (insects) displaces or eradicates another species which is its ecological homologue. Ecological homologues are species that have essentially identical niches (i.e., they fit into an ecosystem in the same way, all or nearly all of their life requirements being the same). For competitive displacement to occur, one of two ecological homologues would have to have a slight edge over the other, which is the most probable situation between any two species. Several instances of this phenomenon have been observed to occur under natural conditions. Practical applications of the principle of competitive displacement have not as yet been extensively developed, but it may potentially be applied in several ways. According to the Subcommittee on Insect Pests (1969), possibilities centering around the importation and colonization of an ecological homologue of a pest species include the following: (1) introduction of a nonbiting species to displace a biting species (e.g., if both competitors shared the same larval niche); (2) introduction of an imported species which, although sharing the same food as a pest species, is unable to survive the winter in the area of application, but which during favorable periods is capable of displacing the original species; and (3) introduction of a potential pest species which is known to be controllable by biological or other means and which is capable of displacing an otherwise uncontrollable pest species.

Sterilization and Genetic Control. A little more than 30 years ago the idea was conceived that if a large enough percentage of the matings of a given population in the field resulted in no offspring, then over a period of generations the population would decrease. Thus if sexually sterilized insects are introduced into or induced in a wild population generation by generation and if the matings of these sterilized individuals exceeds normal matings, the population will decline. If the number of sterile individuals is kept constant (by more releases) for each generation, the ratio between sterile and normal matings will increase rapidly and the rate of population decline will increase correspondingly. The objective of the presence of sterile insects in a population can be attained in basically two ways: rearing, sterilization, and release of large numbers of individuals into wild populations; and exposure of a wild population to substances which induce sterility (chemosterilants).

The earliest application of the sterile-release technique was against the screwworm fly, *Cochliomyia hominivorax*, in the southeastern United States (Knipling, 1959; Baumhover, 1966). During 1958–1959 and following an impressive sequence of research and field trials covering two decades, 3.7 billion sterilized screwworm pupae were reared and released throughout large portions of Florida and Georgia. This resulted in the successful eradication of the fly from this part of the country. Since that time there have been sporadic outbreaks traceable to the movement of infested animals into the territory, but the screwworm has not been a problem since 1959. Baumhover (1966) describes the cost and savings of the screwworm eradication program:

> During the two-year campaign, 3.7 billion screwworm pupae were produced, and 6.3 million lb. of horsemeat and whale meat were used. Twenty light aircraft were used to release flies over a maximum area of 85,000 square miles, and peak employment, including plant personnel, fly distributors, field inspectors, and clerical and administrative help, totalled 500. However, for a research cost of only $250,000 and an eradication-program cost of $10 million, ranchers in the Southeast have experienced $140 million in savings since inception of the program in 1958.

Because of the vast numbers of sterilized individuals which need to be released to effect eradication, the sterile-release method is most practical against insects that occur in relatively small numbers at certain times during the population-density cycle and that are easy to rear in very large numbers. The low numbers can occur naturally or may be brought about by the employment of other methods of insect control. It is obvious that the insects that are released should not cause extreme annoyance or damage. For example, it would seem inappropriate in most instances to apply this method against house flies or cockroaches. The sterile-release method is likely to be impractical against insects that are very prolific and widespread or against insects that appear in large numbers sporadically and unpredictably (e.g., floodwater mosquitoes) because large numbers of artificially reared individuals would have to be maintained at all times (Subcommittee on Insect Pests, 1969).

It is also important that the means used for sterilizing the insects do not interfere to any great extent with their ability to compete with wild individuals for a mate, longevity, vigor, or normal behavioral patterns. Various types of radiation, chemicals, and genetic manipulation have been tried as means of producing sterilization. Since the screwworm-eradication campaign, the use of the sterile-release method against other species has been studied and its application shows promise in some instances. For example, in a pilot program using the sterile-release method in conjunction with a bait spray, the melon fly, *Dacus curcurbitae*, was eradicated from the island of Rota in the South Pacific (Subcommittee on Insect Pests, 1969).

Several chemicals have been discovered in recent years that produce sterility when ingested by insects. Out of approximately 6000 compounds screened, more than 300 show promise as chemosterilants (Subcommittee on Insect Pests, 1969). These chemicals produce sterility in three major ways: (1) they cause insects to fail to produce sperm or ova, (2) they cause the death of sperm or ova after they have been produced, or (3) they produce genetic defects in spermatozoa which prevent zygote development (Subcommittee on Insect Pests, 1969). The last of these mechanisms for the production of sterility is the most desirable because the sterilized males are generally competitive with the unsterilized males in mating with available females.

Chemosterilants are useful in situations where the sterile-release method would be inappropriate (e.g., against species that occur in very large populations and are difficult or impossible to rear in large numbers in the laboratory). Since the best chemosterilants so far known must be ingested to be effective, they are usually applied with baits. Chemosterilants that are effective by contact might be used in association with luring stimuli such as light and sex attractants. Another possibility is the application of a chemosterilant to breeding places. However, since all the promising chemosterilants are mutagenic agents, they present a hazard to other animals, including man.

One of the principal advantages of the use of chemosterilants is that both males and females are rendered sterile. Hence there is an increase in the frequency of matings between sterilized and unsterilized individuals with a resultant decrease in the production of offspring.

In addition to the use of genetic manipulations to produce sterility mentioned above, other genetic approaches to insect control include the use of genetic factors such as the following: "(1) sex-ratio distorters, (2) detrimental genes incorporated into chromosomes that have meiotic drive, (3) chromosome translocations in heterozygous males, and (4) conditional lethal genes that allow the parents to survive in the laboratory but are lethal to their descendants under field conditions" (World Health Organization, 1967).

Resistant Hosts. Under natural conditions, some individuals of a given species possess characteristics that make them more able than others to cope with various environmental stresses (e.g., insect attack). Those individuals which possess such characteristics will tend to be more successful in reproducing and hence passing the hereditary determinants of these characteristics to their progeny. For this reason, over time a population exposed for many generations to a given stress will become composed predominantly of members that are able to cope with this stress. This, in a very general way, is the process of natural selection. In a number of instances, man has applied this process artificially in the laboratory to produce strains or varieties of organisms which are "resistant" or more resistant than members of other strains or varieties to an environmental stress. Another approach has been to in some way encourage the proliferation of some naturally occurring "resistant" strain or variety.

We shall define resistance as the heritable ability of an individual, strain, variety, race, and so on (plant or animal) to repel or withstand the effects of some environmental stress or stresses to a greater degree than other individuals or groups. In the context of our discussion insect pests are obviously "environmental stresses."

A great deal more emphasis has been placed on the studies of insect resistance in plants than in animals. This fact is due at least in part to greater use and lower costs involved in plant breeding versus animal breeding. The search for both resistant plants and animals has unfortunately been neglected with the advent of the more efficient pesticides. Painter (1958) describes three possible components of resistance: (1) nonpreference, (2) antibiosis, and (3) tolerance. Nonpreference is explainable in terms of the host plants lacking a quality (or enough of that quality) which is necessary to attract a given insect to oviposit, feed, or seek shelter. Antibiosis describes the situation in which a resistant plant may have an adverse effect on the biology of an insect. The last component, tolerance, refers to the ability of resistant plants to survive levels of infestation to which susceptible plants would succumb or be severely damaged. Since all these components are under the control of one or more genetic factors, they may occur alone or in combination in a given individual. Obviously, a plant exhibiting a high degree of each would be most desirable. Painter (1958) and Beck (1965) elaborate on the ideas presented here and offer several examples.

Resistance on the part of animals to pest insects may also be considered to involve the abovementioned components (Subcommittee on Insect Pests, 1969).

Environmental Control

Environmental control procedures involve the removal, destruction, modification, or isolation of materials which could potentially favor the survival of an insect pest by affording food or making a site suitable for breeding and/or hibernation.

Removal and Disposal of Material.   One of the most effective ways of controlling many insect pests is the maintenance of high standards of sanitation. Accumulations of trash, garbage, and untreated sewage provide food and breeding sites for numerous annoying and disease-carrying insects (not to mention rats). Thus it is of critical importance that these materials be removed from their source of accumulation and be properly treated and disposed of. Garbage disposals, trash and garbage collection services, sewage collection systems and treatment plants, sanitary land fills, and incineration are among the methods involved in the maintenance of sanitation. Garbage provides a feeding and breeding medium for numerous annoying, disease-carrying insect pests, particularly domestic flies (house flies, blow flies, and so on). Trash materials (e.g., discarded cans, bottles, various other containers, and old tires) may serve as hiding places for many insects and if they are holding water may afford excellent breeding sites for mosquitoes. Raw sewage as well as garbage, in addition to providing an excellent breeding medium for numerous pests, may also be a source of disease organisms transmitted by insects.

Crop rotation (i.e., varying the kind of crops planted in different seasons on a given plot of land), in addition to its beneficial effects on the soil, may also be responsible for the reduction of insects which attack one of the crops being rotated but not the other(s). For example, May beetle larvae (grubs) which feed on the roots of many crops do poorly on certain leguminous crops. Hence the use of these crops in rotation has the effect of reducing the grubs (Subcommittee on Insect Pests, 1969).

Destruction or Modification of Material.   In many instances it is not practical to remove potential insect breeding material. Under these circumstances the material is destroyed or modified at the site of its accumulation. Destruction of crop refuse by burning, turning it under, or allowing farm animals acess to it is an excellent way of destroying the insects harbored by it. Thus the boll weevil, *Anthonomus grandis*, the European corn borer, *Ostrinia nubilalis*, and others are controlled in part by the destruction of infested plants after harvesting. Destruction or removal of slash following cutting operations in a forest are appropriate means for controlling numerous insects that attack trees. The borders of fields used for cropping may contain plants that serve as alternative hosts for pest species or which harbor plant-disease organisms. Cleaning out these borders may aid considerably in the control of certain insects. On the other hand, vegetation on field borders may provide a suitable habitat for beneficial parasitic, predatory, or pollinating insects, and this should be carefully considered before destruction is carried out. Plants that serve as alternative hosts may also grow in uncultivated fields, and in some instances it is appropriate to destroy them. Various methods of tillage have been used at appropriate times to kill susceptible stages of insects in the soil. For example, fall plowing is effective against the pupae of the corn earworm, which overwinters in the soil. Various crop management and treatment practices, such as planting soil-building crops, strip cropping, thinning, topping, preening, and defoliating, have been used in particular instances at least partially for the purpose of insect control.

Naturally or artificially produced aquatic or semiaquatic habitats can produce severe insect problems, particularly in relation to the production of mosquitoes, biting midges, and horse and deer flies. Examples of naturally occurring aquatic or semiaquatic habitats that provide suitable conditions for the breeding of these insects are salt- and freshwater marshes, swamps, and various depressions in which water can collect. Artificially created habitats include the creation of water impoundments for a variety of reasons (e.g., flood control, hydroelectric power, water storage, drinking water for animals, recreational purposes, borrow pits), and the development of water-distribution systems, particularly as they pertain to irrigation. Marshes, swamps, and depressions containing water may be either flooded or drained, depending on a variety of factors.

Along the east coast of Florida, many of the salt marshes, which produce large numbers of salt-marsh mosquitoes, *Aedes taeniorhynchus* and *A. sollicitans*, and biting midges, particularly *Culicoides furens*, have been impounded, a measure that has provided successful control. Careful management of artificially impounded water and water-distribution systems is necessary to avoid insect problems. Practices include (1) proper construction so that water does not stand in any one place for any longer than necessary (e.g., proper land grading for irrigation purposes, appropriate drain construction); (2) deepening or filling very shallow areas; (3) periodical removal of vegetation within the water-level fluctuation boundaries; and (4) periodic removal of vegetation and accumulated debris from drains to avoid clogging, seepage, and overflow.

Isolation of Breeding Material. In many instances, it is inappropriate to remove, destroy, or modify potential breeding material, and hence the only alternative is to isolate it from potential insect pests. This isolation may be temporal or spatial (Rolston and McCoy, 1966). Temporal isolation involves the timing of any practice (e.g., plant harvesting, shearing, dehorning) such that problems with insect attack will be minimal. A classic example is the sowing of winter wheat on "fly-free" dates to avoid infestation by the Hessian fly, *Mayetiola destructor*. The adult flies that have developed on summer wheat live only a few days, and if the winter wheat is planted on a fly-free date, all the flies will be dead by the time it comes through the ground. Fly-free dates, of course, vary with latitude. Dehorning and castration of cattle are done during cool periods of the year to avoid the attacks of screwworm flies. A number of procedures may logically be looked upon as providing spatial isolation of breeding materials. These include (1) planting a given crop as far as possible from an earlier infested one of the same kind; (2) the use of trap crops and trap logs which are intended to divert pests from the main crop; (3) appropriate pasturing of domestic animals to avoid pests [e.g., the late summer use of upland pastures in the western United States to avoid the attacks of tabanid flies, which breed in lowland pastures where there are small streams and abundant trees (Subcommittee on Insect Pests, 1969)]; and (4) the use of properly designed and constructed storage boxes, bins, rooms, and so on, in which cool, dry conditions are usually ideal for the inhibition of pest species of insects.

Chemical Control

Under the heading "chemical control" we shall include those natural or synthetic chemicals that cause directly the death, repulsion, or attraction of insects.

Insecticides. The term *insecticide* means "to kill an insect." Thus the broadest possible definition would be "anything used to kill an insect." We prefer to define an insecticide as a substance or a mixture of substances employed for the purpose of killing insects and related arthropods. The term *pesticide* is more inclusive, referring to insecticides as well as herbicides, rodenticides, and other substances.

Among the advantages afforded by insecticides are the following: (1) they are usually very effective, (2) they generally act within a short period of time, (3) they are effective when applied against large pest populations, and (4) they are readily available for use when needed (Subcommittee on Insect Pests, 1969). The use of insecticides has played an important role in the development of agriculture as we know it today. Both increased yield per unit land area or animal and improved quality of agricultural products are in large part attributable to the successful use of insecticides. Agricultural yield has shown a 54% increase in the 20 years following the general use of synthetic organic insecticides (Subcommittee on Insect Pests, 1969). Undoubtedly other factors have played a role in this impressive increase, but application of insecticides is certainly among the most important. Insecticides have also enabled man to bring several dreaded diseases under control or at least reduce their extent.

On the negative side, insecticides often pose a threat to nontarget organisms, including man. The application of insecticides is commonly very hazardous, and several instances are known of persons who have accidentally come into direct contact with a highly toxic insecticide and have become critically ill and on several occasions succumbed as a result. This is why the careful reading and following of instructions on the labels on pesticide containers and constant concern with safety during their application are so important. Other important considerations are (1) storage of pesticides in well-labeled containers out of reach of children; (2) extreme care in application, including avoidance of spillage on clothing or skin, or inhalation of sprays or dusts; (3) not smoking or eating when working with toxicants; and (4) proper disposal of empty insecticide containers.

Another matter of concern is that of residues in food products, plant and animal, which have been treated with pesticides at some point in their production. Every pesticide that is shipped in interstate commerce in the United States must be registered under the Federal Insecticide, Fungicide, and Rodenticide Act. Presently, about 400 insecticides are registered (Subcommittee on Insect Pests, 1969). The objective of registration is the protection of the consumer, not only in terms of safety, but also with regard to the value of the products. When an insecticide is registered, the amount of residue (tolerance) to be allowed in various food products is established. Some insecticides have such low mammalian toxicities that they are tolerance exempt, whereas others, which are considered dangerous following an extensive series of tests, are subject to a rigid tolerance. If these tolerance levels are exceeded, various degrees of legal action against the violators result. In addition to the strict controls applied by the federal and state governments, the development of a new insecticide is a rigorous, time-consuming process which increases the probability that an insecticide which reaches the market will be effective and reasonably safe if directions are followed carefully. The Subcommittee on Insect Pests (1969) presents an outline (Table 12–1) of the basic steps involved in the development of a new insecticide.

Despite the hazards involved, the safety record of pesticide usage in the United States is not as bad as one might suspect:

In recent years there has been so much publicity about pesticide hazards that much of the general public has developed a distorted impression of the magnitude of the problem. Vital statistics show that pesticides account for only 1 out of every 700 accidental deaths and only 5% of all poison deaths in the United States. As a cause of death, pesticides are far outranked by common drugs and by household agents such as cleaners, polishes, and solvents. For many years, the ubiquitous aspirin tablet has caused about the same number of accidental deaths annually in the United States as all the pesticides combined, and since 1957 aspirin has caused slightly more accidental deaths. As a matter of record, the death rate for all pesticides has remained constant at slightly over one per million of population for the past 25 years.— (Subcommittee on Pest Insects, 1969).

The toxicity of an insecticide is usually expressed in terms of LD_{50} (lethal dosage), which refers to the dosage (milligrams of insecticide per kilogram of body weight of test animal) that is lethal to 50% of a test population. For example, if the LD_{50} dosage for

Table 12–1
Stages in the Development of a New Insecticide

Stage I	Stage II	Stage III	Stage IV	Stage V	Stage VI
Exploration Testing	**Characterization of Performance**	**Advanced Evaluation Tests**	**Extended Field Trials**	**Semicommercial Field Tests**	**Promotion and Sales**
1. Conduct synthesis of new compounds	1. Continue synthesis of related compounds	1. Develop a practical method of synthesis	1. Perfect synthesis methods	1. Conduct large-scale field tests	1. Initiate wide-scale sales activities: public relations, advertising, sales-training schools, technical service
2. Conduct screening tests for biological activity	2. Make laboratory insecticidal evaluation	2. Conduct small-plot field tests	2. Submit product to experiment stations for field testing	2. Obtain an experimental permit for trial sales if desired	
3. Conduct insecti-cidal screening tests	3. Conduct range-finding toxicity tests, such as acute oral skin, eye, skin absorption, and inhalation if necessary	3. Conduct 30- to 90-day toxicity feeding studies	3. Initiate long-term toxicity feeding studies on rats and dogs, and if necessary a three-generation reproductive study	3. Construct production plant	
	4. Improve screening formulations	4. Initiate metabolism studies if required	4. Perfect residue-analysis methods and initiate residue determinations	4. Complete market-potential survey	
	5. Initiate compat-ibility studies	5. Develop analytical procedures	5. Conduct stability, corrosion, and packaging studies	5. Apply for federal and state label registrations	
	6. File for patent coverage	6. Determine stability of formulations	6. Determine effects on wildlife		
		7. Make preliminary market analysis	7. Construct pilot production plant		

a From Subcommittee on Insect Pests (1969).

laboratory rats was given to each individual in a population of these rats, one would expect about 50% of the exposed animals to die. The remaining 50% are less susceptible and larger doses of insecticide would be required to kill them. Theoretically, in a randomly distributed population, the relationship between insecticide (or other toxicant) dosage and percent mortality follows a bell-shaped curve (normal distribution). LD_{50}'s vary depending on the route of entry of the insecticide into a test animal. Thus LD_{50}'s are commonly determined for both the oral and dermal routes. Readers interested in more detailed information regarding insecticides should consult Shepard (1951).

Environmental pollution with insecticides has become a matter of great concern. Highly residual insecticides can pass well beyond their intended targets and may reduce populations of beneficial insects and wildlife. DDT in particular has been attacked in this regard. Two major factors have contributed to its being a problem: (1) it is highly residual, and (2) it becomes concentrated in a stepwise fashion along a food chain based on the simple idea that animals eat many times their own weight in food. Woodwell, Malcolm, and Whittaker (1969) provide a graphic description of this process of "biological concentration":

. . . in the food web in which the herring gull is a scavenger in Lake Michigan, DDT (DDE and DDD) concentrations in the bottom muds at 33–96 feet averaged 0.014 parts per million. In a shrimp *(Pontoporeia affinis)* they were 0.44 ppm, more than ten times higher. Levels increased in fish to the range of a few ppm (3.3 alewife; 4.5 chub; 5.6 whitefish), another tenfold increase, and jumped in the scavenging, omnivorous herring gull to 98.8 ppm, twenty times higher still, and 7000 times as high as in the mud.

Unfortunately, such accumulations of insecticides kill or stop the reproduction of certain animals, birds in particular. Woodwell et al, (1969) point out that many important predatory birds, including the bald eagle, are in danger of becoming extinct in the United States. Although concern for these animals is justified in itself, of far greater concern is the long-term indirect effects pesticide residues may have by disrupting the intricate balance of ecosystems. Rachel Carson in her book *Silent Spring* painted a particularly vivid and frightening picture of what might happen. Graham (1970) discusses what has occurred in this area since the publication of *Silent Spring*.

Insecticides were apparently used before recorded history. The writings of the Greeks, Romans, and Chinese all contain allusions to the use of various substances, such as sulfur, hellebore (a poisonous herb), and arsenic. Prior to the 1940s the insecticidal value of a number of inorganic (e.g., arsenic, mercuric chloride, and carbon disulfide) and natural organic chemicals (e.g., pyrethrum, cube, and nicotine) was known and put to extensive use. The discovery of DDT by Paul Muller in Europe in 1939 revolutionized insect control and marked the beginning of the development and application of synthetic organic insecticides. In 1948 Dr. Muller was awarded the Nobel Prize for medicine for his discovery. Since that time hundreds of compounds of varying insecticidal value have been discovered and thousands of new potential toxicants are being evaluated each year by a long and arduous screening process. The vast majority (about 90%) of currently used pesticides are of the synthetic organic variety.

Insecticides may be classified in different ways. Some are more useful than others for a given stage in the insect life cycle. Thus there are ovicides, larvicides, and adulticides. A useful classification may be developed by using a combination of the primary route or mode of entry and chemical nature of the various insecticides. With regard to the route of entry, they can be grouped into *stomach poisons*, which act via ingestion and absorption from the alimentary canal; *contact poisons*, which are readily absorbed through the cuticle; and *fumigants*, which enter through the spiracles and tracheal system or body wall in the gaseous state. Within each of these categories are a number of different insecticides. These categories are by no means completely exclusive (e.g., many contact insecticides also act quite effectively via the oral route). In the following paragraphs we shall consider many of these. A useful and up-to-date list of commercial and experimental insecticides has been prepared by Kenaga and Allison (1969).

Stomach poisons are useful against insects with chewing mouthparts which ingest the toxicants with their food or when they groom their body parts. They are mostly inorganic compounds and have been used more in the past than they are at present. However, with the development of resistance to synthetic organic insecticides in populations of a number of species, several of these inorganic stomach poisons are coming back into use (e.g., Paris green and sodium fluoride). Paris green has been used very effectively against DDT-resistant mosquito larvae. The arsenicals (arsenic trioxide, Paris green, lead arsenate, sodium arsenite, and so on), fluorine compounds (cryolite, sodium fluorosilicate, sodium fluoride, and so on), boron compounds (sodium tetraborate or borax, and boric acid), phosphorus compounds (yellow phosphorus and zinc phosphide), tartar emetic (an antimony compound), thallium sulfate, and others are examples of stomach poisons that have been applied against a wide variety of insect pests. Most of these form persistent residues and are applied to the natural food plants of pest species, mixed with an attractive substance (poisoned bait) or applied to a surface where pest insects are likely to contaminate their legs and other body parts, later to ingest the poison during cleaning and grooming activities.

A group of synthetic organic insecticides, *systemics*, which are readily absorbed and distributed to the various tissues of plants or animals, thereby rendering them toxic to insects which feed upon them, are appropriately included in the stomach-poison category. This group is, of course, effective against insects with piercing-sucking mouthparts which are able to penetrate host tissues (sucking lice, thrips, aphids) and ingest blood, sap, and so on, as well as those chewing insects which might feed on the same host. In addition, internal parasites of animals (e.g., cattle grubs and screwworm larvae) are susceptible to systemic insecticides. Systemics of plants may be applied to the soil, seeds, foliage, bark of trees, or injected into the inner tissues of woody plants. They are then translocated to the tissues remote from the site of application. Animal systemics are applied as surface sprays which are absorbed dermàlly; as boluses (large pills), which animals are induced to swallow; or in food, water, or salt. Fortunately, these insecticides are gradually broken down by enzymes and after a period of time the animals can be safely used for milk production or slaughtered for meat (Metcalf, Flint, and Metcalf, 1962).

As mentioned previously, contact poisons kill by being absorbed directly through the cuticle. For our purposes, we will divide them into the following three groups: (1) inorganic, (2) natural organic, and (3) synthetic organic. Inorganic contact insecticides include sulfur, mercury, and copper compounds which have been used in killing mites and fungi as well as insects. Many have been replaced today by the synthetic organic compounds. The finely divided silica compounds (e.g., silica aerogel) may also be considered as a part of the inorganic category. However, they are not absorbed into the body, but have their effect by abrading or otherwise disrupting the water-proofing wax layer of the epicuticle, thus causing the subsequent desiccation of an insect. These materials have the advantage of being nontoxic to man and other warm-blooded animals.

The natural organic group of contact insecticides includes the oils and the botanicals. Although oils are used mostly as solvents for insecticides, many have an intrinsic toxicity for insects. For example, kerosene and certain other oils have been long used as mosquito larvicides. The botanicals are as a group probably the safest insecticides. They are prepared either by extraction of active ingredients or by being dried and ground and applied as dusts. Pyrethrum and nicotine are probably the two best-known botanicals. Pyrethrum is prepared from ground-up pyrethrum flowers and has been used as an insecticide for hundreds of years. It is widely used in many "bug bombs" (canned aerosols) because of its rapid knockdown effect on insects and low mammalian toxicity, two features that make it particularly attractive for household use. Nicotine is prepared from tobacco plants. It is no longer widely used. Other botanicals include allethrin, rotenone, ryania, and sabidilla. According to Pratt and Littig (1962), allethrin is probably one of the safest insecticides known.

Since the discovery of DDT in 1939, synthetic organic contact insecticides have become the most widely used group. This group includes the organic thiocyanates, dinitrophenols, several sulfur compounds, chlorinated hydrocarbons, organophosphates, carbamates, and others. The dinitrophenols act to increase oxygen uptake, enhance oxidative metabolism, interfere with the production of ATP, and quickly cause muscle paralysis (Subcommittee on Insect Pests, 1969). The remaining groups act on the nervous systems of insects in a variety of different ways. For example, the organophosphates, carbamates, and organic sulfur compounds are cholinesterase inhibitors. Cholinesterase is an enzyme that breaks down the neurotransmitter acetylcholine, which is liberated at synapses of neurons in response to an impulse. When cholinesterase is blocked by an inhibitor, acetylcholine accumulates, and as a result the insect becomes hyperactive and eventually paralyzed. The synthetic organic insecticides that are probably the most dangerous to man in terms of deaths caused are certain of the chlorinated hydrocarbons and organophosphates. O'Brien (1966) discusses in considerable detail the mode of action of several synthetic organic insecticides as well as botanicals.

Fumigants enter insects via the body wall and tracheal system in the gaseous state. These insecticides are used in situations where the pest insects can be contained in an essentially airtight structure for a period of time. This closed container could range from a small plastic bag to an entire building. Some fumigants, in addition to being toxic to man, are highly flammable (e.g., carbon disulfide) and must be used mixed with nonflammable gases such as carbon dioxide or carbon tetrachloride. Fumigants may be initially in the solid, liquid, or gaseous state. Naphthalene and paradichlorobenzene (PDB) are examples of solids that give off toxic and repellent gases. The latter is the most important, being widely used to protect clothing and insect collections from pest species of insects. Other fumigants include hydrogen cyanide, methyl bromide, carbon disulfide, and carbon tetrachloride. The first two mentioned are commonly used for the control of insects of public health importance and household pests. The others are used mainly in the fumigation of stored foods.

The ideal insecticide would be a highly specific one, killing only the target pest species and leaving the pest's predators and parasites and other beneficial insects unharmed. Unfortunately, to date no such pesticide is known. The systemic insecticides probably come the closest to being specific, in that insects that feed on the plants or animals which contain them are affected by the poison, whereas most other animals in the environment are not even exposed. However, the parasites and predators that feed directly upon the pest species are likely to be poisoned when taking a meal. The biochemical systems of insects (or most animals) are remarkably similar; hence most poisons, whatever their particular mode of action, tend to be toxic to a broad spectrum of animals. Whatever degree of specificity a given insecticide may show depends, at least in part, on its mode and ease of entry. For instance, a sucking insect may fail to ingest a lethal dose of a stomach poison on the surface of foliage since it feeds beneath the surface. Insects with very thick cuticles may resist the absorption of some contact poisons. Another factor that may account for some specificity is resistance to a given insecticide. For example, some insects (e.g., the tomato hornworm, *Manduca quinquemaculata*) are unable to detoxify DDT, whereas other species (e.g., the tobacco hornworm, *Manduca sexta; Melanoplus* grasshoppers; the reduviid bug, *Triatoma protracta*) are able to detoxify this insecticide and hence have a degree of tolerance to it (Subcommittee on Insect Pests, 1969).

Substances used in combination with technical-grade insecticides (the purest commercial forms) which serve as carriers, solvents, wetting agents, and so on, may conveniently be referred to as *adjuvants* since they are used to complement, in some way, the action of an insecticide. Substances such as adjuvants include dusts, granules, solvents, emulsifiers, wetting and spreading agents, adhesives, synergists, and a number of other materials. The mixture of an insecticide and one or more of these adjuvants is called a *formulation*. Dusts are finely divided materials, somewhere in the range 1 to 40 microns in diameter (Fronk, 1962a), which are used as inert carriers and diluents for insecticides when it is necessary or appropriate to apply them in a dry state. Each minute dust particle becomes coated with insecticide, and generally there is a direct relationship between toxicity and particle size. In some instances the insecticides themselves (e.g., sulfur) are ground into dust-sized particles and may be diluted with inert dusts of various types. A wide variety of materials have been used as inert carriers and diluents [e.g., the minerals pyrophyllite (a hydrous aluminum silicate) and talc (a hydrous magnesium silicate) and organic substances such as walnut shell flour]. Granules are similar to dusts except that they are larger in particle size.

Solvents are used to dissolve the basic insecticide and hence serve both as diluents and carriers. Most of the modern synthetic organic pesticides are insoluble in water, and organic solvents such as fuel oil or kerosene are used. However, many organic insecticides can be dissolved in an organic solvent and this solution then diluted with water in the form of an emulsion. This is made possible by the use of an emulsifying agent. Concentrations of insecticide plus organic solvent plus emulsifying agent (emulsifiable concentrates) are the most common versatile formulations (Fronk, 1962a). Wettable powders are similar to emulsifiable concentrates in that they contain a wetting and spreading agent which allows them to be suspended in water and applied as sprays. However, instead of the insecticide being in solution in an organic solvent, it is in the form of a dust.

Synergists are substances that when used with an insecticide cause the mixture to be more toxic than would be expected based on a simple additive effect of the toxicities of the individual components. A good example of this synergistic action is that between piperonyl butoxide (synergist) and pyrethrum. This is a fortunate relationship since the combination of the two is cheaper than would be an equally effective straight pyrethrum formulation (Rolston and McCoy, 1966). Insecticides may also be mixed with attractive baits (poisoned baits), perfumes to nullify some of the disagreeable odors characteristic of certain pesticides, rug shampoos, animal shampoos, fertilizers, and so on. Certain insect toxicants are sometimes prepared as a pill (bolus), which is introduced into the alimentary tracts of animals such as those infested with maggots. Insecticides may also be applied in the form of sprays made up of extremely minute particles (droplet size between 0.1 and 50 microns in diameter; Pratt and Littig, 1962). These sprays are referred to as *aerosols* and are particularly valuable in space spraying (e.g., outdoors against mosquitoes and other flying insects). Smokes are similar to aerosols both in their properties and application, but they are comprised of much smaller particles of insecticide. They are produced by burning or evaporating insecticides. Smokes are not considered to be particularly effective.

A wide variety of mechanical devices (sprayers, aerosol generators, dusters, and so on) exists for the application of insecticides. These devices differ extensively depending upon the type of application for which they are intended and under what circumstances they are used.

The reader desiring more information on this topic should consult Scott and Littig (1960), the report of the Subcommittee on Insect Pests (1969), and the various manufacturers that design and produce insecticide application equipment.

One of the difficulties with the use of many pesticides in insect control has been the development of resistance in a large number of species. According to Hoskins (1963), the World Health Organization defines the phenomenon of insecticide resistance as follows: "the development of an ability in a strain of insects to tolerate doses of toxicants which would prove lethal to a majority of individuals in a normal population of the same species." In more practical terms Hoskins (1963) defines insecticide resistance as "the failure of the customary programs to give practical control, usually even when the dosage was raised and/or frequency of application increased." The failure of a given dosage which previously has been effective in control is due to selection of the individuals in a population which possess the ability to survive this dosage. The traits that enable certain individuals of this normal population to be resistant were present before the introduction of the insecticide; the application of the insecticide merely enabled the already resistant individuals to have a selective (reproductive) advantage relative to the nonresistant ones. In the most common situation the resistant insect possesses a detoxication mechanism which enables it to render a particular insecticide or group of insecticides nontoxic. For example, resistant house flies, *Musca domestica*, produce an enzyme, DDT-dehydrochlorinase, which catalyzes the breakdown of DDT to HCl and DDE, which is nontoxic (Subcommittee on Insect Pests, 1969). *Culex tarsalis*, a mosquito, is able to detoxify the organophosphate insecticide malathion by means of a carboxyesterase enzyme.

Strains of some species have been observed to possess a "behavioral" resistance to certain pesticides. According to Hoskins (1963): "The chief example is a hyperirritability which causes insects to leave a treated surface before picking up a lethal dose. Thus female *Anopheles albimanus* mosquitoes were observed to survive DDT house spraying in certain districts of Panama in much greater numbers than in former years. When confined over treated paper they quickly left the paper and showed signs of restlessness to a greater degree than did mosquitoes from other districts." Another mechanism of resistance which has been discovered is resistant individuals whose cuticles or gut walls are such that they allow much slower penetration of toxicant than the nonresistant individuals. Hoskins (1963) calls this "structural resistance" and cites German cockroaches as examples of insect strains that are resistant to the carbamate Sevin as a result of the slow penetration of the insecticide through the integument.

The first known case of insecticide resistance in the United States was in 1908, when it was found that strains of the San Jose scale, *Aspidiotus perniciosus*, were resistant to lime–sulfur sprays (Subcommittee on Insect Pests, 1969). In 1946, DDT-resistant house flies were first discovered in Sweden. Prior to that time only 9 species of insects and ticks were known to have developed insecticide resistance (Brown, 1968). Resistance in these cases was, of course, to inorganic and botanical insecticides. Since that time insecticide resistance has developed in strains of 224 species of insects and mites and ticks. Of these, 97 are of public health and veterinary importance, and 127 are pests of stored products or field or forest crops (Brown, 1968). One of the means that have been used in attempts to circumvent the resistance problem has been to replace an insecticide to which resistance has developed with one to which it has not. The alternative pesticide has been in a different chemical group since some cross resistance has been detected in certain pesticides that are chemically closely related. This has been successful in some instances but a failure in those in which resistance to both the original and alternate insecticide has developed. The use of insecticides in conjunction with other control procedures (e.g., biological and cultural) in an integrated program has much promise.

Repellents. Repellents may be defined as substances which are mildly toxic or nontoxic to pests, but which prevent damage to plants or animals by causing the pests to make oriented movements away from the source. Various methods for repelling insects have been known for some time. In very early times, smoke from wood fires was used to keep away biting and annoying insects. As early as 1897, oil of citronella was used as a mosquito repellent (Shepard, 1951). Most of the earlier repellent substances were quite odorous and perhaps somewhat repellent to humans as well as insects. However, many of the more modern, synthetic repellents have little, if any, disagreeable odor.

To be appropriate as an insect repellent a substance has to be more than merely repellent to insects. It must be essentially nontoxic, non-irritating, and nonallergenic to man or domestic animals; inoffensive in odor; harmless to fabrics; persistent—not easily removed by perspiration, rubbing, laundering, and so on; and effective against a broad spectrum of pest species. In addition, a good repellent should be cheap and not damaging to plastics, painted surfaces, and the like.

The use of good repellents has a number of advantages: (1) They afford individual protection from insects without the necessity for expensive and time-consuming population eradication; (2) they do not damage or kill beneficial animals or plants; and (3) the ones that are available for use are nontoxic to man. On the other hand, disadvantages include the following: (1) repellents are at best a temporary measure (a few hours at most) and tend to evaporate from or rub off skin or clothing due to perspiration and the like; (2) they commonly have an oily feel and may have a somewhat disagreeable odor; (3) they must be applied in comparatively large doses (in the range of 20 to 40 mg/cm^2 of skin; Smith, 1966); and (4) they may damage certain plastics or painted surfaces.

So far, repellents have been primarily used for the protection of man and animals against attacks from bloodsucking or otherwise annoying insects. The armed forces have over the years had a particular interest in repellents, realizing that protection of troops from disease vectors would greatly increase their military advantage. However, considerable interest has also developed in the use of repellents in plant protection. Metcalf, Flint, and Metcalf (1962) describe three general groups of repellents: those used against (1) crawling insects, (2) the egg laying of insects, and (3) the feeding of insects. Repellents used against crawling insects usually consist of a repellent barrier interposed between an insect and whatever material happens to be attractive to it. For example, creosote has been used as a barrier against the migration of chinch bugs, *Blissus leucopterus,* and trichlorobenzene and other repellent insecticidal chemicals to protect buildings from termite invasion. Creosotes derived from coal and wood tar have been used extensively for the protection of wood against termites, powder post beetles, and rot organisms (Shepard, 1951). Creosote and other oils apparently ". . . smother the natural attraction of the insect to its food or oviposition site" (Shepard, 1951). Several chemicals have been found which are reasonably effective in repelling insects from feeding. Washes containing bordeaux, lime, and other materials are used to repel leafhoppers and some chewing insects, and inert dusts have been useful on cucurbits to repel cucumber beetles (Metcalf, Flint, and Metcalf, 1962). An ideal repellent for plant protection would be one that would somehow block the natural attractants to which pest species respond. Diethyl toluamide, considered to be one of the best general repellents yet discovered, dimethyl phthalate, ethyl hexanediol, dimethyl carbate, powdered sulfur, and many others are examples of repellent chemicals that have been applied to the skin and/or clothing of man.

When a broad spectrum of effectiveness is desired, a mixture of repellents may be appropriate. For example, according to Rolston and McCoy (1966): "The Armed Forces used a repellent containing dimethyl phthalate, ethylhexanediol, and dimethyl carbate for application to skin, and a repellent containing benzyl benzoate, *n*-butylacetanilide, and 2-butyl-2-ethyl-1,3-propanediol for application to clothing. Both mixtures provided protection against medically important mosquitoes, fleas, ticks, and chiggers." Smokes, smudges, and burning of pyrethrum (insecticidal and repellent) are useful repellent measures for outdoors. Chemicals that have been used as repellents against pests of livestock include low concentrations of pyrethrums, butoxypolypropylene glycol, and dibutyl succinate.

In recent years there has been interest in the development of a systemic insect repellent (i.e., one that would be taken orally and following ingestion appear in the skin or skin secretions). Ideally such a systemic would be repellent to a wide variety of arthropods, be of low toxicity, and be effective for a long period of time (12 to 24 hours or more after ingestion; Sherman, 1966). There have been repeated failures in the search for a workable systemic repellent, but research is still being conducted along these lines (Subcommittee on Insect Pests, 1969).

Attractants.   Substances that are strongly attractive are known to be involved in a wide variety of insect activities, ranging from feeding, mating, and oviposition to assembling of large groups. These activities were discussed in Chapter 7 and appropriate references cited. Attractants have a great potential in the area of insect control and can be useful from several standpoints. For an attractant to be useful in these regards, the insect behavior with which it is associated must be understood, it must be extracted and chemically identified, and if possible it should be synthesized. Ideally the synthesized attractant or an analogue would be even more attractive to an insect than the original. Another avenue of approach is to test various chemicals for attractiveness.

As mentioned above, attractants have a variety of possible uses. They may be of considerable value when used in traps in detecting and sampling insect populations to monitor various changes, tracing the movement of marked insects in dispersion and migration studies, basic studies in insect behavior, and in insect control. In surveys, the use of a specific lure in a trap allows selective trapping and hence makes the work easier and much more efficient. The design, color, and location of traps are also important factors. The sex attractant of the gypsy moth, *Porthetria dispar*, is a good example of an attractant that is of value in survey work. Male moths fly into the wind and are attracted to the airborne scent of the nonflying females (Jacobson, 1965). Traps currently used are made of paper and contain a potent synthetic attractant for the gypsy moth, gyplure. These traps contain dental-roll wicks impregnated with gyplure and a liner coated with Tanglefoot (a sticky substance in which insects become mired down). They are hung on tree limbs. According to Jacobson, "Each year immediately before the male moth flight season (July–August), approximately 50,000 traps are placed in the infested New England area." He describes the potency of gyplure: ". . . a single pound of the attractant, which may be made quite inexpensively, is sufficient to last for more than 300 years if used for survey alone."

There are several applications for attractants in insect control. They can be used to attract insects to traps that alone might be effective in reducing some populations. In addition, the traps may contain insecticides, pathogens, or chemosterilants, which further increase the control potential. The use of an attractant with an insecticide avoids the necessity for complete spray coverage and may still be very effective. For example, "The oriental fruit fly, *Dacus dorsalis* Hendel, was eradicated from the Pacific island of Rota by aerial distribution of 5×5 cm fiberboard squares saturated with a bait containing an insecticide, naled, and an attractant, methyleugenol" (Subcommittee on Insect pests, 1969). The methyleugenol is a strong sex attractant for males and its use with an insecticide in eradication of the oriental fruit fly is an example of the male annihilation technique, the basis of which is that "each male removed from the wild fly population by an attractant would represent one unmated female," the assumption being that the two sexes are present in equal numbers and that they mate only once (Christenson, 1963). Although not as yet demonstrated in field studies, the use of attractants with pathogens could theoretically increase the effectiveness of the pathogen by ensuring maximum contact with their insectan hosts (Subcommittee on Insect Pests, 1969). The bases for the use of chemosterilants in insect control were discussed earlier.

Besides luring insects to their sterilization or death by insecticides or pathogens, attractants have possibilities in the perversion of normal insect behavior such that they are "tricked" into destroying themselves. Examples might consist of the following:

(1) treatment of weeds and other undesirable plants with insect attractants, feeding stimulants, and oviposition stimulants to create susceptibility to an insect. Materials need to be highly active in order that the treated plants compete with normal host plants in nature. (2) Treatment of host plants to induce greater susceptibility for purposes of luring insects to specifically treated portions of a crop. This method presents greater possibilities than treatment of weeds, because a preference which already exists may be increased with less effort than that needed to create a preferred host from a nonpreferred plant. (3) Use of chemicals to distort sexual activity, diverting the males or females in their search for mates, or confusing their orientation mechanisms.—(Subcommittee on Insect Pests, 1969)

Other Chemical Controls. Substances that may be considered as potential chemical control agents include antimetabolites, feeding deterrents, and hormones.

Antimetabolites chemically resemble essential nutrients and interfere with metabolism. They are low in mammalian toxocity and thus have the advantage of being quite safe to use (e.g., for insect-proofing of fabrics). They may be effective against insects that have access only to treated food; however, they have limited value against polyphagous insects (Subcommittee on Insect Pests, 1969).

Feeding deterrents or antifeeding compounds may be defined as compounds "which will prevent the feeding of pests on a treated material, without necessarily killing or repelling them" (Wright, 1963). Antifeeding compounds have been used for several years in the mothproofing of fabric. However, the use of these compounds in the protection of crops is a fairly new idea. The use of feeding deterrents is still in the experimental stage. They offer considerable specificity since they would affect only the insects that feed on treated plants and would spare the parasites and predators of these pest species. They also have low mammalian toxicity.

Since growth, development, and sexual maturation are largely regulated by hormones, these substances have potential value in insect control. Ecdysone, juvenile hormone, and various analogues of these compounds have been shown to disrupt the development of an insect if applied at appropriate times and in appropriate doses (Williams, 1970). For example, cyasterone, a substance related to ecdysone, when injected into a diapausing *Cynthia* (moth) pupa in a very low dose (0.2 microgram), stimulates termination of diapause and formation of a normal moth. However, if a high dose (10 micrograms) is injected, the developmental events are accelerated and their sequence disrupted. This results in the premature deposition of cuticle, which literally "locks in" the epidermal tissues before they have completed the developmental changes necessary for survival into the adult stage. Compounds related to ecdysone (i.e., the phytoecdysones) have been found in many different kinds of plants, particularly ferns. In addition to lethal effects, there is evidence that certain of the phytoecdysones may function as feeding deterrents. As with ecdysone and its relatives, juvenile hormone can be applied with lethal effects, either by preventing the transformation of the pupa into an adult or by inhibiting the development of eggs. Again, as with ecdysone, several species of plants have been shown to elaborate compounds that mimic juvenile hormone activity. These plant-synthesized analogues of ecdysone and juvenile hormone are thought to be adaptations that protect the plants from attack by phytophagous insects. Understanding the chemistry and physiological effects of ecdysone and juvenile hormone and their analogues may ultimately provide the key to the synthesis of insecticides with extreme specificity.

Mechanical and Physical Control

Mechanical control involves the use of devices that make possible the direct destruction of insects or act indirectly as barriers, excluders, or collectors. Physical controls include the use of heat, light, electricity, x-rays, and so on, to kill insects directly, reduce their reproductive capacity, or to attract them to something that will kill them.

Mechanical Methods. Mechanical methods include the use of simple manual techniques such as handpicking, swatting and crushing, jarring and shaking, and the use of various kinds of barriers, excluders, and traps. Handpicking is and has been effective where comparatively small numbers of insects are to be dealt with (e.g., the handpicking of bagworms from an ornamental shrub or the handpicking of tomato and tobacco hornworms, extensively used in the past, but now limited to small garden situations). The fly swatter, the bare hand, or any of a wide variety of implements are quite handy against a few insects within a dwelling or on one's person. Jarring and shaking or hand beating of shrubs or trees, especially fruit trees, is sometimes used to remove insects, particularly beetles from their hosts. Sheets or buckets of kerosene or other material have been used beneath the plant being shaken to trap the insects that fall into them. Various sorts of collecting devices have been used against insects. These have ranged from a bucket and paddle to horse-drawn "hopper-dozers" and similar machines which were used for grasshopper control in the western United States in the late 1800s (Subcommittee on Insect Pests, 1969). Unfortunately, these methods were not and are not nearly as effective as the application of pesticides.

Several mechanical means are employed to act as barriers to insect movement. Sticky materials in which insects become hopelessly entangled have been used, for example, in the form of flypaper that traps numerous flying insects or as sticky bands about the trunks of trees to protect them from oviposition damage caused by the periodical cicada. Metal collars around tree trunks have also been used for the same purpose and are effective against any nonflying insects which may try to gain access to the branches and foliage. Metal is also used in the construction of shields around the foundation of houses and buildings to prevent attack by subterranean termites. Screens of metal, cloth, fiberglass, or plastic have been used to cover various openings (doors, windows, vents, and so on) to containers and dwellings to allow the passage of air but exclude most insects. Cloth netting (e.g., cheesecloth) is useful in excluding the periodical cicada from young fruit trees and has been used extensively to protect sleeping persons from mosquito and other disease-carrying biting fly attacks. Other barrier techniques include the protective packaging of food products, low sheet metal fences against Mormon and coulee crickets (Metcalf, Flint, and Metcalf, 1962), plastic sheeting and bags as liners and containers, and the digging of deep furrows around fields being threatened by chinch bug and army-worm attacks (Metcalf, Flint, and Metcalf, 1962).

Traps are used both for control and survey purposes. Control traps are usually used in conjunction with some attractive stimulus (e.g., light, food, or sex attractant) and with some means of killing the insects that enter (e.g., a pesticide or an electrically charged grid). Survey traps are used in the detection of the presence of potential pest species and in the evaluation of the effectiveness of any control procedures that may have been carried out in a given area.

Physical Methods. High and low temperatures are useful in insect destruction and are commonly applied in a variety of situations. Most insects become inactive at temperatures of about 4°C or below, and stored products maintained at such temperatures will suffer little damage, although the insects present would not likely be killed. However, cold can be used to kill drywood termites in furniture in vaults for 4 days at −9°C (Subcommittee on Insect Pests, 1969). High temperatures are more effective for killing insects. Few, if any, insects can survive prolonged exposure to temperatures much above 60 to 66°C. High temperatures have been used against insects that infest stored grain, coffee bean, various seeds, citrus fruits, clothing, bedding, furniture, baled fibrous materials, bulbs, soil, and logs. Whether low or high temperatures are used depends in part on the nature of the product to be protected or disinfested.

Insects such as clothes moths and carpet beetles prefer soiled wool garments. Hence laundering or dry cleaning not only kills any insects that may be present, but also makes clothing less susceptible to attack.

The burning of crop stubble by simple ignition or with special flaming apparatus is highly effective against all life stages of any insects that may be present. Electrically charged grids near windows and doors or in association with an attracting light source have been used to kill insects on contact.

Several possible physical means for insect control are still in the experimental stages of development. These include high-frequency electric fields, particularly against stored-grain species; ionizing radiation such as x-rays, gamma rays, high-energy beta particles against insects attacking materials which would not themselves be harmed by the radiation (see also the section on sterilization of insects, p. 385); laser beams; short-duration light used, for example, to induce overwintering species to break diapause and succumb to unfavorable environmental conditions; light reflection from aluminium foil against aphids; and sounds that destroy, repel, attract, or confuse insects.

Regulatory control involves the enactment and enforcement of laws and regulations that allow quarantines to be imposed for the purpose of excluding or containing a real or potential pest species. Thus quarantines are designed to prevent the entry of possible pest species, to confine them to as small an area as practicable once introduced, or to prevent them from being exported to other countries. Even if quarantine measures only retard the spread of a given species, the resultant amount of money saved may very well justify the cost of maintaining the quarantine. In the definition of regulatory control the phrase "potential pest" is especially significant. It should be borne in mind that because a given species is not a major pest in its native land does not mean that it will not be one in another region where few or none of its natural enemies exist.

Prior to the advent of human means of transport, insect dispersal consisted entirely of natural means [e.g., migration or distribution by air movement (see Chapter 9)]. However, when man developed means of covering long distances in increasingly short periods of time and transported not only himself but a wide variety of materials over these distances, a new parameter was added to the natural means of dispersal already available. The environmental changes that man has brought about have also had a profound influence on the dispersal and distribution of insects. For example, the boll weevil and the harlequin bug have extended their ranges from Mexico into the United States because of the extensive irrigation of large areas of the desert which was once an effective natural barrier (Swain, 1954).

In terms of man's modes of transport serving as means for insect dispersal the insects may either simply be stowaways or actually be infesting some material, living or inanimate. Examples of the introduction of insect pests directly attributable to human transport are numerous. A particularly interesting example is the way in which the gypsy moth, an important forest pest, was introduced into United States. In 1869 an amateur entomologist who was studying silkworms acquired living specimens of the gypsy moth and brought them to Massachusetts (Swan, 1964). Since that time the gypsy moth has spread to several other states in the northeastern United States and poses a constant threat to other areas of the country where susceptible trees are present.

Apparently the first regulatory control legislation was passed in Germany in 1873 and was aimed at the prohibition of the entry into that country of any materials that might harbor the grape phylloxera, *Phylloxera vitifoliae,* from America (Subcommittee on Insect Pests, 1969). Although there was earlier regulatory legislation in the United States, both at the state and national levels, the first major and effective legislation was passed in 1905. This was the Federal Insect Pest Act, which provided for the regulation of importation and interstate movement of potentially injurious insects. The Plant Quarantine Act of 1912 supplemented and extended earlier legislation and gave the Secretary of Agriculture the authority to enforce laws designed to protect the agriculture of the United States from insect pests and plant diseases by the regulation of importation and interstate movement of potential carrier materials. Additional pertinent legislation has included the Postal Terminal Inspection Act of 1915 and subsequent amendments and several more recent legislative actions, both at state and national levels. Under the authority provided by these acts, trained inspectors at key locations are able to examine materials as they cross the international boundaries into the United States or cross boundaries of regions within the United States which are under quarantine for one or more plant diseases or insect pests. These inspectors have the power to prevent movement of infested or infected materials across these boundaries or to render these materials "safe" by appropriate treatment whether it be fumigation, application of liquid insecticides, dry heat, or otherwise. Also the regulatory control legislations of other countries are respected, and potentially pestiferous materials that are to be moved from the United States require inspection and export certification.

There are currently several quarantines being enforced in the United States and in the past there have been many effective and not-so-effective quarantines that have been terminated. The Japanese beetle has been under quarantine regulation since 1919, and as a result there are many areas of this country which, although they provide favorable environments for this insect, are uninfested (Subcommittee on Insect Pests, 1969). Current and past control and quarantine programs have confined the gypsy moth to regions of the northeastern United States. Past domestic quarantines, in combination with other kinds of control, have been in part responsible for the virtual elimination of several pest species of insects and ticks [e.g., the cattle tick, *Boophilus annulatus;* red tick, *Rhipicephalus evertsi,* from Florida; parlatoria date scale, *Parlatoria blanchardii,* from Arizona and California; Mediterranean fruit fly, *Ceratitis capitata,* from Florida and Texas; and many others (Subcommittee on Insect Pests, 1969)]. Quarantines against the European corn borer, the satin moth, the Asiatic garden beetle, and others were terminated because these insects became widespread in spite of the quarantine and other control measures. In addition, the face fly, *Musca autumnalis,* and the cereal leaf beetle, *Oulema melanopus,* have recently been introduced into the United States.

Integrated Control

Although there are many instances in which the application of a single method of insect control has been successful, there have been many failures. During the last twenty years it has become apparent that two or more methods of insect control applied against a given pest situation may be much more effective in population reduction and avoid some of the disadvantages inherent in the use of a single method, particularly insecticides. Earlier the term *integrated control* was used to designate the application of a combination of chemical and biological control. Integrated control is

Applied pest control that combines and integrates biological and chemical measures into a single unified pest-control program. Chemical control is used only where and when necessary, and in a manner that is least disruptive to beneficial regulating factors of the environment. It may make use of naturally occurring insect parasites, predators, and pathogens, as well as those biotic agents artificially increased or introduced.—(Subcommittee on Insect Pests, 1969)

More recently, integrated control has been used in a much broader sense:

Utilization of all suitable techniques to reduce and maintain pest populations at levels below those causing injury of economic importance to agriculture and forestry, or bringing together two or more methods of control into a harmonized system designed to maintain pests at levels below those at which they cause harm—a system that must rest on firm ecological principles and approaches.

This usage makes integrated control synonomous with the concept of *pest management* introduced earlier in this chapter.

A good example of the value of an integrated approach to a pest problem is the history of control of the cotton pests in Canete Valley, Peru (Subcommittee on Insect Pests, 1969). Prior to the development of synthetic organic insecticides, the cotton growers of this valley had relied entirely on inorganic insecticides and hand picking of insects. During the late 1940s, they shifted entirely to the use of synthetic organic insecticides. Although these produced good results at first, resistance began to develop and by 1956 nearly one half the cotton crop was destroyed by insects that survived the chemical treatments. In 1957 an integrated system of pest management was instituted which consisted of the following measures:

(1) Reduction in the planting of ratoon cotton (crop produced from plants cut back and allowed to produce new growth from the crown) to less than 25% of the total area; (2) preparation of soil without irrigation to obtain increased destruction of *Heliothis virescens* pupae; (3) repopulation of predators and parasites by introduction from other valleys or foreign countries, or by artificial rearing for release; (4) establishment of mandatory planting and crop-residue destruction dates; (5) adoption of recommended irrigation schedules; (6) adherence to the various cultural methods considered to be sound agronomic practice for production of uniformly early crops; (7) limiting insecticides used to arsenical and botanical materials only, except under very special circumstances, which allowed application of systemic insecticides at dosage rates one fourth to one half of that recommended by the manufacturer to control *Aphis gossypii*—(Subcommittee on Insect Pests, 1969).

This program has been highly successful and yields of cotton have increased substantially as a result.

Beneficial Insects

Although many insects are pests for man and often cause serious problems, the vast majority are neutral relative to him and in the overall scheme of nature must be considered to be very valuable and important animals. In this section we want to consider several ways in which insects are directly or indirectly beneficial to man.

Insect Products

Honey. Honey is a highly nutritive liquid material prepared from flower nectar by several species of bees. It is rather sticky and viscous, ranges from hyaline to dark amber in color, and is composed mainly of water and several sugars (levulose and dextrose in particular) with small amounts of various other substances, including fatty acids, proteins, vitamins, and minerals. The major insect producer of honey is the honey bee, *Apis mellifera,* which has been domesticated and is maintained in artificial hive containers throughout the world. According to Crane (1963), "The present world production of honey is nearly 500,000 tons, the work of 40 to 45 million colonies of bees in the hands of perhaps 5 million beekeepers . . . ." Honey has been sought out and collected by man for thousands of years and has found use mainly as a food material. Meads, beverages similar to wines, are prepared from honey and were possibly among man's earliest alcoholic drinks (White, 1963). Honey is also widely used in the preparation of bake goods, candies, and ice cream.

Beeswax. Beeswax is a yellowish white solid waxy material which is secreted by specialized epidermal glands between the abdominal sternites of honey bees. These insects use this material to construct the series of hexagonally cross-sectioned cells (honeycomb) in which they store honey and rear their offspring. Beeswax has long been used by man for a variety of purposes and was probably the major wax material of ancient times. In the past seventy or so years, several other wax or waxlike substances have come into prominence, causing the demand for beeswax to decrease somewhat. Nonetheless, it is still widely used in many cosmetics, nearly smokeless church candles, various pharmaceuticals, some polishes, and several other manufactured materials. One of the greatest uses to which beeswax is put is in the preparation of comb foundation, which is affixed to the frames in a commercial beehive. This foundation serves to induce the bees to construct honeycomb in the frames, which in turn makes the hive much easier to manage. Grout (1963) notes that, according to the USDA Crop Reporting Board, the 1961 production of beeswax was 5,092,000 pounds.

Silk. Silk is the product of specialized glands of some lepidopterous larvae which use it in the construction of the cocoons in which they undergo the pupal stage. It is composed of two proteins, fibroin and sericin. Approximately 4,000 years ago in China it was discovered that by boiling a cocoon the filament of silk used to construct it became loose and could be unwound, fortunately in a single strand (Clausen, 1954). When several of these filaments were wound together and woven from this thread, a soft, lustrous, easily dyed fabric was the happy result. From that time on the silk industry (sericulture) developed and flourished in China, and its methods were closely guarded secrets. Finally, in about A.D. 550 silkworm eggs and the secret techniques for preparing silk were smuggled to Constantinople and the silk industry began in Europe (Clausen, 1954).

Several species of silkworms have been cultured for the commercial production of silk, but *Bombyx mori* is and has been the most widely used and has become, through generation after generation of careful selection, a totally domesticated insect. The larvae of this species feed on mulberry leaves, and it requires extreme and continuous attention to rear and obtain silk from them. Many of the steps in silk production require hours of tedious hand labor, a factor that has allowed the silk industry to flourish in the Orient, particularly China and Japan, where such labor is fairly cheap. However, silk has been produced in more than twenty countries throughout the world, including the United States (Yokoyama, 1963). In addition to its use in producing fabric, the gummy material that is secreted by the silk glands and drawn out into the filaments by the silkworm is dissected directly from the glands and artificially drawn into thin threads which still find use in surgical stitching and have in the past been used as fishing leaders (Clausen, 1954). In recent years cheaper and in many ways superior fibers have been prepared synthetically and have threatened to destroy the silk industry.

Lac. Lac is the crude resinous material from which commercial shellac is prepared. It is a natural body secretion of the scale insect *Laccifer lacca,* which inhabits trees in India and Burma. The tiny immature "crawlers" suck the sap from branches of these trees and eventually cover themselves with lac, which serves as a protective shield. Thousands of crawlers feed in close proximity to one another and the lac comes to cover branches almost entirely. When they mature, the females remain wingless and sedentary while the winged males emerge and fertilize the females through tiny openings in the lac. When the fertilized eggs hatch, the young crawlers migrate to uninfested areas and commence to feed, beginning the cycle anew. In areas where lac is collected, workers remove twigs with mature females on them and tie them to uninfested trees. These twigs are referred to as *brood lac* and crawlers emerge from them by the thousands and infest their new host tree (Clausen, 1954). The harvesting of lac is accomplished by removing branches covered with lac (*stick lac*) and grinding them. The resultant *seed lac* is then washed and dried and bleached in the sun and, after drying, heated in cloth bags over open charcoal fires. As the lac is melted, it is squeezed out onto the floor and very quickly pressed and stretched into thin sheets which are then flaked. Shellac is prepared by dissolving this *flake lac* (Metcalf, Flint, and Metcalf, 1962). Lac is a basic ingredient of:

> stiffening agents in the toes and soles of shoes, and in felt, fur, and composition hats; shoe polishes; artificial fruits and flowers; lithographic ink; electrical insulation; protective coverings for wood, paper, fabric, wax emulsions, wood fillers, sealing wax and buttons; glazes on confections; coffee bean burnishing; paints; cements and adhesives; shellac varnishes and moldings; photographic products; phonographic records; playing card finishes;

dental plates; pyrotechnics; foundary work and hair dyes—(Clausen, 1954).

As with beeswax and silk, lac is being replaced by synthetic materials.

Cochineal. Like lac, cochineal is a product of a scale insect, *Coccus cacti*, which lives and feeds on the prickly pear. It is prepared from the dried, ground bodies of these insects and in this form is a red pigment that has been used widely, particularly in the past, as a dye. Metcalf, Flint, and Metcalf (1962) list the current uses as follows: "... as a cosmetic or rouge; for decorating fancy cakes; for coloring beverages and medicines; for dyeing where unusual permanence is desired . . .". The major producers of cochineal are in Mexico, Honduras, and the Canary Islands, and according to Bishopp (1952) approximately 70,000 insects are required to produce a pound of dye. In the past dyes have also been prepared from other species of scale insects.

Insect Galls. Several insect galls have been the source of various pigments used for dyeing wool, skin, hair, leather, and so on, and for the production of permanent inks. Tannic acid, a substance widely used in tanning, dyeing, and preparation of inks, is also derived from insect galls.

Use of Insects in Medicine

Aside from the many purported medicinal uses that occur in ancient literature and folklore, in a few instances insects or insect secretions have proved to be of some value in the treatment of certain human ailments.

Maggot Therapy. During the last three or four centuries the observation has been made on a number of occasions that wounded soldiers on the battlefield fared better if their wounds became infested with the larvae of certain flies (Leclercq, 1969). These larvae were observed to devour necrotic tissue and inhibit infection. In the early 1930s, the use of a strain of *Lucilia sericata* larvae which attack only necrotic tissues was advocated for the treatment of osteomyelitis and chronic open sores. Subsequently it was discovered that one of the nitrogenous wastes, allantoin, excreted by the maggots produced the same inhibition of infection as the whole insects. This substance was then produced commercially. In recent years, the discovery and development of antibiotics has replaced maggot therapy and the use of allantoin.

Cantharidin. Cantharidin is derived from the bodies of blister beetles in the family Meloidae. The best known species of blister beetle is *Lytta vesicatoria,* the Spanish fly, found throughout Europe. Cantharidin is a strong vesicant, which when taken internally acts as a strong irritant of the urogenital system. For this reason it has been used as an aphrodisiac, for the treatment of certain urogenital diseases, and in cattle breeding. It is no longer used in the first two ways because of the extreme danger to human life involved.

Miscellaneous. Several other insect substances have been used as medicaments or have promise for such use. Among these are bee venom, which has been used in the treatment of rheumatisms and other ailments (Leclercq, 1969; Metcalf, Flint, and Metcalf, 1962). Cochineal is purported to relieve pain applied against such things as whooping cough and neuralgia (Metcalf, Flint, and Metcalf, 1962). The Aleppo gall or gallnut formed on several species of oaks in Asia and Europe has been used as a "tonic, astringent, and antidote for certain poisons" (Bishopp, 1952). According to Leclercq (1969), there is evidence that antibiotic substances are present in the hemolymph of, or are secreted by, certain insect species. An ancient technique still applied by primitive peoples in various parts of the world is the use of insects such as ants and carabid beetles, which have well-developed mandibles, in the suturing of a wound (Leclerq, 1969). The insect is induced to bite a wound so that the two edges are brought together, and then it is decapitated; the head remains in this clamped position and provides a suture.

Insects in Biological Research

Insects make ideal organisms for fundamental biological research. They are in general easily collected, easy to rear in large numbers, are small in size and can thus be easily manipulated within a small space, have comparatively short generation times, and as a group display a diversity in form and function nearly unapproached by other groups of animals. *Drosophila melanogaster* has been widely utilized as a research species in genetics. A considerable amount of experimental–developmental endocrine work has been carried out on the reduviid bug, *Rhodnius prolixus.* Other genera that appear often in biological literature include *Tribolium, Ephestia, Calliphora, Musca, Periplaneta,* and *Apis.*

Pollination By Insects

Sexual reproduction in flowering plants is accomplished by the transfer of pollen from the anther of a male flower to the stigma of a female flower. When a pollen grain contacts a stigma, the male germ cell in the pollen grain unites with the female germ cell or egg and a fertile seed develops. When the seed is exposed to the proper conditions in an appropriate substrate, it will give rise to a new plant. The transfer of pollen from a male to a female flower is accomplished primarily by the wind or by the activities of insects that associate with plants. Examples of wind-pollinated plants include cereal plants such as wheat and corn and many species of trees. The flowers of these plants are generally small with weakly developed petals, do not produce nectar, and produce dry pollen grains, which of course are easily picked up by the wind.

Insect-pollinated plants include vegetables such as tomatoes, peas, beans, and onions; most fruit crops; and field crops such as alfalfa, red and white clover, and tobacco. The flowers of insect-pollinated plants are usually brightly colored with well-developed petals, are odorous, secrete nectar, and produce sticky pollen grains.

The flowers of plants and the insects that frequent plants have evolved together. The insects have acquired morphological adaptations and behavioral patterns that enable them to obtain nectar and/or pollen from flowers, and the flowers have developed traits that cause them to be attractive to the insects or allow them to take advantage of the insects (e.g., sticky pollen grains). Insects attracted to flowers are exposed to the pollen grains, which adhere to their bodies and are eventually rubbed off on the stigmas of female flowers. Flowers show many "ingenious" adaptations for ensuring that the insects that visit them come into contact with the pollen. For example, alfalfa female flowers have a tripping mechanism that briefly catches the head of an insect in an ideal position for exposure to the pollen. The nectaries of some plants have become so specialized that only certain insect species are morphologically and behaviorally equipped to remove nectar from them.

Insects involved in the pollination of flowering plants number in the thousands of species and are found mainly in the orders Hymenoptera, Diptera, Lepidoptera, and Coleoptera, although some members of other smaller orders (e.g., Thysanoptera) may also be involved. Several members of the order Hymenoptera, in particular the bees (superfamily Apoidea), are the most important pollinators of commercial crops. In the past, wild, native pollinating insects were sufficient in number to accomplish the pollination of our food crops, but with the intensive land use, clean cultivation, and excessive application of insecticides characteristic of modern agriculture, the populations of many of these species have been so reduced that they can no longer maintain an adequate level of pollination. According to Bohart (1952): "An estimated 80 percent of the insect pollination of our commercial crops is performed by honey bees . . . ." Thus honey bees are no doubt our most valuable pollinators and have the advantage that they can be easily cultured and moved about at will. In fact, "It has been claimed that the value of bees in the pollination of crops is 10 to 20 times the value of the honey and wax which they produce" (Lovell, 1963).

Insects Consumed and as Consumers

As Human Food. Insects actually have a high nutritional value, being quite rich in protein and lipids, and may therefore be a very important supplement of the diets of otherwise vegetarian peoples (Leclercq, 1969). In Africa today, meals are sometimes supplemented with insects as a source of protein, which may partially stave off the protein-deficiency disease of children, kwashiorkor (Leclercq, 1969). Clausen (1954) explains that the Australian aborigines realize that grubs are an essential part of their diet, even though they do not realize that these insect larvae are their sole source of protein. In less acute food situations, various insects have for centuries been valued as delicacies and have even become articles of food commerce. For example, "fusanos"—fried caterpillars, earthworms, and beetle grubs found in agave plants—are exported from Mexico and can be purchased in gourmet shops in the United States (Clausen, 1954). Other insects that have been or are still prepared in some way or eaten raw by various peoples in the world include termites, silkworm pupae, migratory locusts, ants, caterpillars, diving beetles, cicadas, and eggs of waterboatmen.

As Food for Wildlife.   Aside from being eaten by a wide variety of arthropods, including their own kind, insects are important food organisms for many kinds of wildlife, in particular birds and fishes. According to Swan (1964): "It is generally agreed that more than half the food consumed by the more than 1,400 species and subspecies of birds in North America consists of insects." Metcalf, Flint, and Metcalf (1962) cite the work of Forbes, who concluded that two fifths of the food of adult freshwater fishes is insects, the most important of which are bloodworms, mayfly nymphs, and caddisfly larvae.

Insects as Consumers.   The value of the roles insects play as plant eaters, scavengers, predators, and parasites in the total picture of nature is inestimable. They are a part of nearly every terrestrial and freshwater food web. Man's understanding, all too incomplete at times, of their roles as consumers has allowed him in some instances to manipulate populations of insects to control other insects and, in a couple of noteworthy instances, undesirable weeds (e.g., klamath weed and prickly pear). The food habits of insects were discussed in Chapter 7.

Selected References

GENERAL TEXTS
Davidson and Peairs (1966); Graham and Knight (1965); Mallis (1969); Metcalf, Flint, and Metcalf (1962); Pfadt (1962); Rolston and McCoy (1966); Subcommittee on Insect Pests (1969); U.S. Department of Agriculture (1952).

MEDICALLY IMPORTANT INSECTS
Gordon and Lavoipierre (1962); Greenberg (1971); Horsfall (1962); James and Harwood (1969); Leclercq (1969); Oldroyd (1964); Roth and Willis (1957); Zumpt (1965).

BIOLOGICAL CONTROL
Baumhover (1966); Bay (1967); Beck (1965); Burges and Hussey (1971); Christenson (1963); Knipling (1959); Painter (1958); Sweetman (1936a, b); World Health Organisation (1967).

CHEMICAL CONTROL
Beroza (1970); Brown (1968); Hoskins (1963); Jacobson (1965); Jacobson and Crosby (1970); Kenaga and Allison (1969); O'Brien (1966, 1967); Pratt and Littig (1962); Scott and Littig (1960); Shepard (1951); Williams (1970); Wood, Silverstein, and Nakajima (1970); Wright (1963).

PESTICIDES IN THE ENVIRONMENT
Cope (1971); Gilette (1970); Graham (1970); Miller and Berg (1969); Woodwell, Malcolm, and Whittaker (1969).

BENEFICIAL INSECTS
Bishopp (1952); Bohart (1952); Crane (1963); Free (1970); Grout (1963); Lovell (1963); Swan (1964); White (1963); Yokoyama (1962).

References Cited

Adkisson, P. L. 1966. Internal clocks and insect diapause. Science, 154:234–41.

Agrell, I. 1964. Physiological and biochemical changes during insect development, In *The Physiology of Insecta,* Vol. 1, pp. 91–148, M. Rockstein, ed. New York, Academic Press, Inc.

Alexander, R. D., and D. J. Borror. 1956. The songs of insects, a phonograph record. Ithaca, N.Y., Cornell University Press.

Alexander, R. D., and T. E. Moore. 1962. The evolutionary relationships of 17- and 13-year cicadas, and three new species (Homoptera, Cicadidae, *Magicicada*). University of Michigan, Museum of Zoology, Misc. Publs. 121.

Allee, W. C., A. E. Emerson, O. Park, T. Park, and K. P. Schmidt. 1949. *Principles of Animal Ecology.* Philadelphia, W. B. Saunders Company.

Alloway, T. M. 1972. Learning and memory in insects. Ann. Rev. Entomol., 17:43–56.

Andrewartha, H. G., and L. C. Birch. 1954. *The Distribution and Abundance of Animals.* Chicago, University of Chicago Press.

Andrewes, Sir Christopher. 1970. *The Lives of Wasps and Bees.* New York, American Elsevier Publishing Company, Inc.

Arnett, R. H., Jr. 1968. *The Beetles of the United States.* Ann Arbor, Mich., The American Entomological Institute.

———. 1970. *Entomological Information Storage and Retrieval.* Baltimore, The Bio-Rand Foundation, Inc.

Arnold, J. W. 1964. Blood circulation in insect wings. Entomological Society of Canada, Mem. 38.

Askew, R. R. 1971. *Parasitic Insects.* London, William Heinemann Ltd.

Baker, E. W., T. M. Evans, D. J. Gould, W. B. Hull, and H. L. Keegan. 1956. *A Manual of Parasitic Mites of Medical or Economic Importance.* New York, National Pest Control Association, Inc.

Baldus, K. 1926. Experimentelle Untersuchungen über die Entfernungslokalisation der Libellen (*Aeschna cyanea*). Z. vergl. Physiol., 3:375–505.

Barth, R. 1937. Muskulatur und Bewegungsart der Raupen. Zool. Jahrb., 62:507–66.

Batra, S. W. T., and L. R. Batra. 1967. The fungus gardens of insects. Sci. Amer., 217(5):112–20.

Baumhover, A. H. 1966. Eradication of the screwworm fly. J. Amer. Med. Assoc., 196(3):150–58.

Bay, E. C. 1967. Mosquito control by fish: a present day appraisal. W. H. O. Chronicle, 21(10):415–23.

Beament, J. W. L. 1958. The effect of temperature on the waterproofing mechanism of an insect. J. Exp. Biol., 35:494–519.

Beck, S. D. 1965. Resistance of plants to insects. Ann. Rev. Entomol., 10:207–32.

———. 1968. *Insect Photoperiodism.* New York, Academic Press, Inc.

Beroza, M. (ed.). 1970. *Chemicals Controlling Insect Behavior.* New York, Academic Press, Inc.

Birukow, G. 1966. Orientation behavior in insects and factors which influence it. In *Insect Behavior*, P. T. Haskell, ed. Royal Entomological Society of London, Symp. 3, pp. 2–12.

Bishopp, F. C. 1952. Insects as helpers. In *Yearbook of Agriculture,* pp. 79–87. Washington, D.C., U.S. Department of Agriculture.

Blackwelder, R. E. 1967. *Taxonomy: A Text and Reference Book.* New York, John Wiley & Sons, Inc.

——— and A. A. Boyden. 1952. The nature of systematics. Syst. Zool., 1:26–33.

References Cited

Blatchley, W. S. 1920. *Orthoptera of Northeastern America.* Indianapolis, The Nature Publishing Co.

Blum, M. S. 1969. Alarm pheromones. Ann. Rev. Entomol., 14:57–80.

Bodenstein, D. 1971. *Milestones in Developmental Physiology of Insects.* New York, Appleton-Century Crofts.

Bohart, G. E. 1952. Pollination by native insects. In *Yearbook of Agriculture,* pp. 107–21. Washington, D.C., U.S. Department of Agriculture.

Borror, D. J., and D. M. DeLong. 1971. *An Introduction to the Study of Insects,* 3rd ed. New York, Holt, Rinehart and Winston, Inc.

———— and D. M. DeLong. 1964. *An Introduction to the Study of Insects,* rev. ed. New York, Holt, Rinehart and Winston, Inc.

———— and R. E. White. 1970. *A Field Guide to the Insects of America North of Mexico.* Boston, Houghton Mifflin Company.

Brammer, J. D., and R. H. White. 1969. Vitamin A deficiency: effect on mosquito eye ultrastructure. Science, 163:821–23.

Breland, O. P., C. D. Eddleman, and J. J. Biesele. 1968. Studies of insect spermatozoa I. Entomol. News, 79(8):197–216.

Brian, M. V. 1965. *Social Insect Populations.* New York, Academic Press, Inc.

Briggs, J. D. 1964. Immunological responses. In *The Physiology of Insecta,* Vol. 3, pp. 259–85, M. Rockstein, ed. New York, Academic Press, Inc.

Britton, W. E. (ed.). 1923. *The Hemiptera or Sucking Insects of Connecticut.* Connecticut State Geological and Natural History Survey, Bull. 34

Brown, A. W. A. 1968. Insecticide resistance comes of age. Bull. Entomol. Soc. Amer., 14(1):3–9.

Brues, C. T. 1946. *Insect Dietary.* Cambridge, Mass., Harvard University Press.

———— A. L. Melander, and F. M. Carpenter. 1954. *Classification of Insects.* Harvard University, Museum of Comparative Zoology, Bull. 108

Buck, J. B. 1948. The anatomy and physiology of the light organ in fireflies. Ann. N.Y. Acad. Sci., 49:397–483.

————. 1953. Physical properties and chemical composition of insect blood. In *Insect Physiology,* pp. 147–90, K. Roeder, ed., New York, John Wiley & Sons, Inc.

————. 1962. Some physical aspects of insect respiration. Ann. Rev. Entomol., 7:27–56.

Bullock, T. H., and G. A. Horridge. 1965. *Structure and Function in the Nervous Systems of Invertebrates,* Vols. 1 and 2. San Francisco, W. H. Freeman and Company.

Burges, H. D., and N. W. Hussey. 1971. *Microbial Control of Insects and Mites.* New York, Academic Press, Inc.

Burks, B. O. 1953. *The Mayflies or Ephemeroptera of Illinois.* Illinois Natural History Survey, Bull. 26(1).

Bursell, E. 1964. Environmental aspects: temperature. In *The Physiology of Insecta,* Vol. 1, pp. 284–323, M. Rockstein, ed. New York, Academic Press, Inc.

————. 1971. *An Introduction to Insect Physiology.* New York, Academic Press, Inc.

Butler, C. G. 1963. The honey bee colony—life history. *The Hive and the Honey Bee,* Chap. 3, R. A. Grout, ed. Hamilton, Ill., Dadant & Sons, Inc.

Butt, F. H. 1960. Head development in the arthropods. Biol. Rev., 35:43–91

Carpenter, F. M. 1931. The biology of the Mecoptera. Psyche, 38(2):41–55.

————. 1953. The geological history and evolution of insects. Amer. Sci., 41:256–70.

Carson, R. L. 1962. *Silent Spring.* Boston, Houghton Mifflin Company.

Carthy, J. D. 1958. *An Introduction to the Behavior of Invertebrates.* London, George Allen & Unwin Ltd.

References Cited

————. 1965. *The Behavior of Arthropods.* San Francisco, W. H. Freeman and Company.

Caudell, A. N. 1918. Zoraptera not an apterous order. Proc. Entomol. Soc. Wash., 22:84–97.

Chadwick, L. E. 1953. The motion of the wings. In *Insect Physiology,* pp. 577–614, K. Roeder, ed. New York, John Wiley & Sons, Inc.

Chamberlin, W. J. 1952. *Entomological Nomenclature and Literature,* 3rd ed. Dubuque, Iowa, William C. Brown Company.

Chandler, H. P. 1956. Megaloptera. In *Aquatic Insects of California,* pp. 229–33, R. L. Usinger, ed. Berkeley, University of California Press.

Chauvin, R. 1967. *The World of an Insect.* London, World University Library, New York, McGraw-Hill Book Company.

Christenson, L. D. 1963. The male annihilation technique in the control of fruit flies. In *New Approaches to Pest Control and Eradication.* Advances in Chemistry Series 41, pp. 31–35. Washington, D.C., American Chemical Society.

Christiansen, K. 1964. Bionomics of Collembola. Ann. Rev. Entomol., 9:147–78.

Chu, H. F. 1949. *How to Know the Immature Insects.* Dubuque, Iowa, William C. Brown Company.

Claassen, P. W. 1931. *Plecoptera Nymphs of America (North of Mexico).* The Thomas Say Foundation, Publ. 3.

Clark, A. M., and M. Rockstein. 1964. Aging in insects. In *The Physiology of Insecta,* pp. 227–81, Vol. 1, M. Rockstein, ed., New York, Academic Press, Inc.

Clark, L. R., P. W. Geier, R. D. Hughes, and R. F. Morris. 1967. *The Ecology of Insect Populations in Theory and Practice.* London, Methuen & Company Ltd.

Clausen, L. W. 1954. *Insect Fact and Folklore.* New York, Macmillan Inc. (Collier Books).

Clements, A. N. 1963. *The Physiology of Mosquitoes.* New York, Macmillan Publishing Co., Inc.

Cloudsley-Thompson, J. L. 1962. Microclimates and the distribution of terrestrial arthropods. Ann. Rev. Entomol. 7:199–222.

Cole, F. R. (with the collaboration of E. I. Schlinger). 1970. *The Flies of Western North America.* Berkeley, University of California Press.

Commonwealth Scientific and Industrial Research Organization. 1970. *The Insects of Australia.* Canberra, Australia.

Comstock, J. H. 1918. *The Wings of Insects.* Ithaca, N.Y., Comstock Publishing Company, Inc.

————1940. *An Introduction to Entomology,* 9th ed. Ithaca, N.Y., Cornell University Press.

Cope, O. B. 1971. Interaction between pesticides and wildlife. Ann. Rev. Entomol., 16:325–64.

Corbet, P. S. 1963. *A Biology of Dragonflies.* Chicago, Quadrangle Books, Inc.

————. 1966. The role of rhythms in insect behavior. In *Insect Behavior,* pp. 13–28. P. T. Haskell, ed., Royal Entomological Society of London, Symp. 3.

Cornwell, P. B. 1968. *The Cockroach,* Vol. 1. London, Hutchinson Publishing Group Ltd.

Cott, H. B. 1940. *Adaptive Coloration in Animals.* London, Methuen & Company Ltd.

Counce, S. J. 1961. The analysis of insect embryogenesis. Ann. Rev. Entomol., 6:295–312.

Craig, R. 1960. Physiology of insect excretion. Ann. Rev. Entomol., 5:53–68.

Crane, E. 1963. The world's beekeeping—past and present. In *The Hive and the Honey Bee,* rev. ed., pp. 1–18, R. A. Grout, ed., Hamilton, Ill., Dadant & Sons, Inc.

References Cited

Cummins, K. W., L. D. Miller, N. A. Smith, and R. M. Fox. 1965. *Experimental Entomology*. New York, Van Nostrand Reinhold Company.

Curran, C. H. 1965. *The Families and Genera of North America Diptera*, 2nd rev. ed. Woodhaven, N.Y., Henry Tripp.

Cushing. E. C. 1957. *History of Entomology in World War II*. Smithsonian Institution, Publ. 4294.

Danilevsky, A. S., N. I. Goryshin, and V. P. Tyshchenko. 1970. Biological rhythms in terrestrial arthropods. Ann. Rev. Entomol., 15:201–44.

Davey, D. G. 1965. *Reproduction in the Insects*. Edinburgh, Oliver & Boyd Ltd.

Davidson, R. H., and L. M. Peairs. 1966. *Insect Pests of Farm, Garden, and Orchard*, 6th ed. New York, John Wiley & Sons, Inc.

Davis, D. E. 1966. *Integral Animal Behavior*. New York, Macmillan Publishing Co., Inc.

Day, M. F. 1954. The mechanism of food distribution to the midgut or diverticula in the mosquito. Aust. J. Biol. Sci., 7:515–24.

———— and D. F. Waterhouse. 1953a. Structure of the alimentary system. In *Insect Physiology*, pp. 273–98, K. Roeder, ed. New York, John Wiley & Sons, Inc.

———— and D. F. Waterhouse. 1953b. Functions of the alimentary system. In *Insect Physiology*, pp. 299–310, K. Roeder, ed. New York, John Wiley & Sons, Inc.

———— and D. F. Waterhouse. 1953c. The mechanism of digestion. In *Insect Physiology*, pp. 311–30, K. Roeder, ed. New York, John Wiley & Sons, Inc.

Day, W. C. 1956. Ephemeroptera. In *Aquatic Insects of California*, pp. 79–105. R. L. Usinger, ed. Berkeley, University of California Press.

Demerec, M. 1950. *Biology of Drosophila*. New York, John Wiley & Sons, Inc.

Denning, D. G. 1956. Trichoptera, *Aquatic Insects of California*, pp. 237–70, R. L. Usinger, ed. Berkeley, University of California Press.

Dethier, V. G. 1953. Chemoreception. In *Insect Physiology*, pp. 544–76, K. Roeder, ed. New York, John Wiley & Sons, Inc.

————. 1955. The physiology and histology of the contact chemoreceptors of the blow fly. Quart. Rev. Biol., 30:348–71.

————. 1963. *The Physiology of Insect Senses*. New York, John Wiley & Sons, Inc.

————. 1966. Feeding behavior. In *Insect Behavior*, pp. 46–58, P. T. Haskell, ed. Royal Entomological Society of London, Symp. 3.

Dethier, V. G., and E. Stellar. 1964. *Animal Behavior—Its Evolutionary and Neurological Basis*, 2nd ed. Englewood Cliffs, N.J., Prentice-Hall, Inc.

DeWilde, J. 1964a. Reproduction. In *The Physiology of Insecta*, Vol. 1, pp. 9–58, M. Rockstein, ed. New York, Academic Press, Inc.

————. 1964b. Reproduction—endocrine control. In *The Physiology of Insecta*, Vol. 1, pp. 59–90, M. Rockstein, ed. New York, Academic Press, Inc.

Dillon, E. S. and L. S. Dillon. 1961. *A Manual of Common Beetles of Eastern North America*. New York, Harper & Row, Inc.

Dodson, E. O. 1960. *Evolution: Process and Product*, rev. ed. New York, Van Nostrand Reinhold Company.

Dunn, J. A. 1960. The natural enemies of the lettuce root aphid, *Pemphigus bursarius* (L.). Ent. Res., Bull. 51:271–278.

DuPorte, E. M. 1961. *Manual of Insect Morphology*. New York, Van Nostrand Reinhold Company.

Ebeling, W. 1964. The permeability of insect cuticle. In *The Physiology of Insecta*, Vol. 3, pp. 507–56. M. Rockstein, ed. New York, Academic Press, Inc.

Edmondson, W. T. (ed.). 1959. Freshwater Biology. New York, John Wiley & Sons, Inc.

References Cited

Edmunds, G. F., Jr. 1959. Ephemeroptera, in *Ward and Whipple's Freshwater Biology,* pp. 908–16, W. T. Edmondson, ed. New York, John Wiley & Sons, Inc.

———— and R. K. Allen. 1963. An annotated key to the nymphs of the families and subfamilies of mayflies (Ephemeroptera). Univ. Utah Biol. Ser. 13(1):1–49.

Edwards, G. A. 1953. Respiratory mechanisms. In *Insect Physiology,* pp. 55–95, K. Roeder, ed. New York, John Wiley & Sons, Inc.

Ehrlich, P. R., and A. H. Ehrlich. 1961. *How to Know the Butterflies.* Dubuque, Iowa, William C. Brown Company.

———— and R. W. Holm. 1963. *The Process of Evolution.* New York, McGraw-Hill Book Company.

Eisenstein, E. M. and M. J. Cohen. 1966. Learning in an isolated insect ganglion. Anim. Behav., 13:104–108.

Eisner, T. 1970. Chemical defense against predation in arthropods. In *Chemical Ecology,* pp. 157–217, E. Sondheimer and J. B. Simeone, eds. New York, Academic Press, Inc.

———— and Y. C. Meinwald. 1965. Defensive secretion of a caterpillar. Science, 150:1733–35.

———— and Y. C. Meinwald. 1966. Defensive secretions of arthropods. Science, 153:1341–50.

Eisner, T., E. van Tassel, and J. E. Carrel. 1967. Defensive use of a "fecal shield" by a beetle larva. Science, 158; 1471–73.

Emerson, K. C. 1964. *Checklist of the Mallophaga of North America* (North of Mexico). Proving Ground, Dugway, Utah. I, Suborder Ischnocera; II, Suborder Amblycera.

Engelmann, F. 1968. Endocrine control of reproduction in insects. Ann. Rev. Entomol., 13:1–26.

————. 1970. *The Physiology of Insect Reproduction.* New York, Pergamon Press, Inc.

Essig, E. O. 1931. *A History of Entomology.* New York, Macmillan Publishing Co., Inc.

————. 1942. *College Entomology.* New York, Macmillan Publishing Co., Inc.

————. 1958. *Insects and Mites of Western North America.* New York, Macmillan Publishing Co., Inc.

Evans, H. E. 1957. *Studies on the Comparative Ethology of Digger Wasps of the Genus Bembix.* Ithaca, N.Y., Comstock Publishing Associates.

————. 1968. Life on a Little-Known Planet. New York, E. P. Dutton & Company, Inc.

———— and M. T. W. Eberhard. 1970. *The Wasps.* Ann Arbor, Mich., University of Michigan Press.

Ewing, H. E. 1940. *The Protura of North America.* Ann. Entomol. Soc. Amer., 33(3):495–551.

———— and I. Fox. 1943. *The Fleas of North America.* U.S. Department of Agriculture, Misc. Publ. 500.

Ferris, G. F. 1928. *The Principals of Systematic Entomology.* Stanford, Calif., Stanford University Press.

————. 1951. *The Sucking Lice.* Pacific Coast Entomological Society, No. 1.

Florkin, M., and C. Jeuniaux. 1964. Hemolymph: composition. In *The Physiology of Insecta,* Vol. 3. pp. 109–52, M. Rockstein, ed. New York, Academic Press, Inc.

———— and E. Schoffeniels. 1969. *Molecular Approaches to Ecology.* New York, Academic Press, Inc.

Folsom, J. W., and R. A. Wardle. 1934. *Entomology, with Special Reference to Its Ecological Aspects,* 4th ed. Philadelphia, P. Blakiston's Son & Company.

References Cited

Fox, R. M., and J. W. Fox. 1964. *Introduction to Comparative Entomology.* New York, Van Nostrand Reinhold Company.

Fraenkel, G. S. 1932. Untersuchungen über die Koordination von Reflexen und automatisch-nervösen Rhythmen bei Insekten III. Z. vergl. Physiol., 16:394–460.

———— and D. L. Gunn. 1961. *The Orientation of Animals.* New York, Dover Publications, Inc.

Free, J. B. (originally published 1940). 1970. *Insect Pollination of Crops.* New York, Academic Press, Inc.

von Frisch, K. 1950. *Bees, Their Vision, Chemical Senses, and Language.* Ithaca, N.Y., Cornell University Press.

————. 1967. *The Dance, Language, and Orientation of Bees,* translated by L. E. Chadwick. Cambridge, Mass., Harvard University Press, 1967.

————. 1971. *Bees, Their Vision, Chemical Senses, and Language,* 2nd ed. Ithaca, N.Y., Cornell University Press.

Frison, T. H. 1937. *Descriptions of Plecoptera.* Illinois Natural History Survey, Bull. 21(3), pp. 78–99.

Froeschner, R. C. 1947. Notes and keys to the Neuroptera of Missouri. Ann. Entomol. Soc. Amer., 40(1):123–36.

Fronk, W. D. 1962a. Chemical control. In *Fundamentals of Applied Entomology,* pp. 159–90, R. E. Pfadt, ed. New York, Macmillan Publishing Co., Inc.

————. 1962b. Vegetable crop insects. In *Fundamentals of Applied Entomology,* pp. 346–75, R. E. Pfadt, ed. New York, Macmillan Publishing Co., Inc.

Frost, S. W. 1959. *Insect Life and Insect Natural History,* 2nd rev. ed. New York, Dover Publications, Inc.

Garman, P. A. 1927. *The Odonata of Connecticut.* Connecticut State Geological and Natural History Survey, Bull. 39.

Gelperin, A. 1971. Regulation of feeding. Ann. Rev. Entomol., 16:365–78.

Gilbert, L. I. 1964. Physiology of growth and development: endocrine aspects. In *The Physiology of Insecta,* Vol. 1, pp. 149–225, M. Rockstein, ed. New York, Academic Press, Inc.

Gilette, J. W. 1970. *The Biological Impact of Pesticides in the Environment.* Corvallis, Ore., Oregon State University Press.

Gillett, J. D., and V. B. Wigglesworth. 1932. The climbing organ of an insect, *Rhodnius prolixus* (Hemiptera: Reduviidae). Proc. Roy. Soc. London, Ser. B, 111:364–76.

Gilmour, D. 1965. *The Metabolism of Insects.* San Francisco, W. H. Freeman and Company.

Gloyd, L. K. and M. Wright. 1959. Odonata. In *Freshwater Biology,* pp. 917–40, W. T. Edmondson, ed. New York, John Wiley & Sons, Inc.

Gochnauer, T. A. 1963. Diseases and enemies of the honey bee. In *The Hive and the Honey Bee,* pp. 477–511, R. A. Grout, ed. Hamilton, Ill., Dadant & Sons, Inc.

Goetsch, W. 1957. *The Ants.* Ann Arbor, Mich., University of Michigan Press.

Goodwin, T. W. (ed.), 1966. *Aspects of Insect Biochemistry.* New York, Academic Press, Inc.

Gordon, R. M., and M. M. T. Lavoipierre. 1962. *Entomology for Students of Medicine.* Oxford, Blackwell Scientific Publications Ltd.

Graham, F., Jr. 1970. *Since Silent Spring.* Boston, Houghton Mifflin Company.

Graham, S. A., and F. B. Knight. 1965. *Principles of Forest Entomology,* 4th ed. New York, McGraw-Hill Book Company.

Gray, P. (ed.). 1970. *Encyclopedia of the Biological Sciences.* New York, Van Nostrand Reinhold Company.

Greenberg, B. 1971. *Flies and Disease,* Vol. I, *Ecology, Classification, and Biotic Associations,* Princeton, N.J., Princeton University Press.

References Cited

Grégoire, C. 1964. Hemolymph coagulation. In *The Physiology of Insecta,* pp. 153–88, M. Rockstein, ed., New York, Academic Press, Inc.

Grout, R. A. 1963. The production and uses of beeswax. In *The Hive and the Honey Bee,* pp. 425–36, R. A. Grout, ed. Hamilton, Ill., Dadant & Sons, Inc.

Gupta, A. P. 1969. Studies of the blood of Meloidae (Coleoptera) I. The haemocytes of *Epicauta cinerea* (Forster), and a synonymy of haemocyte terminologies. Cytologia, 34(2): 311–44.

———— and D. J. Sutherland. 1966. *In vitro* transformations of the insect plasmatocyte in some insects. J. Ins. Physiol., 12:1369–75.

Gurney, A. B. 1938. A synopsis of the order Zoraptera, with notes on the biology of *Zorotypus hubbardi.* Proc. Entomol. Soc. Wash., 40(3):57–87.

———— and S. Parfin. 1959. Neuroptera. In *Freshwater Biology,* pp. 973–80, W. T. Edmundson, ed., New York, John Wiley & Sons, Inc.

Guthrie, D. M. and A. R. Tindall. 1968. *The Biology of the Cockroach.* New York, St. Martin's Press, Inc.

Hackman, R. H. 1964. Chemistry of the insect cuticle. In *The Physiology of Insecta,* Vol. 3, pp. 471–506, M. Rockstein, ed. New York, Academic Press, Inc.

Hagen, H. R. 1951. *Embryology of the Viviparous Insects.* New York, The Ronald Press Company.

Hammack, G. M. 1970. *The Serial Literature of Entomology—A Descriptive Study.* College Park, Md., The Entomological Society of America.

Harcourt, D. G. 1969. The development and use of life tables in the study of natural insect populations. Ann. Rev. Entomol., 14:175–96.

———— and E. J. Leroux. 1967. Population regulation in insects and man. Amer. Scientist, 55(4):400–15.

Haskell, P. T. 1961. *Insect Sounds.* Chicago, Quadrangle Books, Inc.

————. 1964. Sound production. In *The Physiology of Insecta,* Vol. 1, pp. 563–608, M. Rockstein, ed. New York, Academic Press, Inc.

Hebard, M. 1934. *The Dermaptera and Orthoptera of Illinois.* Illinois Natural History Survey, Bull. 20(3).

Helfer, J. R. 1963. *How to Know the Grasshoppers, Cockroaches, and Their Allies.* Dubuque, Iowa, William C. Brown Company.

Hennig. W. 1966. *Phylogenetic Systematics.* Chicago, University of Chicago Press.

Herms, W. B., and M. T. James. 1961. *Medical Entomology,* 5th ed. New York, Macmillan Publishing Co., Inc.

Hertz, M. 1929. Die Organization des optischen Feldes bei der Biene I. Z. vergl. Physiol., 8:693–748.

Hess, W. N. 1917. The chordotonal organs and pleural discs of cerambycid larvae. Ann. Entomol. Soc. Amer., 10:63–78.

Highnam, K. C. 1964. Endocrine relationships in insect reproduction. In *Insect Reproduction,* pp. 26–42, K. C. Highnam, ed. Royal Entomological Society of London, Symp. 3.

———— and L. Hill. 1970. *The Comparative Endocrinology of the Invertebrates.* New York: American Elsevier Publishing Company, Inc.

Hinde, R. A. 1969. *Animal Behavior: A Synthesis of Ethology and Comparative Psychology,* 2nd ed. New York, McGraw-Hill Book Company.

Hinton, H. E. 1946. A new classification of insect pupae. Proc. Zool. Soc. London, 116:282–328.

————. 1948. On the origin and function of the pupal stage. Trans. Roy. Entomol. Soc. London, 99:395–409.

————. 1951. A new Chironomid from Africa, the larva of which can be dehydrated without injury. Proc. Zool. Soc. London, 121:371–380.

————. 1958. The phylogeny of the Panorpid orders. Ann. Rev. Ent., 3:181–206.

References Cited

————. 1960. Cryptobiosis in the larva of *Polypedilum vanderplanki* Hint. (Chironomidae). J. Ins. Physiol., 5:286–300.

————. 1963. The origin and function of the pupal stage. Proc. Roy. Entomol. Soc. London, Ser. A, 50:96–113.

————. 1964. Sperm transfer in insects and the evolution of haemocoelic insemination. In *Insect Reproduction,* pp. 95–107, K. C. Higham, ed. Royal Entomological Society of London, Symp. 3.

————. 1966. Apolysis in arthropod moulting cycles. Nature, 211(5051):871.

————. 1969. Respiratory systems of insect egg shells. Ann. Rev. Entomol., 14:343–68.

Hodgson, E. S. 1964. Chemoreception. In *The Physiology of Insecta,* Vol. 1, pp. 363–96, M. Rockstein, ed. New York, Academic Press, Inc.

Holland, W. J. 1903. *The Moth Book.* New York, Dover Publications, Inc. (originally published by Doubleday & Company, Inc., 1903).

Honigberg, B. M. (chm.), W. Balamuth, E. C. Bovee, J. O. Corliss, M. Gojdics, R. P. Hall, R. R. Kudo, N. D. Levine, A. R. Loeblich, Jr., J. Weiser, and B. H. Wenrich. 1964. A revised classification of the phylum Protozoa. J. Protozool., 11(1):7–20.

Horridge, G. A. 1962a. Learning of leg position by the ventral nerve cord of headless insects. Proc. Roy. Soc. London, Ser. B, 157:33–52.

————. 1962b. Learning of leg position by headless insects. Nature (London), 193:697–98.

Horsfall, W. R. 1962. *Medical Entomology—Arthropods and Human Disease.* New York, The Ronald Press Company.

Hoskins, W. M. 1963. Resistance to insecticides. Internat. Rev. Trop. Med., 2:119–74.

House, H. L. 1961. Insect nutrition. Ann. Rev. Entomol., 6:13–26.

————. 1965a. Insect nutrition. In *The Physiology of Insecta,* Vol. 2, pp. 769–814, M. Rockstein, ed. New York, Academic Press, Inc.

————. 1965b. Digestion. In *The Physiology of Insecta,* Vol. 2, pp. 815–62, M. Rockstein, ed. New York, Academic Press, Inc.

Howard, L. O. 1930. A history of applied entomology. Smithsonian Institution, Misc. Coll. 84.

Howse, P. E. 1970. *Termites: A Study in Social Behavior.* New York, Hillary House Publishers Ltd.

Hoyle, G. 1965. Neural control of skeletal muscle. In *The Physiology of Insecta,* Vol. 2, pp. 407–449, M. Rockstein, ed. New York, Academic Press, Inc.

Huber, F. 1965. Neural integration (central nervous system). In *The Physiology of Insecta,* Vol. 2, pp. 333–406, M. Rockstein, ed. New York, Academic Press, Inc.

Hughes, G. M. 1965. Locomotion: terrestrial. In *The Physiology of Insecta,* Vol. 2, pp. 227–54, M. Rockstein, ed. New York, Academic Press, Inc.

Hungerford, H. B. 1959. Hemiptera. In *Freshwater Biology,* pp. 958–72, W. T. Edmondson, ed. New York, John Wiley & Sons, Inc.

Huxley, H. E. 1965. The contraction of muscle. In *The Living Cell,* Readings from Scientific American, pp. 279–89. San Francisco, W. H. Freeman and Company.

Hynes, H. B. N. 1970. The ecology of stream insects. Ann. Rev. Entomol., 15:25–42.

Jacobson, M. 1965. *Insect Sex Attractants.* New York, Interscience Publishers, A Division of John Wiley & Sons, Inc.

Jacobson, M., and D. G. Crosby. 1970. *Naturally Occurring Insecticides.* New York, Marcel Dekker, Inc.

Jacques, H. E. 1947. *How to Know the Insects.* Dubuque, Iowa, William C. Brown Company.

References Cited

James, M. T., and R. F. Harwood. 1969. *Herm's Medical Entomology*. New York, Macmillan Publishing Co., Inc.

Jander, R. 1963. Insect orientation. Ann. Rev. Entomol., 8:94–114.

Janzen, D. H. 1967. Interaction of the bull's-horn acacia (*Acacia cornigera* L.) with an ant inhabitant (*Pseudomyrmex ferruginea* F. Smith) in eastern Mexico. Univ. of Kansas Sci. Bull., 47(6):315–558.

Jenkin, P. M., and H. E. Hinton. 1966. Apolysis in arthropod molting cycles. Nature, 211(5051):871.

Jewett, S. G., Jr. 1956. Plecoptera, *Aquatic Insects of California*, pp. 155–81, R. L. Usinger, ed. Berkeley, University of California Press.

Johannsen, O. A., and F. H. Butt. 1941. *Embryology of Insects and Myriapods*. New York, McGraw-Hill Book Company.

Johansen, C. 1962. Insect pests of tree fruits. In *Fundamentals of Applied Entomology*, pp. 376–407, R. E. Pftadt, ed. New York, Macmillan Publishing Co., Inc.

Johnsgard, P. A. 1967. *Animal Behavior*. Dubuque, Iowa, William C. Brown Company.

Johnson, C. G. 1963. The aerial migration of insects. Sci. Amer. 209(6):132–38

———. 1965. Migration. In *The Physiology of Insecta*, Vol. 2, pp. 187–226, M. Rockstein, ed. New York, Academic Press, Inc.

———. 1966. A functional system of adaptive dispersal by flight. Ann. Rev. Entomol., 11:233–60.

———. 1969. *Migration and Dispersal of Insects by Flight*. London, Methuen & Company Ltd.

Jones, J. C. 1962. Current concepts concerning insect hemocytes. Amer. Zool., 2:209–46.

———. 1964. The circulatory system of insects. In *The Physiology of Insecta*, Vol. 3, pp. 1–107, M. Rockstein, ed. New York, Academic Press, Inc.

Kenaga, E. E., and W. E. Allison. 1969. Commercial and experimental organic insecticides. Bull. Entomol. Soc. Amer., 15(2):85–148.

Kendeigh, S. C. 1961. *Animal Ecology*. Englewood Cliffs, N.J., Prentice-Hall, Inc.

Kennedy, J. S. (ed.) 1961. *Insect Polymorphism*, Royal Entomological Society of London, Symp. 1.

Kevan, D. K. M. 1963. *Soil Animals*. London, H. F. & G. Witherby Limited.

King, R. C. 1970. *Ovarian Development in Drosophila melanogaster*. New York, Academic Press, Inc.

Klots, A. B. 1951. *A Field Guide to the Butterflies*. Boston, Houghton Mifflin Company.

Knipling, E. F. 1959. Screwworm eradication: concepts and research leading to the sterile-male method. Smithsonian Institution, Publ. 4365.

Knudsen, J. W. 1966. *Biological Techniques*. New York, Harper & Row, Inc.

Kormondy, E. J. 1969. *Concepts of Ecology*. Englewood Cliffs, N. J., Prentice-Hall, Inc.

Krishna, K., and F. M. Weesner (eds.). 1969. *Biology of Termites*, Vol. 1. New York, Academic Press, Inc.

——— and F. M. Weesner (eds.). 1970. *Biology of Termites*, Vol. 2. New York, Academic Press, Inc.

Lai-Fook, J. 1967. The structure of developing muscle insertions in insects. J. Morph. 123:503–28.

Lawson, F. A. 1959. Identification of the nymphs of families of Hemiptera. J. Kans. Entomol. Soc., 32(2):88–92.

Leclercq, M. L. 1969. *Entomological Parasitology*. New York, Pergamon Press, Inc.

Lees, A. D., 1955. *The Physiology of Diapause in Arthropods*. Cambridge, Cambridge University Press.

Lewis, T., and L. R. Taylor. 1967. *Introduction to Experimental Ecology*. New York, Academic Press, Inc.

Lindauer, M. 1965. Social behavior and mutual communication. In *The Physiology of Insecta*, Vol. 2, pp. 124–87, M. Rockstein, ed. New York, Academic Press, Inc.

Linsley, E. G., and R. L. Usinger. 1961. Taxonomy, In *The Encyclopedia of the Biological Sciences*, P. Gray, ed. New York, Van Nostrand Reinhold Company.

Lloyd, J. L. 1971. Bioluminescent communication in insects. Ann. Rev. Entomol., 16:97–122.

Locke, M. 1964. The structure and formation of the integument in insects. In *The Physiology of Insecta*, Vol. 3, pp. 379–470, M. Rockstein, ed. New York, Academic Press, Inc.

————. 1965. Permeability of the insect cuticle to water and lipids, Science, 147:295–98.

Lovell, H. B. 1963. The honey bee as a pollinating agent. In *The Hive and the Honey Bee*, pp. 463–76, R. A. Grout, ed. Hamilton, Ill., Dadant & Sons, Inc.

Lüscher, M. 1961a. Air-conditioned termite nests. Sci. Amer., 205:138–45.

————. 1961b. Social control of polymorphism in termites. In *Insect Polymorphism*, pp. 57–67, J. S. Kennedy, ed. Royal Entomological Society of London, Symp. 1.

Macan, T. T. 1962. Ecology of aquatic insects. Ann. Rev. Entomol., 7:261–88.

————. 1963. *Freshwater Ecology*. London, Longman Group Ltd.

McCann, F. V. 1970. Physiology of insect hearts. Ann. Rev. Entomol., 15:173–200.

McConnell, E. and A. G. Richards. 1955. How fast can a cockroach run? Bull. Brooklyn Entomol. Soc., 50:36–43.

McElroy, W. D. 1964. Insect bioluminescence. In *The Physiology of Insecta*, Vol. 1, pp. 463–508, M. Rockstein, ed. New York, Academic Press, Inc.

———— and H. H. Seliger. 1962. Biological luminescence. In *The Living Cell*. Readings from the Scientific American, pp. 122–34. San Francisco, W. H. Freeman and Company.

McGill, T. E. 1965. *Readings in Animal Behavior*. New York, Holt, Rinehart and Winston, Inc.

Maier, N. R. F., and T. C. Schneirla. 1935. *Principles of Animal Psychology*. New York, McGraw-Hill Book Company.

Mallis, A. 1969. *Handbook of Pest Control*, 5th ed. New York, McNair Dorland Co.

Mangan, A. 1934. *La Locomotion chez les animaux, I: Le Volume des insectes*. Paris, Hermann & Cie.

Manning, A. 1966. Sexual behavior. In *Insect Behavior*, pp. 59–68, P. T. Haskell, ed. Royal Entomological Society of London, Symp. 3.

Mansingh, A. 1971. Physiological classification of dormancies in insects. Can. Entomol. 103:983–1009.

Markl, H. 1962. Schweresinnesorgane bei Ameisen und anderen Hymenopteren. Z. vergl. Physiol., 44:475–569.

———— and M. Lindauer. 1965. Physiology of insect behavior. In *The Physiology of Insecta*, Vol. 2, pp. 3–122, M. Rockstein, ed. New York, Academic Press, Inc.

Marler, P., and W. J. Hamilton, III, 1966. *Mechanisms of Animal Behavior*. New York, John Wiley & Sons, Inc.

Maruyama, K. 1965. The biochemistry of the contractile elements of insect muscle. In *The Physiology of Insecta*, Vol. 2, pp. 451–82, M. Rockstein, ed. New York, Academic Press, Inc.

Masek Fialla, K. 1941. Die Korpertemperatur Poikilothermes Thiere im Abhangigkeit vom Kleinklima. Z. wiss. Zool., 154:170–247.

References Cited

Matheson, R. 1944. *Handbook of the Mosquitoes of North America,* 2nd ed. Ithaca, N.Y., Cornell University Press.

Matsuda, R. 1970. Morphology and evolution of the insect thorax. Entomological Society of Canada, Mem. 76.

Maynard, E. A. 1951. *The Collembola of New York State.* Ithaca, N.Y., Comstock Publishing Company, Inc.

Mayr, E. 1963. *Animal Species and Evolution.* Cambridge, Mass., Harvard University Press.

———. 1969. *Principles of Systematic Zoology.* New York, McGraw-Hill Book Company.

———. 1970. *Populations, Species, and Evolution.* Cambridge, Mass., Harvard University Press.

——— E. G. Linsley, and R. L. Usinger. 1953. *Methods and Principles of Systematic Zoology.* New York, McGraw-Hill Book Company.

Mazokhin-Porshnyakov, G. A. 1969. *Insect Vision.* New York, Plenum Publishing Corporation.

Meglitsch, R. A. 1967. *Invertebrate Zoology.* London, Oxford University Press.

Metcalf, C. L., W. P. Flint, and R. L. Metcalf. 1962. *Destructive and Useful Insects.* New York, McGraw-Hill Book Company.

Metcalf, Z. P. 1954–1963. *General Catalogue of the Homoptera.* Raleigh, N.C., University of North Carolina at Raleigh.

———. 1962–1967. *General Catalogue of the Homoptera.* U.S. Department of Agriculture, Agr. Res. Ser.

——— and C. L. Metcalf. 1928. *A Key to the Principal Orders and Families of Insects.* Published by the authors.

Miller, M. W., and G. Berg (eds.). 1969. *Chemical Fallout—Current Research on Persistent Pesticides.* Springfield, Ill., Charles C Thomas, Publisher.

Miller, N. C. E. 1956. *Biology of the Heteroptera.* London, Methuen & Company Ltd.

Miller, P. L. 1964. Respiration—aerial gas transport. In *The Physiology of Insecta,* Vol. 3, pp. 557–615, M. Rockstein, ed. New York, Academic Press, Inc.

Miller, W. H., C. D. Bernard, and J. L. Allen. 1968. The optics of insect compound eyes. Science, 162: 760–67.

Mitchell, R. T., and H. Zim. 1962. *Butterflies and moths* (A Golden Nature Guide). New York, Golden Press (Western Publishing Company, Inc.).

Mockford, E. L. 1951. The Psocoptera of Indiana. Proc. Ind. Acad. Sci., 60:192–204.

Muirhead-Thomson, R. C. 1968. *Ecology of Insect Vector Populations.* New York, Academic Press, Inc.

Nachtigall, W. 1960. Über Kinematik, Dynamik und Energetik des schwimmens einheimischen Dytisciden. Z. vergl. Physiol., 43:48–118.

———. 1963. Zur Lokomotionsmechanik schwimmender Dipterenlarven. Z. vergl. Physiol., 46:449–66.

———. 1965. Locomotion: swimming (hydrodynamics) of aquatic insects. In *The Physiology of Insecta,* Vol. 2, pp. 255–81, M. Rockstein, ed. New York, Academic Press, Inc.

Needham, J. G., and P. W. Claassen. 1925. *A Monograph of the Plecoptera or Stoneflies of America North of Mexico.* The Thomas Say Foundation, Publ. 2.

——— (chm.), P. S. Galtsoff, F. E. Lutz, and P. S. Welch. 1937. *Culture Methods for Invertebrate Animals.* New York, Dover Publications, Inc.

——— and H. B. Heywood. 1929. *A Handbook of the Dragonflies of North America.* Springfield, Ill., Charles C Thomas, Publisher.

———, J. R. Traver, and Y. Hsu. 1935. *The Biology of Mayflies.* Ithaca, N.Y., Comstock Publishing Company, Inc.

———— and M. J. Westfall, Jr. 1955. *A Manual of the Dragonflies of North America (Anisoptera).* Berkeley, University of California Press.

Neville, A. C. 1970. Cuticle ultrastructure in relation to the whole insect. In *Insect Ultrastructure,* pp. 1–16, A. C. Neville, ed. Royal Entomological Society of London, Symp. 5.

O'Brien, R. D. 1966. Mode of action of insecticides. Ann. Rev. Entomol. 11:369–402.

————. 1967. *Insecticides Action and Metabolism.* New York, Academic Press, Inc.

Odum, E. P., and H. T. Odum. 1959. *Fundamentals of Ecology,* 2nd ed. Philadelphia, W. B. Saunders Company.

Oldroyd, H. 1964. *The Natural History of Flies.* New York, W. W. Norton & Company, Inc.

————. 1968. *Elements of Entomology.* London, George Weidenfeld & Nicolson Ltd.

Osborn, H. 1937. *Fragments of Entomological History.* Columbus, Ohio, published by the author.

Packard, A. S. 1898, *A Text-book of Entomology.* New York, Macmillan Publishing Co., Inc.

Painter, R. H. 1958. Resistance of plants to insects. Ann. Rev. Entomol., 3:267–90.

Patton, R. L. 1953. Insect excretion. In *Insect Physiology,* pp. 387–403, K. Roeder, ed. New York, John Wiley & Sons, Inc.

————. 1963. *Introductory Insect Physiology.* Philadelphia, W. B. Saunders Company.

Pennak, R. W. 1953. *Freshwater Invertebrates of the United States.* New York, The Ronald Press Company.

Peterson, A. 1948. *Larvae of Insects,* part I, Lepidoptera and plant infesting Hymenoptera. Columbus, Ohio, published by the author.

————. 1951. *Larvae of Insects,* part II, Coleoptera, Diptera, Neuroptera, Siphonaptera, Mecoptera, Trichoptera. Columbus, Ohio, published by the author.

————. 1959. *Entomological Techniques.* Ann Arbor, Mich., Edwards Brothers, Inc.

Pfadt, R. E. (ed.). 1962. *Fundamentals of Applied Entomology.* New York, Macmillan Publishing Co., Inc.

Pielou, E. C. 1969. *An Introduction to Mathematical Ecology.* New York, John Wiley & Sons, Inc.

Pomerantz, C. 1959. Arthropods and psychic disturbances, Bull. Entomol. Soc. Amer., 5:65–7.

Popham, E. J. 1965. A key to Dermaptera subfamilies. Entomologist, 98: 126–36.

Powell, J. A., and R. A. Mackie. 1966. *Biological Interrelationships of Moths and Yucca whipplei (Lepidoptera: Gelechiidae, Blastobasidae, Prodoxidae).* Berkeley, University of California Press.

Pratt, H. D., and K. S. Littig. 1962. *Insecticides for the Control of Insects of Public Health Importance.* U.S. Public Health Service, Publ. 772.

Pringle, J. W. S. 1957. *Insect Flight.* Cambridge, Cambridge University Press.

————. 1965. Locomotion: flight. In *The Physiology of Insecta,* Vol. 2, pp. 283–329, M. Rockstein, ed. New York, Academic Press, Inc.

Prosser, C. L., and F. A. Brown. 1961. *Comparative Animal Physiology,* 2nd ed. Philadelphia, W. B. Saunders Company.

Rabe, W. 1953. Beitrage zum Orientierungsproblem der Wasserwanzen. Z. vergl. Physiol., 35:300–25.

Rehn, J. A. G., and H. J. Grant, Jr. 1961. A monograph of the Orthoptera of North America (north of Mexico), Vol. 1. Philadelphia Academy of Natural Science, Monogr. 12.

References Cited

Reichstein, T., J. von Euw, J. A. Parsons, and M. Rothschild. 1968. Heart poisons in the monarch butterfly. Science, 161:861–65.

Richards, A. G. 1951. *The Integument of Arthropods*. Minneapolis, University of Minnesota Press.

Richards, O. W. 1953. *The Social Insects*. New York, Harper & Row, Inc.

———. 1961. An introduction to the study of polymorphism in insects. In *Insect Polymorphism,* pp. 2–10, J. S. Kennedy, ed. Royal Entomological Society of London, Symp. 1.

——— and R. G. Davies. 1957. *Imm's A General Textbook of Entomology*. London, Methuen & Co. Ltd.

Riley, C. V. 1888. The habits of *Thalessa* and *Tremex*. U.S. Div. Ent. Ins. Life, 1:168–79.

Roe, A., and G. G. Simpson (eds.). 1958. *Behavior and Evolution*. New Haven, Conn., Yale University Press.

Roeder, K. D. 1959. A physiological approach to the relation between prey and predator, *Studies in Invertebrate Morphology,* pp. 287–306. Smithsonian Institution, Misc. Coll. 157.

———. 1963. *Nerve Cells and Insect Behavior*. Cambridge, Mass., Harvard University Press.

———. 1965. Moths and ultrasound. Sci. Amer., 212(4):94–102.

———. 1966. Auditory system of noctuid moths. Science. 154:1515–21.

Rolston, L. H., and C. E. McCoy, 1966. *Introduction to Applied Entomology*. New York, The Ronald Press Company.

Ross, E. S. 1970. *Biosystematics of the Embioptera,* Ann. Rev. Entomol., 15:157–72.

Ross, H. H. 1938. Descriptions of nearctic caddisflies. Illinois National History Survey, Bull. 21(4).

———. 1959. Trichoptera. In *Freshwater Biology,* pp. 1024–49, W. T. Edmondson, ed. New York, John Wiley & Sons, Inc.

———. 1962a. *How to Collect and Preserve Insects*. Illinois Natural History Survey, Circ. 39.

———. 1962b. *A Synthesis of Evolutionary Theory*. Englewood Cliffs, N.J., Prentice-Hall, Inc.

———. 1965. *A Textbook of Entomology,* 3rd ed. New York, John Wiley & Sons, Inc.

Roth, L. M., and T. Eisner. 1962. Chemical defenses of arthropods. Ann. Rev. Entomol., 7:107–36.

——— and E. R. Willis. 1957. The medical and veterinary importance of cockroaches. Smithsonian Institution, Misc. Coll. 134(10).

——— and E. R. Willis. 1960. The biotic associations of cockroaches. Smithsonian Institution, Misc. Coll. 141.

Sacktor, B. 1965. Energetics and respiratory metabolism of muscular contraction. In *The Physiology of Insecta,* Vol. 2, pp. 483–580, M. Rockstein, ed. New York, Academic Press, Inc.

Salt, G. 1970. *The Cellular Defense Reactions of Insects*. New York, Cambridge University Press.

Salt, R. W. 1961. Principles of insect cold-hardiness. Ann. Rev. Entomol., 6:55–74.

Schaller, F. 1971. Indirect sperm transfer by soil arthropods. Ann. Rev. Entomol., 16:407–6.

Schildknecht, H., and K. Holoubek. 1961. Die Bombardierkäfer und ihre Explosionschemie V. Mitteilung über insekten Abwehrstoffe. Angew. Chem., 73(1):1–6.

Schneider, F. 1962. Dispersal and migration. Ann. Rev. Entomol., 7:223–42.

Schneiderman, H. A., and L. I. Gilbert. 1964. Control of growth and form in insects. Science, 143:325–33.

References Cited

—— and C. M. Williams. 1955. An experimental analysis of the discontinuous respiration of the cecropia silkworm. Biol. Bull., 109:123–43.

Schneirla, T. C. 1953. Insect behavior in relation to its setting. In *Insect Physiology,* pp. 685–722, K. D. Roeder, ed. New York, John Wiley & Sons, Inc.

Schön, A. 1911. Bau und Entwicklung des tibialen Chordotonalorgane bei der Honigbiene und bei Ameisen. Zool. Jahrb. Anat., 31:439–72.

Schwabe, J. 1906. *Beiträge zur Morphologie und Histologie der tympanalen Sinnesapparate der Orthopteren.* Zoologica, Stuttgart.

Scott, H. G. 1961. Collembolla: pictorial keys to the nearctic genera. Ann. Entomol. Soc. Amer., 54(1):104–13.

—— and K. S. Littig. 1960. *Insecticidal Equipment for the Control of Insects of Public Health Importance.* U.S. Public Health Service, Publ. 774.

Scudder, G. G. E. 1971. Comparative morphology of insect genitalia. Ann. Rev. Entomol., 16:379–406.

Sharov, A. G. 1966. *Basic Arthropodan Stock with Special Reference to Insects.* New York, Pergamon Press, Inc.

Shepard, H. H. 1951. *The Chemistry and Action of Insecticides.* New York, McGraw-Hill Book Company.

Sherman, J. L. 1966. Development of a systemic insect repellent. J. Amer. Med. Assoc., 196(3):166–68.

Simpson, G. G. 1961. *Principles of Animal Taxonomy.* New York, Columbia University Press.

Siverly, R. E. 1962. *Rearing Insects in Schools.* Dubuque, Iowa, William C. Brown Company.

Skaife, S. H. 1961. *Dwellers in Darkness.* London, Longman Group Ltd., New York, Doubleday & Company, Inc.

Slabaugh, R. E. 1940. A new thysanuran, and a key to the domestic species of Lepismatidae (Thysanura) found in the United States. Entomol. News, 51(4):95–8.

Slifer, E. H. 1970. The structure of arthropod chemoreceptors. Ann. Rev. Entomol., 15:121–42.

Smart, J. 1959. Notes on the mesothoracic musculature of Diptera. In *Studies in Invertebrate Morphology,* pp. 331–64. Smithsonian Institution, Misc. Coll. 157.

Smith, C. N. 1966. Personal protection from blood sucking arthropods. J. Amer. Med. Assoc., 196(3):146–49.

Smith, D. S. 1968. *Insect Cells: Their Structure and Function.* Edinburgh, Oliver & Boyd Ltd.

Smith, E. L. 1969. Evolutionary morphology of external insect genitalia, I. Origin and relationships to other appendages. Ann. Entomol. Soc. Amer. 62(5):1051–79.

Smith, L. M. 1960. The family Projapygidae and Anajapygidae (Diplura) in North America. Ann. Entomol. Soc. Amer., 55(5):575–83.

Smith, R. C. 1958. *Guide to the Literature of the Zoological Sciences,* 5th ed. Minneapolis, Burgess Publishing Company.

—— and R. H. Painter. 1966. *Guide to the Literature of the Zoological Sciences,* 7th ed. Minneapolis, Burgess Publishing Company.

Smith, R. F., and A. E. Pritchard. 1956. Odonata. In *Aquatic Insects of California,* pp. 106–53, R. L. Usinger, ed. Berkeley, University of California Press.

Snodgrass, R. E. 1925. *Anatomy and Physiology of the Honey Bee.* New York, McGraw-Hill Book Company.

——. 1935. *Principles of Insect Morphology.* New York, McGraw-Hill Book Company.

——. 1950. Comparative studies on the jaws of mandibulate arthropods. Smithsonian Institution, Misc. Coll. 116(1), Publ. 4018.

References Cited

———. 1952. *A Textbook of Arthropod Anatomy*. Ithaca, N.Y., Comstock Publishing Associates.

———. 1956. *The Anatomy of the Honeybee*. Ithaca, N.Y., Cornell University Press.

———. 1954. *The Dragonfly Larva*. Smithsonian Institution, Misc. Coll. 123(2).

———. 1957. *A Revised Interpretation of the External Reproductive Organs of Male Insects*. Smithsonian Institution, Misc. Coll. 135(6), pp. 1–60.

———. 1958. *Evolution of Arthropod Mechanisms*. Smithsonian Institution, Misc. Coll. 138(2), pp. 1–77.

———. 1959. *The Anatomical Life of the Mosquito*. Smithsonian Institution, Misc. Coll. 138(8).

———. 1960. *Facts and Theories Concerning the Insect Head*. Smithsonian Institution, Misc. Coll. 142(1), pp. 1–61.

———. 1961. *The Caterpillar and the Butterfly*. Smithsonian Institution, Misc. Coll. 143(6).

———. 1963a. *A Contribution toward an Encyclopedia of Insect Anatomy*. Smithsonian Institution, Misc. Coll. 146(2), pp. 1–48.

———. 1963b. The anatomy of the honey bee. In *The Hive and the Honey Bee,* pp. 141–90, R. A. Grout, ed. Hamilton, Ill., Dadant & Sons, Inc.

Snyder, T. E. 1954. *Order Isoptera—The Termites of the United States and Canada*. New York, National Pest Control Association, Inc.

Sokal, R. R., and P. H. A. Sneath. 1963. *Principles of Numerical Taxonomy*. San Francisco, W. H. Freeman and Company.

Southwood, T. R. E. 1966. *Ecological Methods with Particular Reference to the Study of Insect Populations*. London, Methuen & Company Ltd.

Stannard, L. J., Jr. 1968. *The Thrips or Thysanoptera of Illinois*. Illinois National History Survey, Bull. 29(4), pp. 215–552.

Steinhaus, E. A. 1949. *Principles of Insect Pathology,* New York, McGraw-Hill Book Company.

——— (ed.). 1963. *Insect Pathology: An Advanced Treatise,* 2 vols. New York, Academic Press, Inc.

Stellwaag, F. 1916. Wie steuern die Insekten während des Fluges. Biol. Zentrbl., 36:30–44.

Stevens, C. F. 1966. *Neurophysiology: A Primer*. New York, John Wiley & Sons, Inc.

Steyskal, G. C. 1965. Trend curves of the rate of species description in zoology. Science, 149:880–82.

Stiles, K. A., R. W. Hegner, and R. Boolootian. 1969. *College Zoology,* 8th ed. New York, Macmillan Publishing Co., Inc.

Stobbart, R. H., and J. Shaw. 1964. Salt and water balance: excretion. In *The Physiology of Insecta,* Vol. 3, pp. 190–258, M. Rockstein, ed. New York, Academic Press, Inc.

Stone, A., et al. 1965. *A Catalog of the Diptera of America, North of Mexico*. U.S. Department of Agriculture, Agricultural Handbook 276.

Sturm, H. 1956. Die Paarung beim Silberfischen, *Lepisma saccharina*. Z. Tierpsychol., 13:1–2.

Subcommittee on Insect Pests. 1969. *Insect-Pest Management and Control*. Washington, D.C., National Academy of Science.

Swain, R. B. 1952. How insects gain entry. In *Insects—The Yearbook of Agriculture,* pp. 350–55. Washington, D.C., U.S. Department of Agriculture.

———. 1957. *The Insect Guide*. Garden City, N.Y., Doubleday & Company, Inc.

Swan, L. A. 1964. *Beneficial Insects*. New York, Harper & Row, Inc.

Swartzkopff, J. 1964. Mechanoreception. In *The Physiology of Insecta,* Vol. 1, pp. 509–61, M. Rockstein, ed. New York, Academic Press, Inc.

References Cited

Sweetman, H. L. 1936a. *The Biological Control of Insects.* Ithaca, N.Y., Comstock Publishing Company, Inc.

———. 1936b. *The Principles of Biological Control.* Dubuque, Iowa, William C. Brown Company.

Thorpe, W. H. 1963. *Learning and Instinct in Animals,* 2nd ed. London, Methuen & Company Ltd.

Tietz, H. M. 1963. *Illustrated Keys to the Families of North American Insects.* Minneapolis, Burgess Publishing Company.

Trager, W. 1953. Nutrition. In *Insect Physiology,* pp. 350–86, K. D. Roeder, ed. New York, John Wiley & Sons, Inc.

Treherne, J. E., and J. W. L. Beament (eds.). 1965. *The Physiology of the Insect Nervous System.* New York, Academic Press, Inc.

Tuxen, S. L. 1964. *The Protura. A revision of the species of the world with keys for determination.* Paris, Hermann & Cie.

——— (ed.). 1970. *Taxonomist's Glossary of Genitalia in Insects,* 2nd ed. rev. Copenhagen, Munksgaard.

Ulrich, W. 1966. Evolution and classification of the Strepsiptera. Proc. 1st Internat. Cong. Parasitol., 1:609–11.

U.S. Department of Agriculture. 1952. *Insects—The Yearbook of Agriculture,* Washington, D.C., U.S. Department of Agriculture.

Usinger, R. L. (ed.). 1956a. *Aquatic Insects of California, with Keys to North American Genera and California Species.* Berkeley, University of California Press.

———. 1956b. Aquatic Hemiptera. In *Aquatic Insects of California,* pp. 182–228, R. L. Usinger (ed.). Berkeley, University of California Press.

———. 1966. Monograph of Cimicidae (Hemiptera-Heteroptera). The Thomas Say Foundation, Vol. 7, pp. 1–585.

Uvarov, Sir B. 1966. *Grasshoppers and Locusts,* Vol. 1, Anatomy; Physiology; Development; Phase Polymorphism; Introduction to Taxonomy. Cambridge, Cambridge University Press.

Vogel, R. 1921. Zur Kenntnis des Baues und der Funktion des Stachels und des Vorderdarmes der Kleiderlaus. Zool. Jahrb., Anat., 42:229–58.

Warburton, F. E. 1967. The purposes of classification. Syst. Zool. 16(3):241–45.

Waterhouse, D., and M. F. Day. 1953. Functions of the gut in absorption, excretion, and intermediary metabolism. In *Insect Physiology,* pp. 331–49, K. D. Roeder, ed. New York, John Wiley & Sons, Inc.

Way, J. J. 1963. Mutualism between ants and honeydew-producing Homoptera. Ann. Rev. Entomol., 13:307–44.

Weber, N. A. 1966. Fungus-growing ants. Science, 153(3736):587–609.

Welch, H. E. 1965. Entomophilic nematodes. Ann. Rev. Entomol., 10:275–302.

West, L. S. 1951. *The Housefly: Its Natural History, Medical Importance, and Control.* Ithaca, N.Y., Comstock Publishing Company, Inc.

Weygoldt, P. 1969. *The Biology of Pseudoscorpions.* Cambridge, Mass., Harvard University Press.

White, J. W. 1963. Honey. In *The Hive and the Honey Bee,* pp. 369–406, R. A. Grout, ed. Hamilton, Ill., Dadant & Sons, Inc.

White, M. J. D. 1964. Cytogenetic mechanisms in insect reproduction. In *Insect Reproduction,* pp. 1–12, K. C. Highnam, ed. Royal Entomological Society of London, Symp. 2.

Whittaker, R. H. 1970. *Communities and Ecosystems.* New York, Macmillan Publishing Co., Inc.

Wickler, W. 1968. *Mimicry in Plants and Animals.* London, World University Library.

References Cited

Wigglesworth, Sir V. B. 1930. The formation of the peritrophic membrane in insects with special reference to the larvae of mosquitoes. Quart. J. Microscop. Sci., 73:593–616.

———. 1941. The sensory physiology of the human louse, *Pediculus humanus corporis* DeGeer (Anoplura). Parasitology, 33:67–109.

———. 1945. Transpiration through the cuticle of insects. J. Exp. Biol., 21:97–114.

———. 1954. *The Physiology of Insect Metamorphosis.* Cambridge, Cambridge University Press.

———. 1959a. *The Control of Growth and Form: A Study of the Epidermal Cell in an Insect.* Ithaca, N.Y., Cornell University Press.

———. 1959b. Insect blood cells. Ann. Rev. Entomol., 4:1016.

———. 1964a. *The Life of Insects.* New York, The New American Library, Inc.

———. 1964b. The hormonal regulation of growth and reproduction in insects. Ann. Rev. Ins. Physiol., 2:247–336.

———. 1965. *The Principles of Insect Physiology,* 6th ed. London, Methuen & Company Ltd.

———. 1970. *Insect Hormones.* Edinburgh, Oliver & Boyd Ltd.

Wilbur, D. A. 1962. Stored grain insects. In *Fundamentals of Applied Entomology,* pp. 466–94, R. E. Pfadt, ed. New York, Macmillan Publishing Co., Inc.

Williams, C. M. 1970. Hormonal interactions between plants and insects. In *Chemical Ecology,* E. Sondheimer and J. B. Simeone, eds. New York, Academic Press, Inc.

Wilson, D. M. 1966. Insect walking. Ann. Rev. Entomol., 11:103–22.

Wilson, E. O. 1965. Chemical communication in the social insects. Science, 149(3688):1064–71.

———. 1970. Chemical communication within animal species. In *Chemical Ecology,* pp. 133–55, E. Sondheimer and J. B. Simeone, eds. New York, Academic Press, Inc.

———. 1971. *The Insect Societies.* Cambridge, Mass., Harvard University Press.

Wolglum, R. S., and F. A. McGregor. 1958. Observations on the life history and morphology of *Agulla bractea* Carpenter (Neuroptera: Raphidiodea: Raphidiidae). Ann. Entomol. Soc. Amer., 51:129–41.

Wood, D. L., R. M. Silverstein, and M. Nakajima (eds.), 1970. *Control of Insect Behavior by Natural Products.* New York, Academic Press, Inc.

Woodwell, G. M., W. M. Malcolm, and R. H. Whittaker. 1969. A-Bombs, bugbombs, and us. In *The Subversive Science—Essays Toward an Ecology of Man,* pp. 230–41, P. Shepard and D. McKinley, eds. Boston, Houghton Mifflin Company.

World Health Organization. 1967. The genetic control of insects. WHO Chronicle, 21(12):517–24.

Wright, D. P., Jr. 1963. Antifeeding compounds for insect control. In *New Approaches to Pest Control and Eradication,* Advances in Chemistry Series 41. Washington, D.C., American Chemical Society.

Wyatt, G. R. 1961. The biochemistry of insect hemolymph. Ann. Rev. Entomol., 6:75–102.

Yeager, J. F. 1938. Mechanographic method of recording insect cardiac activity, with reference to effect of nicotine on isolated heart preparations of *Periplaneta americana.* J. Agr. Res., 56:267–82.

Yokoyama, T. 1963. Sericulture. Ann. Rev. Entomol., 8:287–306.

Zumpt, F. 1965. *Myiasis in Man and Animals in the Old World.* London, Butterworth & Company (Publishers) Ltd.

Index